BRITISH GEOLOGICAL SURVEY

C R BRISTOW,
C M BARTON,
E C FRESHNEY,
C J WOOD,
D J EVANS,
B M COX,
H C IVIMEY-COOK and
R T TAYLOR

Geology of the country around Shaftesbury

Memoir for 1:50 000 geological sheet 313
(England and Wales)

CONTRIBUTORS

Biostratigraphy
D K Graham
H G Owen
I P Wilkinson
M A Woods

Stratigraphy
P M Allen
G K Lott

Water supply
M E Lewis

Geophysics
J D Cornwell
C P Royles
S Self

Engineering geology
T P Gostelow

LONDON: HMSO 1995

First published 1995

ISBN 011 884505 5

Bibliographical reference

BRISTOW, C R, and BARTON, C M, FRESHNEY, E C, WOOD, C J, EVANS, D J, COX, B M, IVIMEY-COOK, H C, and TAYLOR, R T. 1995. Geology of the country around Shaftesbury. *Memoir of the British Geological Survey*, Sheet 313 (England and Wales).

Authors

C R Bristow, BSc, PhD
C M Barton, BSc, PhD
E C Freshney, BSc, PhD
British Geological Survey, Exeter

D J Evans, BSc, PhD
B M Cox, BSc, PhD
H C Ivimey-Cook, BSc, PhD
British Geological Survey, Keyworth

D K Graham, BA
R T Taylor, BSc, PhD
C J Wood, BSc
formerly *British Geological Survey*

Contributors

P M Allen, BSc, PhD, J D Cornwell, BSc, PhD, T P Gostelow, BSc, PhD, G K Lott, BSc, PhD, C P Royles, BSc, S J Self, M A Woods, BSc, I P Wilkinson, BSc, PhD
British Geological Survey, Keyworth

M A Lewis, BA, MSc
British Geological Survey, Wallingford

H G Owen, BSc, PhD
Natural History Museum, London

Printed in the UK for HMSO
Dd 292048 C8 03/95

Other publications of the Survey dealing with this and adjoining districts

BOOKS

British Regional Geology
The Hampshire Basin and adjoining areas, 4th edition

Memoirs
Salisbury (sheet 298), 1903*
Bridport and Yeovil (sheets 312 and 327), 1958
Ringwood (sheet 314), 1902*
Dorchester (sheet 328), 1899*
Bournemouth (sheet 329), 1991

Water supply
Wells and springs of Dorset, 1916*

MAPS

1:584 000
Tectonic map of Great Britain and Northern Ireland

1:625 000
Solid geology (south sheet)
Quaternary geology (south sheet)
Aeromagnetic map (south sheet)

1:250 000
Portland (solid geology)
Portland (gravity)
Portland (aeromagnetic)
Bristol Channel (solid geology)
Bristol Channel (gravity)
Bristol Channel (aeromagnetic)

1:100 000
Hydrogeological map of the Chalk and associated minor aquifers of Wessex

1:50 000
Sheet 297, Glastonbury, 1973
Sheet 298, Wincanton, 1977
Sheet 299, Salisbury, 1903
Sheet 312, Yeovil, 1958
Sheet 313, Shaftesbury, 1993
Sheet 314, Ringwood, 1902
Sheet 327, Bridport, 1940
Sheet 328, Dorchester, 1898
Sheet 329, Bournemouth, 1991

* Out of print

Geology of the country around Shaftesbury

The Shaftesbury district in north Dorset lies at the western end of the Wessex Basin. The district is mostly rural, with the small towns of Shaftesbury and Blandford Forum the principal urban areas. The contrasting scenery reflects the underlying geology. In the centre of the area is the low-lying ground of the Blackmoor Vale, the setting of several of Thomas Hardy's novels; in the south and east, the skyline is dominated by the Chalk escarpment, with, in the north-east, a secondary escarpment formed by the Upper Greensand. Prominent hills (e.g. Hod and Hambledon hills, Shaftesbury) along these escarpments were chosen as sites for fortified settlements by the Romans and, later, the Saxons; extensive earthworks remain at several of the sites, which now form popular tourist venues.

The geological sequence ranges from the upper part of the Lower Jurassic to the Upper Cretaceous Chalk. Details of the outcrops of these strata are presented; they incorporate much new stratigraphical and palaeontological data. Deep boreholes and geophysical data have been interpreted to provide descriptions of the concealed formations and also to elucidate the structure. Extensive areas of landslip have been mapped beneath the escarpments of the Corallian Group and the Upper Greensand.

In the tabular geological succession opposite, the columns show, from left to right, system, group, formation and lithological summary, member/bed and lithological summary (with thickness), and generalised thickness of formations. Formations below the Bridport Sands are not exposed in the district, and are encountered only in boreholes. The thicknesses for the formations below the Bridport Sands are those proved in the Fifehead Magdalen and Mappowder boreholes (see Figure 3).

Cover photograph

Gold Hill, Shaftesbury. The steep hill descends the scarp face of the Upper Greensand Shaftesbury Sandstone. The open fields in the middle distance overlie landslipped Gault, with the low escarpment of the Shaftesbury Sandstone visible to the left. The skyline is formed by the Chalk escarpment (A15212).

Aerial view of landslips, Shaftesbury, viewed from the west.

CONTENTS

FIGURES

PLATES

TABLES

NOTES

The word 'district' used in this memoir means the area included in 1:50 000 Geological Sheet 313 (Shaftesbury).

Figures in square brackets are National Grid references; places within the Shaftesbury district lie within the 100 km square ST.

The authorship of fossil species is given in the index of fossils.

Numbers preceded by the letter A refer to photographs in the Geological Survey collections.

PREFACE

The Shaftesbury area is a particularly attractive part of England that has long been a source of inspiration to writers and poets. It has also been an area of great interest to geologists since the early nineteenth century.

The 1:10 000-scale geological survey of the Shaftesbury (313) Geological Sheet, previously mapped at a scale of one inch to the mile, forms part of the Wessex Basin Project and has involved a comprehensive biostratigraphical and structural synthesis of the Jurassic, Cretaceous and Quaternary sedimentary rocks at the western end of the Wessex Basin. The results will have an important bearing on the scientific and economic appraisal of the basin as a whole.

Major revisions of the Corallian, Upper Greensand and Chalk sequences have been undertaken, and extensive landslips have been recognised and delineated for the first time. The detailed maps will be particularly relevant to the needs of the extractive industry, the water-supply authorities, planners and civil engineers. In particular, the recently recognised major landslips, especially well developed around the expanding town of Shaftesbury, form a significant constraint to development in certain areas.

Exploratory scientific, engineering and mineral-resource boreholes have been drilled for BGS as part of the study. They provide a wealth of data in an area of poorly exposed strata, for which there was little previous information. Geophysical logs from these and other shallow boreholes provide a good correlation with commercially drilled and logged water and hydrocarbon boreholes.

Structural syntheses, which combine evidence from field mapping and seismic data generated by the various oil companies, provide evidence for both the small- and large-scale structures that have controlled the evolution of the Wessex Basin. They are directly relevant to the study of hydrocarbon generation and entrapment in this area which lies on the edge of Britain's most important onshore petroleum-producing basin.

I hope that this new and fascinating account of the geology of the Shaftesbury area is read not only by geologists, but also by the many people who are interested in better understanding the evolution of the 'Hardy County'.

Peter J Cook, DSc
Director

British Geological Survey
Kingsley Dunham Centre
Keyworth
Nottingham
NG12 5GG

November 1994

OPEN-FILE REPORTS

Open-file reports, which are available for the whole of the Shaftesbury district, are shown on the diagram below, together with the dates of publication and the authors' names. They contain more detail than appears in the memoir. Copies of these reports are available for purchase from the British Geological Survey, Keyworth, Nottingham NG12 5GG and at 30 Pennsylvania Road, Exeter, Devon EX4 6BX.

ST 72 NE and SE (Bristow, 1990a)

Boundary of Shaftesbury district

ST 62 SE (Taylor, 1990)

ST 72 SW (Taylor, 1991)

ST 82 SW and SE (Bristow, 1989a)

ST 92 SW (Bristow, 1990b)

ST 61 NE (Barton, 1989)

ST 71 NW (Barton, 1990)

ST 71 NE and SE (Bristow, 1989b)

ST 81 (Bristow, 1992a)

ST 91 NW and SW (Bristow, 1991)

ST 71 SW (Barton and Freshney, 1990)

ST 61 SE and ST 60 NE (Freshney, 1992)

ST 70 NW and NE (Freshney, 1989)

ST 80 NW and SW (Bristow, 1992b)

ST 80 NE, SE and 90 NW (Barton, 1991)

ST 60 SE, 70 SW and SE (Freshney, 1993)

ST 90 SW and SE (Bristow, 1987)

ACKNOWLEDGEMENTS

The authors thank the following for help during fieldwork: Mr M J Abbot (Drillers) for allowing us to log geophysically one of their boreholes; Messrs R Brand and H C Prudden, Mrs M Keats (née Samuel), Dr H Torrens and Professor M R House for details of temporary sections; Carless Exploration Ltd and Gas Council (Exploration) Ltd for permission to use unpublished oil-company data; Ambrit Resources Plc, Industrial Scotland Energy Limited, Kelt UK Limited, Monument Oil and Gas Plc and Teredo Petroleum Plc for permission to reproduce parts of seismic reflection lines and borehole information; British Petroleum and Shell UK for permission to publish certain parts of the structure contour and isopach maps; Mr R D Clark of Bristol Museum for access to specimens; Mr T Cosgrove of Wessex Water for permission to log geophysically several water-production wells; Dorset County Council and Mr P Smith (Frank Graham & Partners) for permission to deepen and log geophysically one of their boreholes; Mr P E Enson and Miss C M Hebditch for the loan of specimens from the Dorset County Museum, J H Hibbitts (Wessex Water Wells Ltd) for a copy of the gamma-ray log of the Quarleston Borehole; the late S C A Holmes for his pre-war notes and palaeontological collections from the Chalk of the district; Mrs L Light of Gillingham Museum for the loan of specimens; Dr W J Kennedy of Oxford University, Professor R N Mortimore of the University of Brighton and Dr R G Bromley of the University of Copenhagen for making available notes, sections and specimens from Chalk pits in the Shaftesbury area; Mr C J R Mitchell of Manor Farm, Iwerne Courtney, and Fontmell Magna Primary School for the loan of specimens from the Melbury Sandstone; Mr H P Powell of the University Museum, Oxford, for the loan of Mr F H A Engleheart's unpublished typescript on the Forest Marble; Mr P Simmons of Shillingstone Lime and Stone Co. Ltd for access to his company's quarry; Mr A Watts for repeated access to the exposures along his recently constructed track up Okeford Hill; Dr C W Wright for palaeontological help, and Dr J K Wright for copies of his field maps of the Corallian of north Dorset.

The following land owners allowed BGS to drill stratigraphical boreholes on their ground: Mr J C Down of Manor Farm, East Stour (East Stour Borehole), Mr A K Loveridge of Spiders Farm, Sturminster Newton (Knackers Hole Borehole), Mr E Waltham of Church Farm, Purse Caundle (Purse Caundle Borehole), Mrs C V Wilcox, Vale Farm, Combe Throop (Combe Throop Borehole), Mr P J Rowland of Hazelbury Bryan (Hazelbury Bryan Borehole), Mr J D Dennison (Cannings Court Borehole) and Mr N Mullins of Church Farm, Shaftesbury (Church Farm Nos. 1–3). The last has kindly given us permission to reproduce an aerial photograph (Plate 11) of his farm.

The memoir was compiled by Dr C R Bristow and edited by Dr R W Gallois and Mr R G Wyatt.

HISTORY OF SURVEY OF THE SHAFTESBURY SHEET

The district was first geologically surveyed on the one-inch (1:63 360) scale by H W Bristow and published on Old Series sheets 15 and 18, in 1856 and 1850 respectively. H B Woodward and W A Ussher carried out minor revisions of the Jurassic strata on Sheet 18, and the map was reissued in 1875. No memoir accompanied these sheets, but details relevant to the district were incorporated in two parts of the five-volume regional stratigraphical memoir covering the Jurassic system (Woodward, 1894, 1895).

Revision of the Cretaceous mapping on the six-inch (1:10 560) scale was carried out between 1893 and 1899 by A J Jukes-Browne, F J Bennett, C Reid and W Whitaker. Details from this area were incorporated in Jukes-Browne and Hill's classic three-volume memoir 'The Cretaceous Rocks of Britain' (1900, 1903, 1904). Drift deposits were surveyed by H G Dines and S E Hollingworth in 1921 and 1922; at the same time, minor modifications were made to the boundaries of the Gault and Lower Greensand. The results were incorporated in New Series One-inch Geological Sheet 313 (Shaftesbury), published in 1923; a descriptive memoir by H J O White was also published in 1923. The sheet was reissued without revision at the 1:50 000 scale in 1977. The revised version at this scale was published in 1993.

Memoirs dealing with the water supply of the district on a county basis have been written by Whitaker and Edmunds (1925) for Wiltshire, Whitaker and Edwards (1926) for Dorset, and Richardson (1928a) for Somerset.

The component 1:10 000-scale National Grid sheets of Geological Sheet 313 are shown here, together with initials of the geological surveyors and the dates of survey. The surveying officers were C R Bristow, E C Freshney, C M Barton, R T Taylor and P M Allen.

ST 62		ST 72		ST 82	ST 92
SE RTT 1986/1988	SW RTT 1989/1990	SE CRB/PMA 1986, 1987 1988	SW CRB 1987	SE CRB 1988	SW CRB 1988
NE CMB 1987/1988	NW CMB 1988/1989	NE CRB 1986	NW CRB 1986	NE CRB 1989	NW CRB 1989
ST 61		ST 71		ST 81	ST 91
SE ECF 1988	SW ECF/CMB 1989	SE CRB 1986/1987	SW CRB 1987	SE CRB 1990	SW CRB 1989
NE ECF 1988	NW ECF 1987	NE ECF 1982/1986	NW CRB 1990	NE CMB 1990	NW CMB 1989
ST60		ST 70		ST 80	ST 90
SE ECF 1989	SW ECF 1990	SE ECF 1987	SW CRB 1990	SE CMB 1990	SW CRB 1985

Uncoloured dyeline copies of the maps are available for purchase from the British Geological Survey, Keyworth, Nottingham NG12 5GG. The fair-drawn maps can be inspected in the Exeter Office of the British Geological Survey.

ONE

Introduction

GEOGRAPHICAL SETTING

The Shaftesbury district falls mostly in the county of Dorset, with small tracts in the north-west and north-east falling in Somerset and Wiltshire respectively (Figure 1). The district takes its name from the Saxon hill-top fort of Shaftesbury (present population 6200), situated at 218 m above Ordnance Datum (OD) on the Upper Greensand escarpment. In the south-east, the market and brewing town of Blandford Forum (population 8800), with its satellite army camp, is the largest town within the district. Sturminster Newton, known both for its cattle market and as the one-time home of the Dorset poet Thomas Barnes, is the third largest town (population 2600). Other small towns and large villages have a fairly even scatter across the district (Figure 1).

The principal drainage is to the south-east by the River Stour and its tributaries, the rivers Lydden, Divelish and Cale, and the Fontmell, Manston, Chivrick's, Caundle and Bow brooks (Figure 1). In the north-east, northward-draining streams combine to form the River Nadder, which joins the River Avon at Salisbury. Streams draining the Milborne Port area in the north-west unite to form the River Yeo near Sherborne, which flows north-westwards into Bridgewater Bay.

The geological deposits preserved within the district, which lies close to the western end of the Wessex Basin, are listed on the inside front cover of this memoir. The

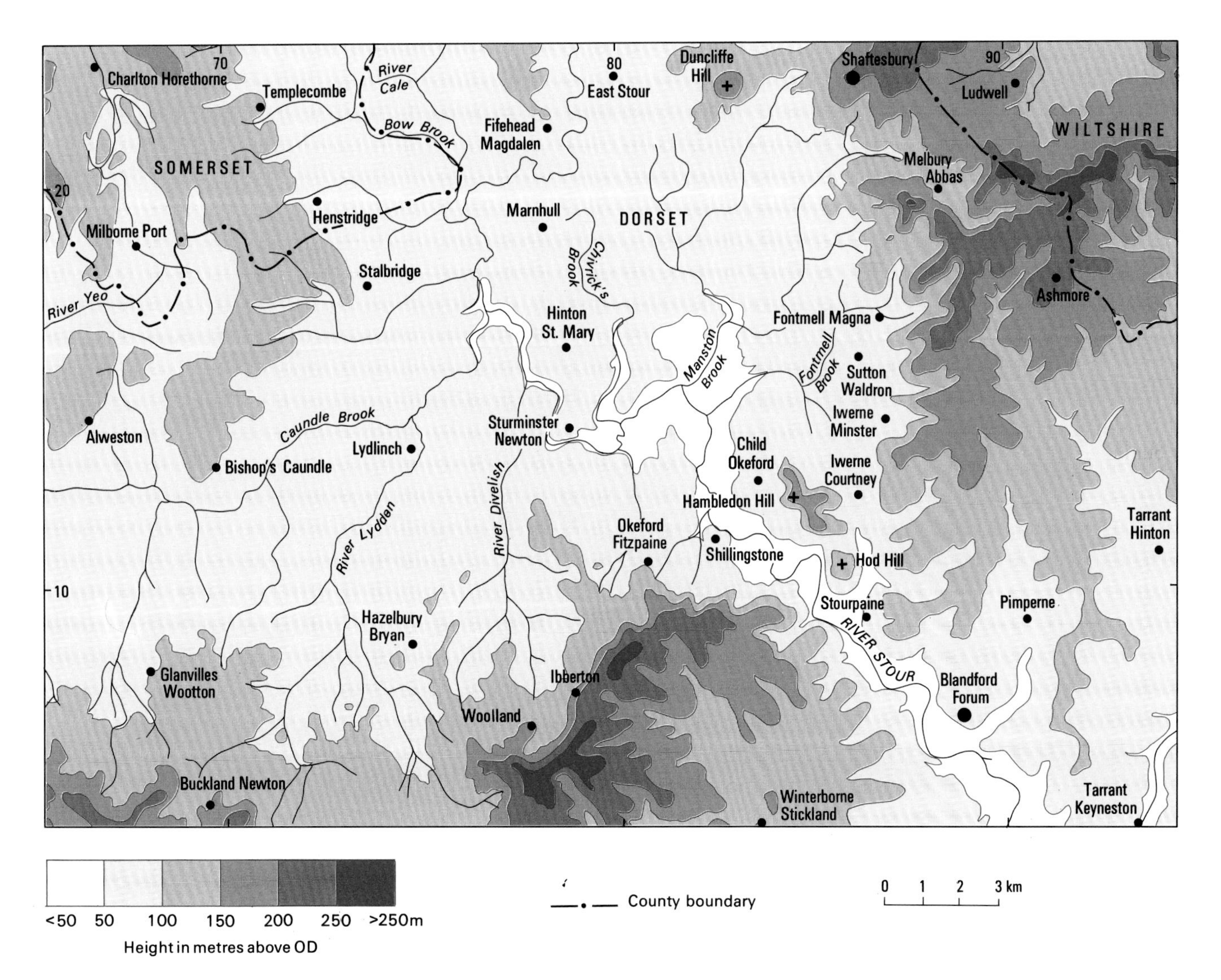

Figure 1 Sketch map illustrating the physical features of the district.

oldest strata to crop out, those of the Jurassic Bridport Sands, occur over a limited area near Charlton Horethorne in the north-west; younger Jurassic strata crop out successively from west to east, culminating in the Kimmeridge Clay. Cretaceous deposits comprise the Lower Greensand, which is only locally developed, Gault, Upper Greensand and Chalk (Figure 2). At depth, older Jurassic, Triassic, Permian and Palaeozoic formations have been proved in boreholes drilled for hydrocarbon exploration. For the most part, drift deposits are not extensive, although spreads of terrace and alluvial deposits obscure some areas of solid geology to the east of Henstridge. Clay-with-flints is locally widespread on some of the dip slopes of the Upper Chalk.

The topography is varied and directly related to the geology. In the western and northern parts of the district, prominent ridges, in part fault-controlled, of resistant Jurassic limestones rise above the clay vales. The Blackmoor (or Blackmore) Vale, the broadest of these, is developed principally on the Oxford Clay, although the name is also loosely applied to the broad vale of the Kimmeridge Clay. This low-lying area, drained by numerous, in places anastomosing, streams and rivers, was formerly a swampy area devoid of permanent settlement. It forms part of the setting for several of Thomas Hardy's novels (*Tess of the D'Urbevilles, Far from the Madding Crowd* and *The Woodlanders*). Drainage works have improved the boggy conditions, but it is still only sparsely populated. All the principal settlements in this north-western part of the district are sited on the well-drained limestone formations. In the south and east, the Chalk forms a prominent escarpment rising from the clay vale to a maximum height of 277 m above OD at Win Green Clump [925 207]. In the north-east, the well-developed Upper Greensand forms an equally prominent feature to the west of the Chalk escarpment. Most of the principal settlements in the eastern and southern parts of the district are sited on the free-draining sands of the Upper Greensand, but a number of villages and towns, including Blandford Forum, are located along the valleys which cross the Chalk outcrop.

The economy is largely based on agriculture and associated support industries, but there are, in addition to the old-established tannery at Milborne Port and the Blandford Brewery, a number of light industries around

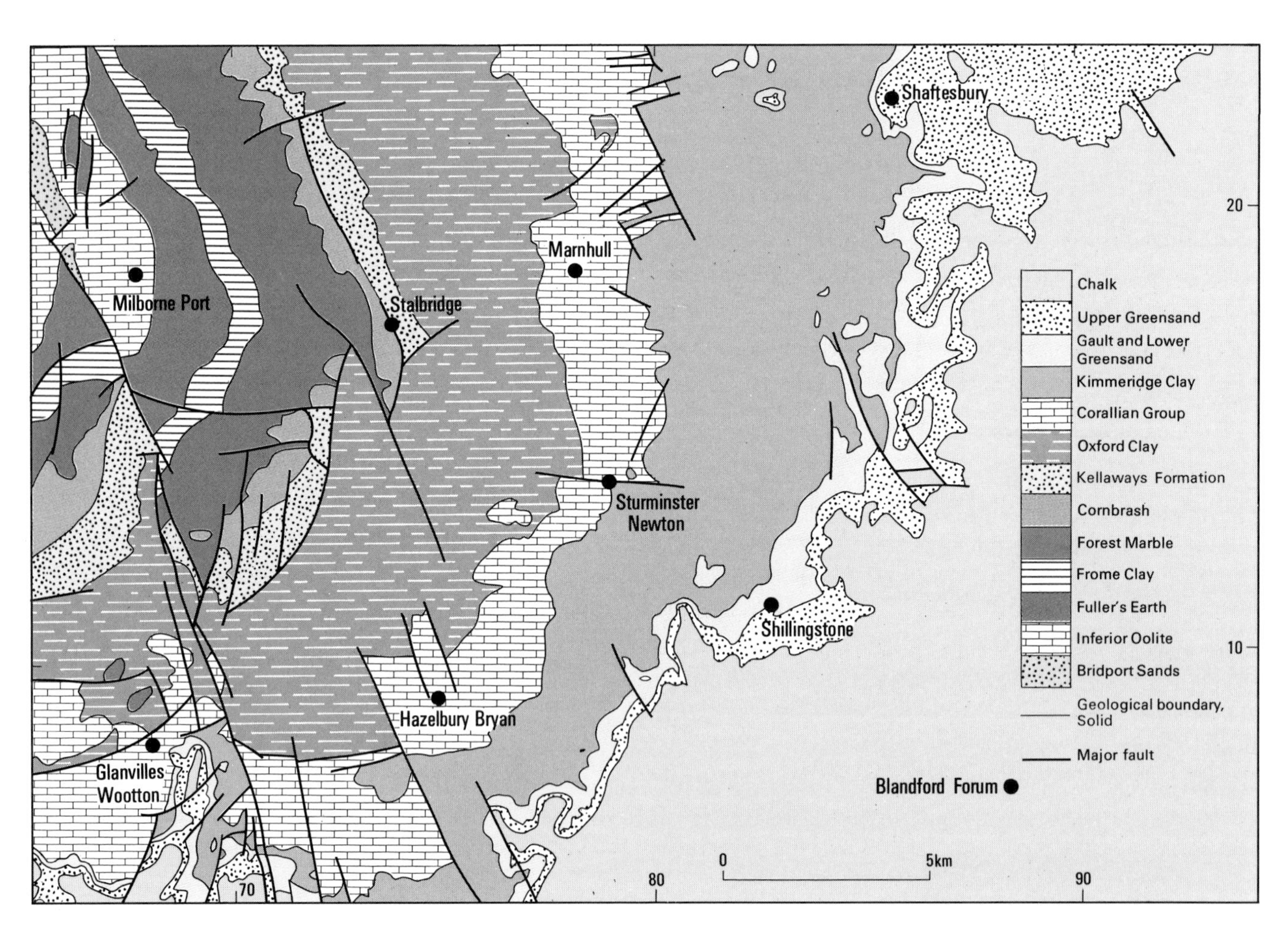

Figure 2 Sketch map of the Solid geology of the district.

the towns and larger villages. There is a mix of arable and pasture farmland, with the latter predominating on the heavy clay soils. There are scattered broad-leaved and coniferous woodlands across the district. The most extensive are those on the Chalk, particularly on the steep scarp face and the sides of some of the coombes.

GEOLOGICAL HISTORY

The nature of the Palaeozoic rocks which form the pre-Permian basement varies across the district from folded and faulted Devonian phyllites in the south, through folded and faulted, but only weakly metamorphosed Devonian sedimentary rocks in the central tract, to Carboniferous platform carbonates in the north. A period of crustal subsidence, which began in the early Permian, initiated the formation of the Wessex Basin (Chadwick, 1985). The basin is crossed by several east–west-trending fault zones which subdivide the basin into smaller basins and highs, of which the Winterborne Kingston Trough in the south, the Cranborne–Fordingbridge High in the centre, and the Mere Basin in the north (Figure 4), had a profound effect on depositional history in the Permian and Triassic periods, but had little effect in the Jurassic and Cretaceous.

Permian deposits occur only in the Winterborne Kingston Trough, where they rest with marked unconformity on Devonian slates and phyllites. They consist of reddened breccias and sandstones deposited in an alluvial or aeolian environment, but with some fluvial input, during a hot, semiarid or desert regime. Continental red-bed fluvial and alluvial conditions continued throughout most of the Triassic Period, with the deposition of the Sherwood Sandstone and Mercia Mudstone Group, but in an extended basin. At the close of the Triassic Period, marine conditions returned with deposition of the Westbury Formation shales and continued throughout much of the Mesozoic era.

Throughout Jurassic times, fully marine conditions existed across the whole of the district. Deposition was more-or-less continuous throughout the Lias, but an important break has been recognised in the Middle Lias. Non-sequences within the shallow-water deposits of the Inferior Oolite are probably related to penecontemporaneous fault movements close to the basin margin. The succeeding Great Oolite Group is represented in the district mostly by mudstone, which was laid down in a quiet low-energy environment. The Cornbrash is a widespread marine limestone divisible into lower (Bathonian) and upper (Callovian) units, generally with a non-sequence separating the two. The succeeding sandy Kellaways Formation passes up into the shallow-water, marine-shelf mudstones of the Oxford Clay. In turn, the Oxford Clay is overlain by the mudstones, sandstones and limestones of the Corallian Group, which form a sequence of clastic/carbonate shallowing-upward cycles. The sandstones and limestones probably represent nearshore deposits of restricted areal extent. The succeeding Kimmeridge Clay formed during a period of high global sea level in low-energy depositional environments that varied from aerobic to anaerobic.

Towards the end of Kimmeridgian times, the Jurassic sea shallowed, and shallow-water marine deposits were laid down during the Portlandian. Further shallowing, accompanied by a probable north-eastwards retreat of the sea during the late Portlandian and early Cretaceous (Berriasian), led to the establishment of a lagoonal environment in the area east of Shaftesbury, in which the Purbeck Beds were deposited.

After Berriasian times, there was uplift and erosion of the western part of the Wessex Basin. Erosion was not uniform across the Shaftesbury district, but was most severe over the southern part of the Cranborne–Fordingbridge High. There, any Berriasian and Portlandian deposits, all of the Upper Kimmeridge Clay and some of the Lower Kimmeridge Clay were removed. This period of erosion manifests itself in the 'late-Cimmerian Unconformity', a series of widespread stratigraphical breaks in the late Jurassic and earliest Cretaceous throughout north-west Europe.

The deposition of the Aptian Lower Greensand (Child Okeford Sands Member) occurred during the first pulse of a series of marine transgressions that successively overstepped westwards across the eroded Jurassic surface. The Lower Albian part of the Lower Greensand (Bedchester Sands Member) overlaps the Child Okeford Sands, and is itself overlapped by the Middle Albian Gault. The Gault was the first of the Cretaceous formations to extend with fairly uniform lithology across the district. Albian uniformity was, however, short lived, because Upper Albian and Cenomanian sedimentation was affected by an important north-west-trending high, the Mid-Dorset Swell, in the south-west of the district. Deposition during the rest of the Cretaceous Period was fairly uniform, but hardgrounds and glauconitised nodular beds in the Chalk indicate possible periods of local shallowing and erosion.

At the close of the Cretaceous Period, uplift and minor folding were followed by erosion. There is no undoubted Palaeogene deposit in the district, but the common occurrence of well-rounded flints, characteristic of the Tertiary deposits, in the Clay-with-flints, which may, in part, be a remanié Palaeogene deposit, indicates the former more widespread extent of Tertiary beds. At the end of the Palaeogene Period, the Alpine Orogeny, a widespread phase of folding and faulting, culminated in the inversion of the major Mesozoic basins and highs. Thus, the eastern part of the Cranborne–Fordingbridge High inverted to form the syncline underlying the Hampshire 'Basin'. It is probable that the last movements along the major faults in the Shaftesbury district took place during the Miocene. No Neogene deposit is known.

The Clay-with-flints was probably formed during the Pleistocene Period. The formation of Head is thought to have commenced under freeze/thaw conditions, also in the Pleistocene; some accretion continues by colluvial process at the present day. In the late Pleistocene, probably during the Devensian cold stage, meltwater from semipermanent snow caps on the higher ground transported large quantities of sand and gravel and deposited them as river terrace deposits. Landslips were probably initiated at that time; some movement still takes place. The major sedimentary process in the district during the Holocene, and continuing to the present day, was the deposition of alluvium.

HISTORY OF RESEARCH

In addition to work by Survey officers, aspects of the geology of the district have attracted the attention of many workers for almost two centuries from the time of Pulteney's (1813) observations on the Upper Greensand near Shaftesbury. J Buckman, and his son S S Buckman, who lived nearby at Bradford Abbas, made detailed stratigraphical and palaeontological observations in the Lower and Middle Jurassic strata (e.g. 1875, 1879, 1889, 1893). Richardson (1916, 1919, 1932) made detailed observations of the Inferior Oolite of Dorset, Somerset and Gloucestershire. Several sections in the Shaftesbury area were recorded by Douglas and Arkell (1928) during their study of the Cornbrash. The Corallian rocks of the district formed part of Blake and Hudleston's (1877) wider study of that group in England. Arkell reviewed the published literature, and visited many exposures when compiling 'The Jurassic System in Great Britain' (1933); the references to the district were further updated in Arkell (1956). More recent work on the Jurassic system by Wright (1981) on the Corallian rocks of north Dorset, and a review of the Cornbrash and Kellaways Formation of England by Page (1989), include localities in the Shaftesbury district. A regional survey of Bathonian sedimentation was carried out by Martin (1967).

By comparison, much less attention has been paid to the Cretaceous strata of the district. Tresise (1960, 1961) wrote about the lithology of the sediments and the origin of the chert in the Upper Greensand of the Wessex Basin. Parallel research by Drummond (1970) and Kennedy (1970) described the effects of the Mid-Dorset Swell on Albian–Cenomanian deposition.

The resurvey of the Shaftesbury Sheet at the 1:10 000 scale was carried out by Drs C R Bristow, E C Freshney, C M Barton, R T Taylor and P M Allen during 1986–1990. The 1:50 000 geological map of the district was published in 1993.

TWO

Concealed formations

The survey of the concealed geology has been facilitated by seismic reflection data and calibrated by deep oil-exploration boreholes in and adjacent to the district (Figures 3 and 4; Table 1). Two deep boreholes, at Fifehead Magdalen and Mappowder, were drilled within the district, and others, notably at Winterborne Kingston, Bruton, Ryme Intrinseca, Spetisbury, Bere Regis and Cranborne, provide data from adjacent districts. Together, these data suggest that there are important variations, both lateral and vertical, in the concealed geology across the district.

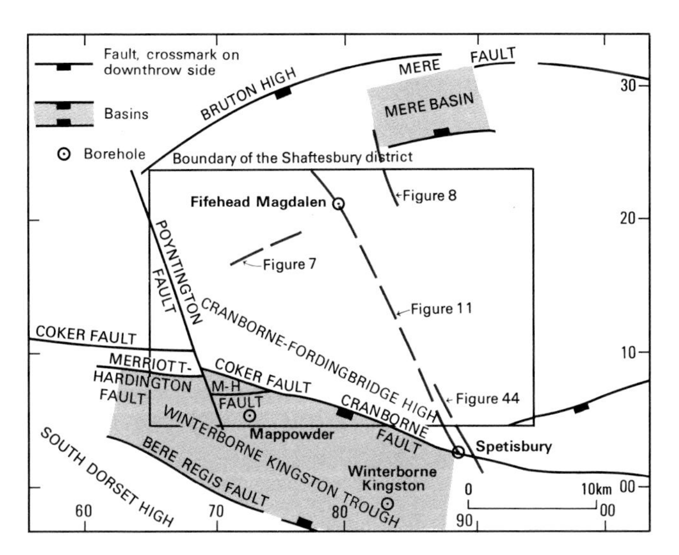

Figure 4 Sketch map showing the major structural features of the Shaftesbury and adjacent districts and the location of figured seismic and cross-sections.

CRUSTAL STRUCTURE BENEATH THE DISTRICT

Seismic reflection profiles, recorded to 10 or 12 seconds two-way travel time (TWTT), show that the crust beneath southern England can be divided vertically into three zones, based upon their reflection character (Chadwick et al., 1983, 1989; Whittaker and Chadwick, 1984). The three zones vary in thickness from place to place and are probably tectonic units encompassing rocks of widely different ages (Figure 5). The Shaftesbury district lies in a region where the character of the lowest crustal zone is transitional between two types, when compared with areas to the north and south.

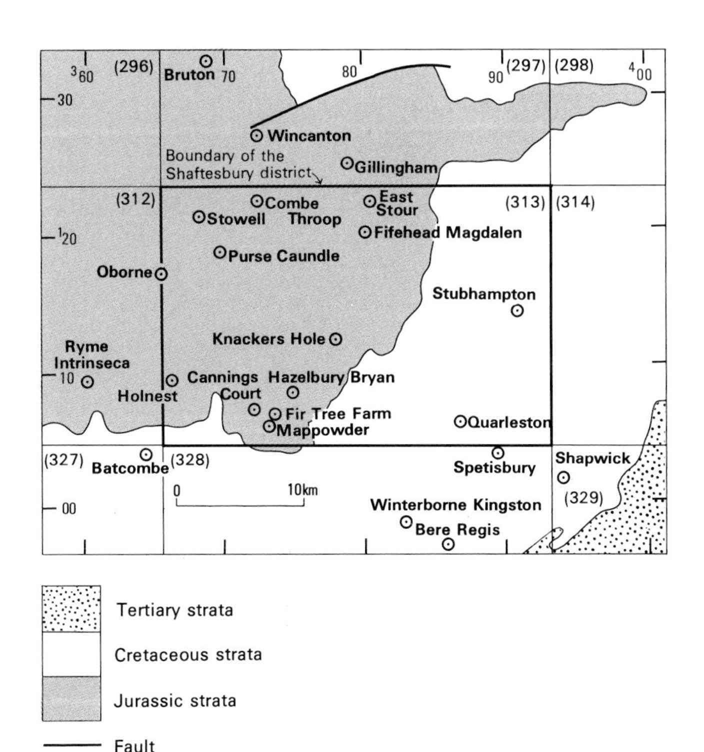

Figure 3 Geological sketch map of of the Shaftesbury and surrounding districts showing the position of important boreholes.

In the district, the uppermost crustal zone (zone 1), which corresponds to the post-Variscan sedimentary cover rocks (see Concealed Permian and Mesozoic sequences, below), is typified by reflections from flat-lying or gently folded strata down to 1 to 2 seconds TWTT (2 to 2.5 km depth).

A middle crustal zone (zone 2) is typically seismically featureless and corresponds to the Variscan foldbelt. Analogy with sequences exposed in Devon suggests that it comprises a sequence of Devonian and Carboniferous slates and phyllites, highly deformed by polyphase folding and thrusting. It extends down to between 5 and 6 seconds TWTT (c.15 km depth), with a few low-amplitude, discontinuous reflections locally crosscut by southward dipping reflections interpreted as thrusts. These reflections do not extend into the underlying zone 3. The Variscan rocks in southern Britain are probably the lateral equivalent of the Rheno-Hercynian foldbelt seen in northern Europe, and represent the most northerly part of the Variscan foldbelt.

The deepest crustal seismic zone (zone 3), probably consisting of Precambrian metamorphic lower crust, is largely poorly reflective, its seismic character resembling the 'foreland-type' lower crust present beneath the Midlands Microcraton, to the north (e.g. Chadwick et al., 1989). South of the district, the lower crust is generally highly reflective, typifying the reflective 'orogenic' crust of Chadwick et al. (1989). The northern margin of re-

Table 1 List of important boreholes in the district and surrounding area (see Figure 3), with abbreviations used on figures, and grid reference.

Name	Symbol	Grid reference	Purpose	Drilled by
Arne G1	(A)	SY 9575 8704	Hydrocarbon exploration	Gas Council
Arreton*	(Ar)	SZ 5320 8580	Hydrocarbon exploration	Gas Council
Batcombe	(Ba)	ST 6112 0314	Hydrocarbon exploration	Carless
Bransgore	(Be)	SZ 1958 9505	Hydrocarbon exploration	BP
Bruton	(Br)	ST 6896 3284	Stratigraphical test	BGS
Bushey Farm*	(BF)	SY 9693 8306	Hydrocarbon exploration	Gas Council
Bere Regis	(BR)	SY 8644 9563	Hydrocarbon exploration	BP
Cannings Court	(CC)	ST 7187 0734	Stratigraphical test	BGS
Chaldon Herring*	(CH)	SY 7837 8402	Hydrocarbon exploration	D'Arcy
Chilworth*	(Ch)	SU 3927 1798	Hydrocarbon exploration	Omoco
Combe Keynes*	(CK	SY 8240 8412	Hydrocarbon exploration	Gas Council
Combe Throop	(CT)	ST 7260 2350	Stratigraphical test	BGS
Cranborne	(Ce)	SU 0341 0907	Hydrocarbon exploration	BP
East Stour	(ES)	ST 8013 2297	Stratigraphical test	BGS
Encombe*	(E)	SY 9412 7832	Hydrocarbon exploration	BP
Farley*	(Fa)	SU 3927 1798	Hydrocarbon exploration	Esso
Fifehead Magdalen	(FM)	ST 7985 2100	Hydrocarbon exploration	Carless
Fir Tree Farm	(FT)	ST 7360 0710	Water supply	
Fordingbridge	(F)	SU 1876 1180	Hydrocarbon exploration	BP
Frith Farm	(Fr)	ST 7070 1750	Water supply	
Gillingham	(G)	ST 7959 2661	Water supply	
Hazelbury Bryan	(HB)	ST 7515 0810	Stratigraphical test	BGS
Holnest	(Ho)	ST 6647 0972	Stratigraphical test	Dorset CC/BGS
Hurn	(H)	SU 0999 0071	Hydrocarbon exploration	BP
Kimmeridge K5*	(K5)	SY 9043 7933	Hydrocarbon exploration	BP
Knackers Hole	(KH)	ST 7791 1188	Stratigraphical test	BGS
Langton Herring*	(LH)	SY 6060 8170	Hydrocarbon exploration	BP
Lulworth Banks*	(Lu)	SY 7851 7710	Hydrocarbon exploration	
Mappowder	(Ma)	ST 7288 0580	Hydrocarbon exploration	Carless
Marchwood*	(Mc)	SZ 3991 1118	Geothermal exploration	BGS
Martinstown	(M)	SY 6481 8702	Hydrocarbon exploration	BP
Marshwood*	(Md)	SY 3885 9880	Hydrocarbon exploration	Cangeo Warner
Nettlecombe	(N)	SY 5052 9544	Hydrocarbon exploration	BP
Norton Ferris	(NF)	ST 7820 3700	Hydrocarbon exploration	Carless
Oborne	(Ob)	ST 6520 1830	Water supply	
Poxwell*	(Po)	SY 7479 8362	Hydrocarbon exploration	D'Arcy Ex. Co.
Purse Caundle	(PC)	ST 7012 1826	Stratigraphical test	BGS
Quarleston	(Q)	ST 8594 0565	Water supply	
Ryme Intrinsica	(RI)	ST 5747 0968	Hydrocarbon exploration	Carless
Sandhills*	(Sa)	SZ 4570 9085	Hydrocarbon exploration	Gas Council
Seaborough	(Se)	ST 4348 0620	Hydrocarbon exploration	Berkeley
Shapwick	(Sk)	ST 9428 0134	Hydrocarbon exploration	BP
Shrewton	(Sh)	SU 0314 4199	Stratigraphical test	BGS
Southampton*	(So)	SZ 4156 1202	Geothermal exploration	BGS
Spetisbury	(Sp)	ST 8881 0269	Hydrocarbon exploration	Gas Council
Stoborough No. 2	(S2)	SY 9126 8659	Hydrocarbon exploration	Gas Council
Stowell	(St)	ST 6849 2180	Water supply	
Stubhampton	(S)	ST 9146 1413	Water supply	
Waddock Cross	(WC)	SY 8035 9125	Hydrocarbon exploration	Gas Council
Wareham No. 6	(W6)	SY 9059 8721	Hydrocarbon exploration	Gas Council
Wincanton	(W)	ST 7155 2840	Water supply	
Winterborne Kingston	(WK)	SY 8470 9796	Stratigraphical test	BGS
Woodlands	(Wo)	SU 0659 0627	Hydrocarbon exploration	BP
Wytch Farm No. 3	(WF)	SY 9276 8538	Hydrocarbon exploration	Gas Council
Wytch Farm X14	(WX)	SY 9804 8526	Hydrocarbon exploration	Gas Council

* indicates that the borehole lies outside the area of Figure 3.

flective lower crust probably lies beneath the southern part of the district, most of which is therefore underlain by poorly reflective lower crust of the Variscan Foreland.

The boundary between the Variscan foldbelt and Variscan Foreland is believed to correspond to the zone of Variscan frontal thrusts, which incrop below Permian cover rocks to the north near Calne (Kenolty et al., 1981; Chadwick et al., 1983). This interpretation is supported by gravity data in which the gravity effect of the Mesozoic rocks has been removed (Figure 13). The resultant northward decrease in residual Bouguer gravity anomaly values, reflecting density variations in the base-

Figure 5 North–south depth sections showing a) line drawing of principal reflectors identified from seismic sections, b) geological interpretation of reflection events (modified after Chadwick et al., 1983).

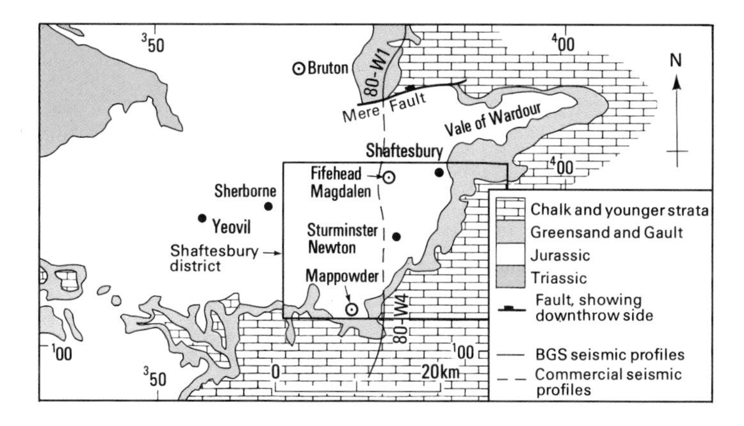

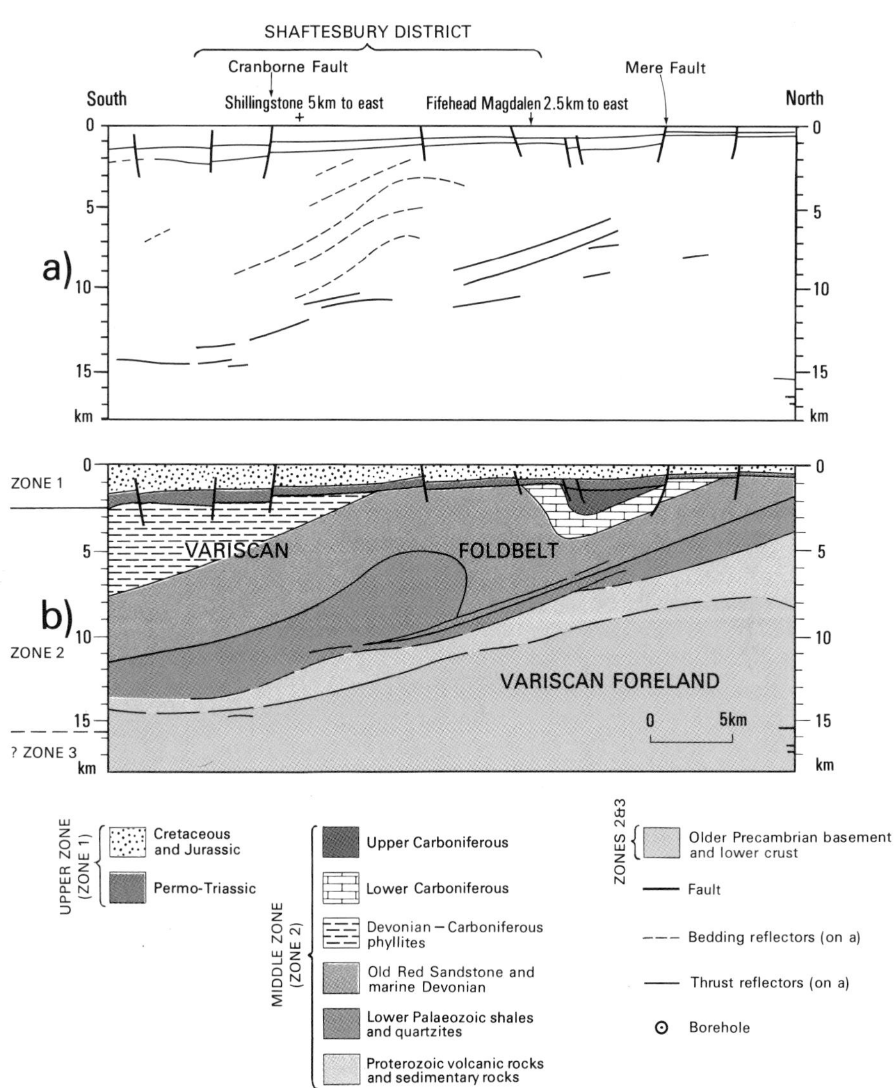

ment rocks only, is consistent with the view that higher density rocks of the Variscan foldbelt have been thrust over lower density foreland rocks.

The base of seismic zone 3 coincides with well-developed reflections at about 11 seconds TWTT (about 30 km depth), from the Mohorovičić Discontinuity (Moho) at the base of the crust.

PRE-PERMIAN (UPPER PALAEOZOIC) FORMATIONS

Borehole, geophysical and outcrop data together suggest that the rocks of the Variscan foldbelt beneath the Permian and Mesozoic cover can be divided geographically into three tracts of differing character (Figure 6).

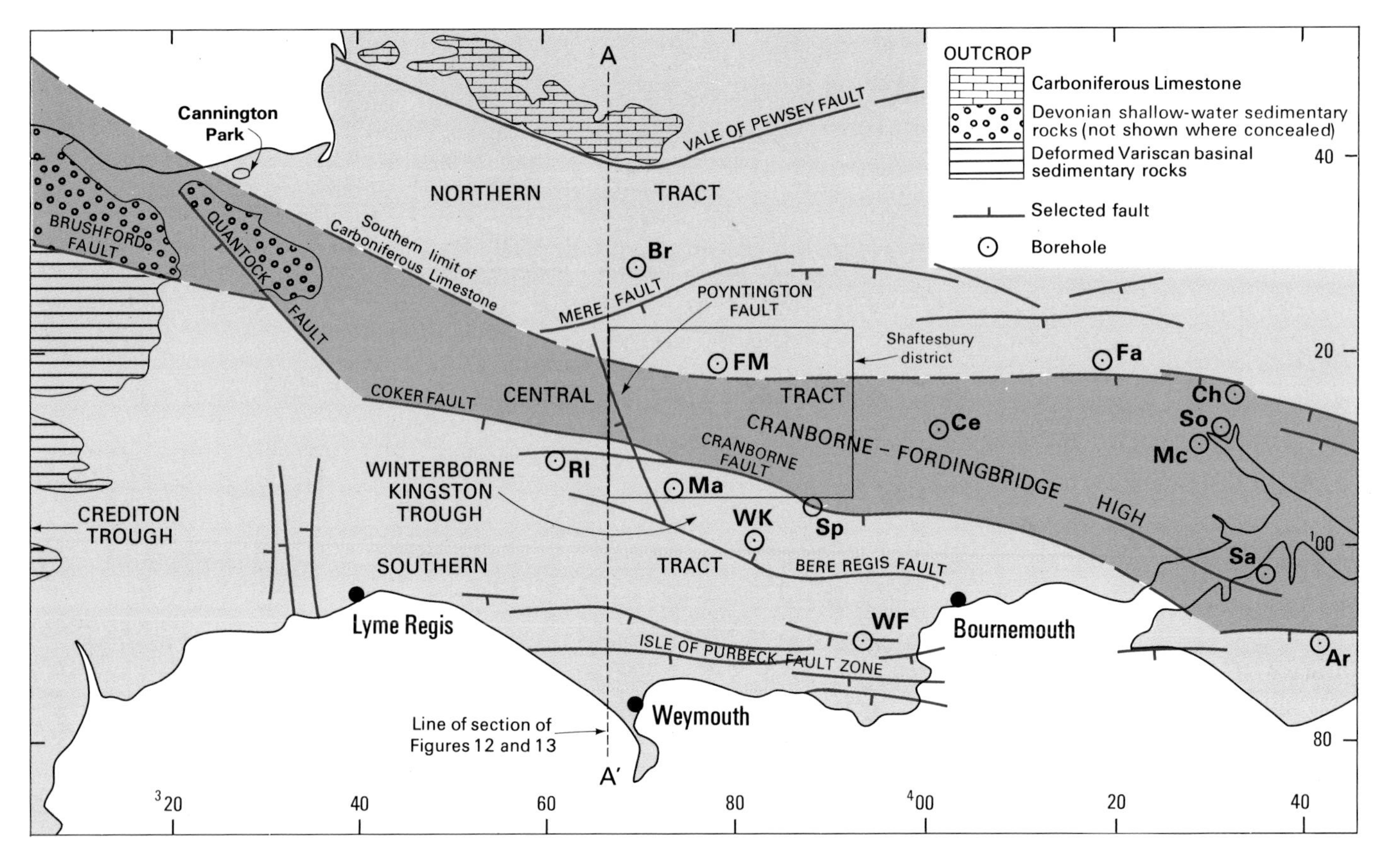

Figure 6 Subdivisions of the Variscan Foldbelt in the Shaftesbury and adjacent districts (see Table 1 for key to boreholes).

A northern tract comprises Carboniferous Limestone and low-grade metamorphic or unmetamorphosed Devonian strata of continental (Old Red Sandstone) type, and is overlain by Triassic sandstones. There, the base of zone 1 is marked by the Top Variscan Basement reflector, a high-amplitude event associated with the truncation of Palaeozoic strata (Figure 7) and the onlap of the Permo-Triassic sequences (Figure 8). Truncated reflections within the basement are interpreted as northward-dipping Carboniferous strata.

Central and southern tracts are identified partly from regional studies, geophysical data and borehole information. In those areas, the Top Variscan Basement seismic event loses its character and is difficult to pick on seismic reflection profiles. The basement rocks in the central tract comprise cleaved Devonian mudstones, sandstones and subordinate limestones transitional between continental and marine types. The southern tract comprises Devonian phyllites and slates, overlain by a thick Permo-Triassic succession; the base of this was proved at Mappowder at a depth of 2129 m below OD, at Ryme Intrinseca at 1608 m below OD and at Spetisbury at 1912 m below OD. Phyllites of presumed similar age in the Wytch Farm X14 Borehole yielded radiometric ages of 337 ± 5 to 357 ± 5 Ma, indicating phases of mid to late Devonian metamorphism (Colter and Havard, 1981).

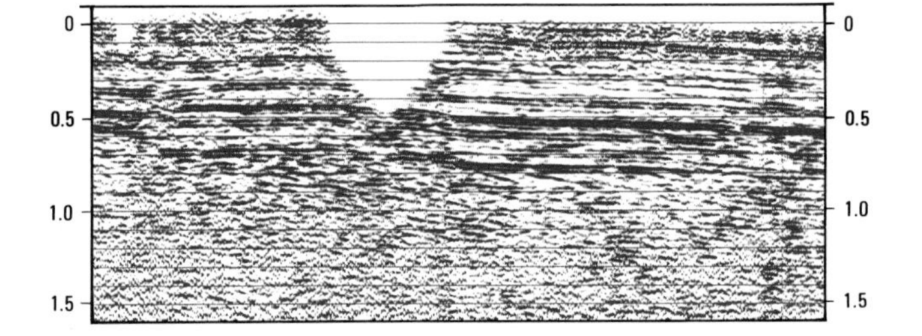

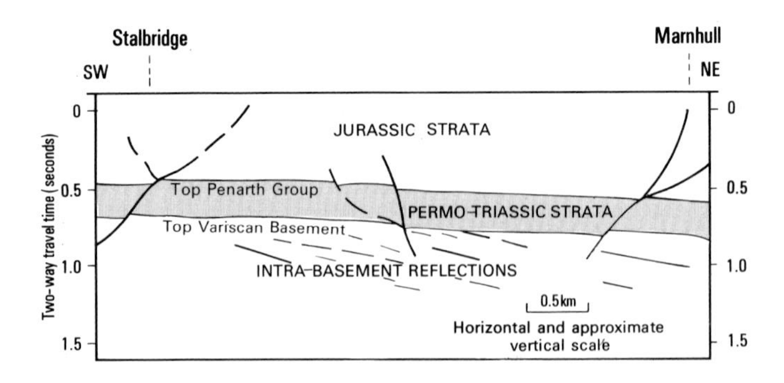

Figure 7 Seismic section showing the basal Permian and Triassic unconformity truncating northward-dipping sequences of Carboniferous strata (see Figure 4 for location).

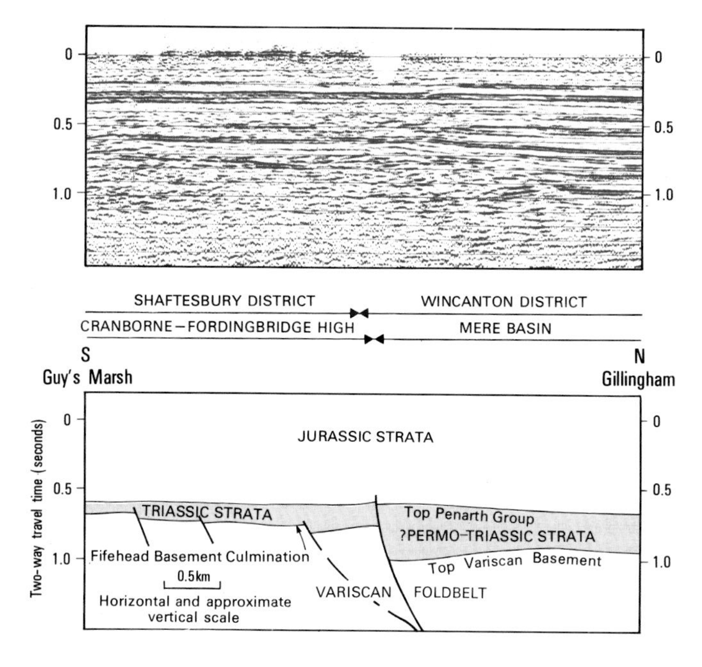

Figure 8 Seismic section showing thinning of Triassic strata onto the basement high at Fifehead Magdalen, and the southern margin of the Mere Basin (see Figure 4 for location).

Northern tract

Within this tract, strata of Devonian and Carboniferous age have been proved by the Fifehead Magdalen Borehole. The borehole was drilled in an area where basement forms a faulted dome-like structure, with a top approximately 1 km below OD (Figures 9 and 10). The Permian and Mesozoic cover sequences show attenuation above the structure (Figure 11). An aeromagnetic profile across the district (Figure 12) shows that the northern tract is associated with a low-amplitude anomaly low as far north as the Vale of Pewsey, where well-defined magnetic anomalies are interpreted to represent Silurian andesitic lavas.

Devonian rocks

In the Fifehead Magdalen Borehole, the composite log records 25 m of medium grey, patchily red-stained, hard, blocky, finely micaceous claystones between 1344 m and the terminal depth (TD) at 1369 m. These are believed to be the Shirehampton Beds, of Devonian age.

Carboniferous rocks

Carboniferous Limestone occurs at depth in the north of the district. It cannot be distinguished satisfactorily from the Shirehampton Beds on seismic reflection data, so that its subcrop position cannot be accurately determined. It probably extends northwards and westwards from Fifehead Magdalen to the sheet margins. The thickness of the Carboniferous Limestone was proved to be 352 m (992 to 1344 m) in the Fifehead Magdalen Borehole; 92 m of younger Carboniferous Limestone were proved in the Bruton Borehole (see below).

The composite log of the Fifehead Magdalen Borehole shows an ascending sequence of Lower Limestone

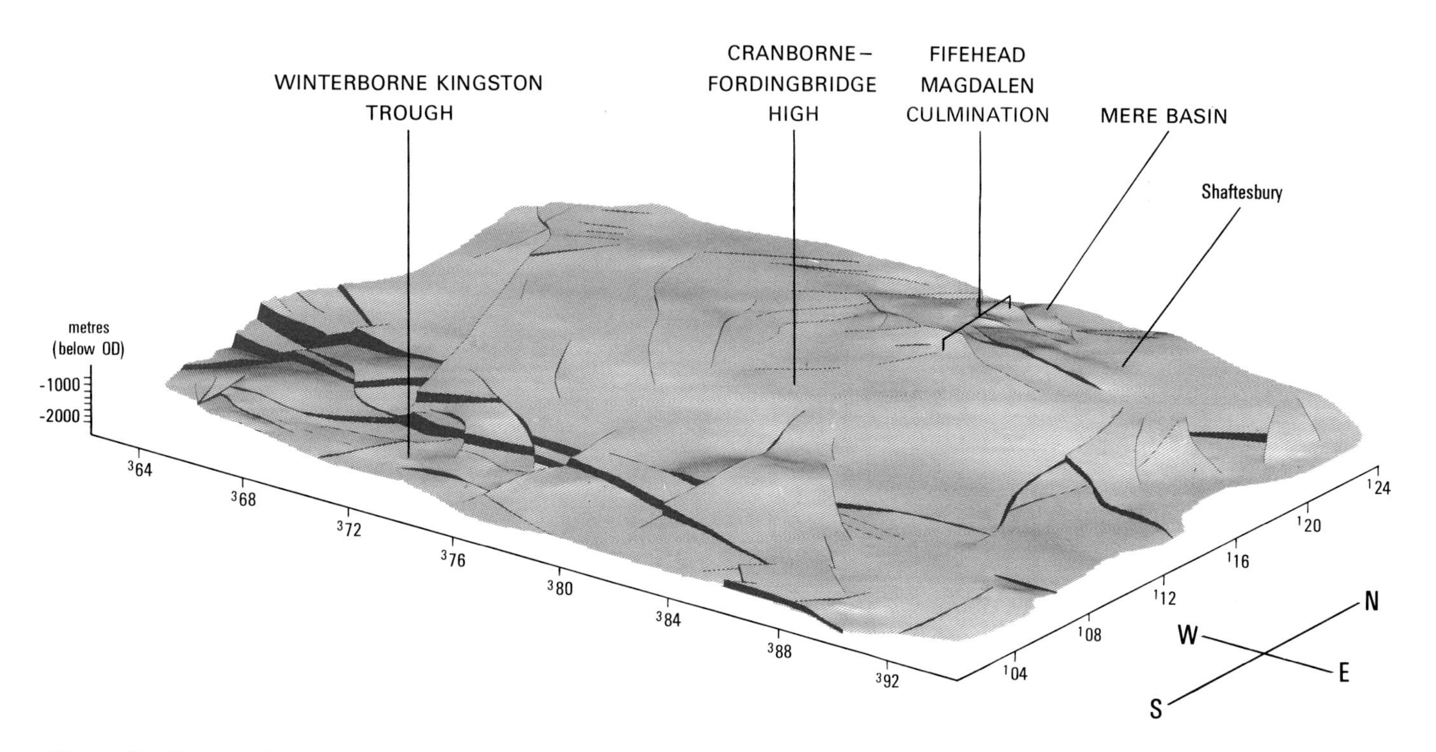

Figure 9 Perspective view of the top surface of the Variscan Basement of the district, viewed from the south-east.

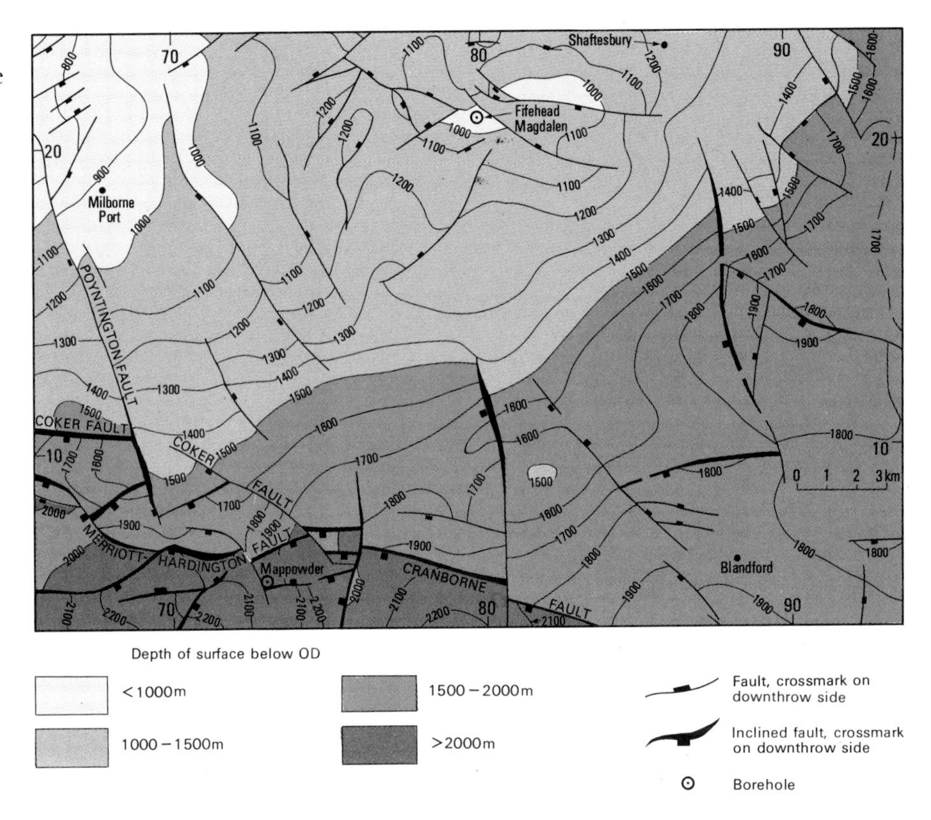

Figure 10 Contour map of the top surface of the Variscan Basement in the Shaftesbury district (contours at 100 m intervals below OD).

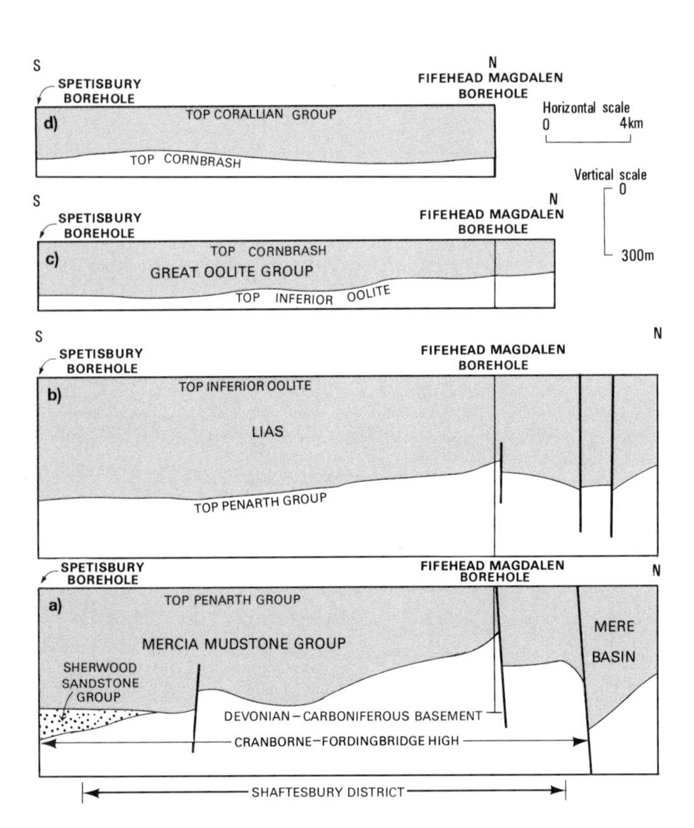

Figure 11 Restored cross-sections across the district, showing thickness changes in a) Triassic rocks, b) Lias, c) Great Oolite Group, and d) Kellaways Formation, Oxford Clay and Corallian Group (see Figures 4 and 15 for line of section).

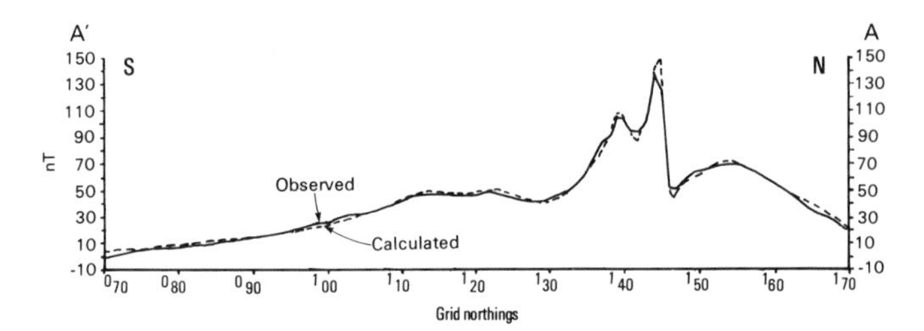

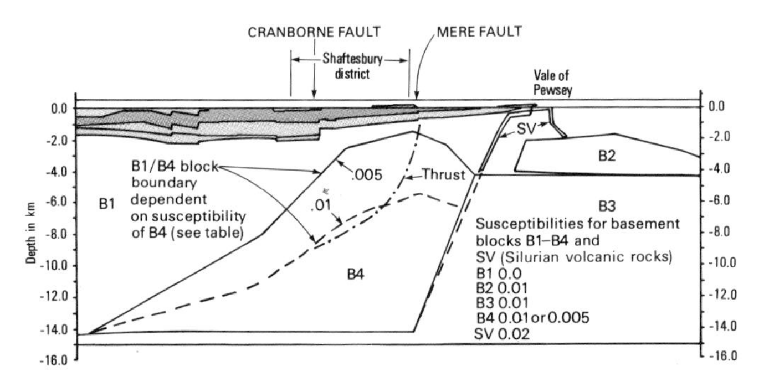

Figure 12 Aeromagnetic profile (AA') and calculated profile for the model shown (for location of section see Figure 6)

Shale, Blackrock Group and possible Clifton Down Group. The sequence ranges in age from early to late Tournaisian. The Lower Limestone Shale, 137 m thick (1207 to 1344 m), consists of grey, locally red, purplish grey and greenish grey, hard claystones and metaclaystones, with subordinate beds of patchily dolomitic limestone. The lower two-thirds (approximately 100 m) of the succeeding Blackrock Group (1027–1207 m) consist dominantly of dolomitic limestone, and the upper third consists of calcium carbonate limestone. The limestones referred to the Clifton Down Group (992 to 1027 m) are mainly crystalline carbonate, with some algal micrites.

In the Bruton Borehole, a sequence of younger strata was proved between 293 m and the TD at 385 m. The dominant lithologies are hard, pale to medium grey, patchily red and green stained, calcitic mudstones and siltstones with scattered bioclastic debris including brachiopods, corals and crinoids. Beds of dolomite up to 5.1 m thick, and red and green-mottled mudstones, siltstones and fine-grained sandstones up to 9.2 m thick also occur. The foraminiferida *Archaediscus pseudomoelleri, Nodosarchaediscus* sp. and *Rectodiscus* sp., and the coral *Palaeosmilia* sp., indicate strata of either Arundian or Asbian age (A R E Strank and M Mitchell *in* Holloway and Chadwick, 1984).

Central tract

This tract (Figure 6) is associated with an elongate, positive magnetic anomaly, which can be traced east-south-eastwards from Exmoor and the Quantock Hills towards the north-western corner of the district (Figure 14). In the Exmoor area, there is evidence that the magnetic source rocks occur in the cores of regional-scale anticlines, and magnetic continuity suggests that the fold structures extend east into the present district (see also Figure 5). The eastern end of the magnetic anomaly swings into an east-north-easterly trend in the north-west corner of the district and is possibly truncated by faulting.

Boreholes east of the district at Cranborne [SU 0341 0907], and in the Southampton–Isle of Wight area, define a 15 to 30 km wide belt in which Permian rocks rest on over 800 m of sandstones, siltstones and sandy mudstones, red-stained in part, commonly recrystallised and exhibiting a weakly developed cleavage. A sparse, poorly preserved fauna from a pre-Permian basement sequence over 370 m thick at Cranborne suggests a mid to late Devonian age.

Southern tract

Bouguer gravity anomaly values increase southwards across the district into the southern tract, where a low magnetic anomaly is observed. The latter indicates a source at considerable depth or a lack of strongly magnetised rocks.

The Mappowder Borehole proved thick Permian and Triassic sequences above 49 m of greyish green, cleaved phyllites veined by white crystalline quartz of probable Devonian age (depth from 2231 to 2280 m). Slate and phyllite of probable Devonian age have also been proved in the Spetisbury, Ryme Intrinseca and Wytch Farm X14 boreholes (Figure 3; Table 1). The phyllite in the latter borehole yielded radiometric ages of 337 ± 5 to 357 ± 5 Ma (mid to late Devonian; Colter and Havard, 1980).

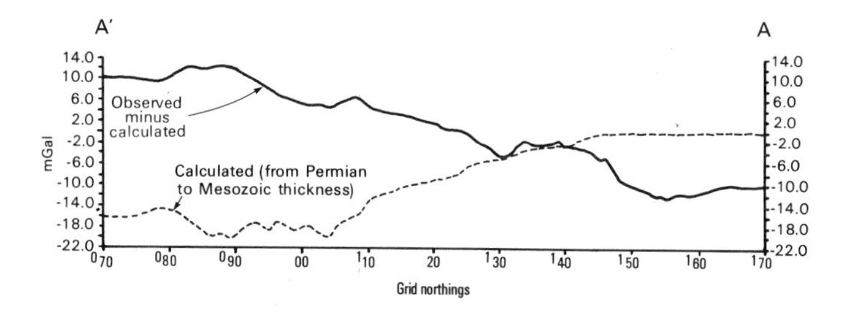

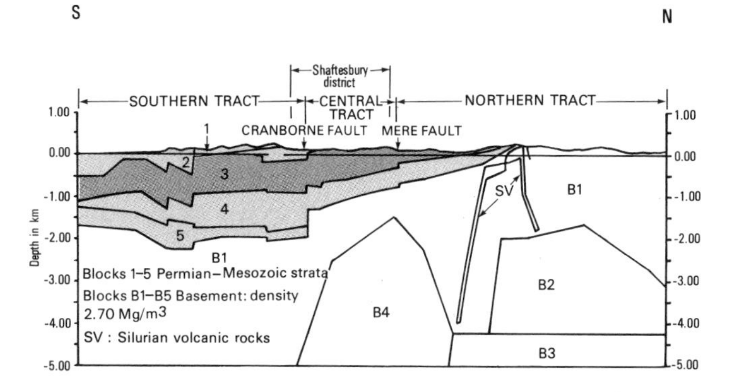

Figure 13 Bouguer anomaly profile (AA') obtained by stripping the anomaly due to the Permian-Mesozoic strata from the observed profile (for location of profile see Figure 6).

CONCEALED PERMIAN AND MESOZOIC SEQUENCE

The uppermost seismic reflection zone of the crust (zone 1) shows well-developed, subhorizontal, high-amplitude reflections, representing a cover sequence of post-Variscan age.

The sub-Permian surface shows major structural features which affect the nature and distribution of some of the cover sequence. In the south of the district, east–west-trending faults form the northern boundary of a thick Permo-Triassic basin, the Winterborne Kingston Trough (Figure 10); its southern boundary is formed by the South Dorset High (Figure 4). North of the trough, Permian deposits are absent, and an attenuated Triassic sequence overlies the Cranborne–Fordingbridge High, the northern limit of which occurs just north of the district and forms the southern margin of the Mere Basin.

The Permian and Triassic sequences give rise to variable-amplitude seismic relections, which show onlap from both the north (Figure 8) and south onto the Cranborne–Fordingbridge High. The Permian and Triassic deposits are over 1200 m thick in the Winterborne Kingston Trough, but thin rapidly across the Cranborne–Fordingbridge High, such that an incomplete Triassic succession is only 182 m thick in the Fifehead Magdalen Borehole (Figure 15). The thinning over the high is not uniform, because faulting caused local thickening of the Triassic strata (Figure 8). The sequence thickens northwards over a growth fault beyond the district into the Mere Basin (Figure 8). The reflection associated with the top of the Penarth Group is continuous across both the 'high' and the synsedimentary faults, showing that, after the Triassic Period, these major structural features had much less effect (Figure 8). Thus, the Jurassic and Cretaceous deposits extend across the former Cranborne–Fordingbridge High and Winterborne Kingston Trough with, in general, only a gradual southerly increase in thickness. The Corallian Beds are an exception, having their thickest development approximately over the Winterborne Kingston Trough. The Mere Basin (Figure 4), a longer-lasting, more subdued structure, was most active during the Permian and Triassic periods, but also affected sedimentation throughout much of the Jurassic.

The local basement culminations or 'highs' recognised in the Fifehead Magdalen (Figure 8) and Bruton (Holloway and Chadwick, 1984) areas appear to be located on footwall blocks adjacent to major synsedimentary faults.

?Permian

Deposits of possible Permian age are present only beneath the south-west of the district in the Winterborne Kingston Trough, where the Wytch Farm Breccias rest with angular unconformity on Variscan Basement rocks. The breccias are succeeded by a thick sequence of mudstones. These two units traditionally have been assigned to the Permian System because of their lithological similarity to the Permian sequence at outcrop in Devon (Colter and Havard, 1981). However, as the Aylesbeare Mudstone Group in Devon is now placed in the Triassic (Warrington and Scrivener, 1990), it is possible that the Wytch Farm Breccias are also of Triassic age.

Wytch Farm Breccias

The presumed correlative of the Wytch Farm Breccias in the Mappowder Borehole is 16 m thick and consists of clasts of quartz and phyllite set in a matrix of mottled red and yellowish brown to light greyish green, calcareous, sandy siltstone. In the Ryme Intrinseca Borehole, in

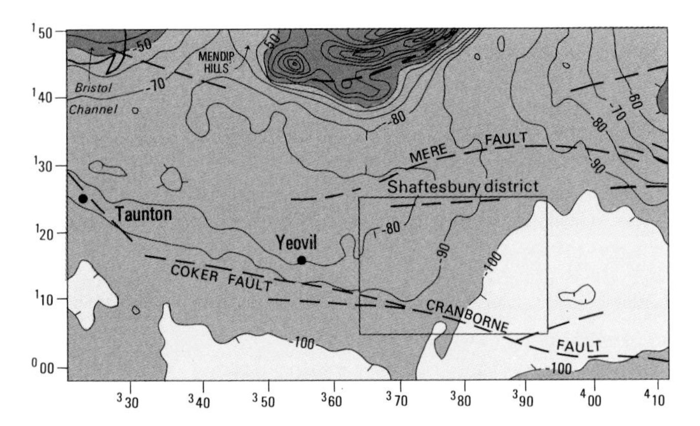

Figure 14 Reduced-to-pole aeromagnetic map of Shaftesbury and adjacent districts, with contours at 10 nT intervals. Locations of major growth faults (at level of basement surface).

the Yeovil district, a basal breccia to the Permian and Triassic sequence, some 40 m thick, similarly consists of clasts of phyllite and quartz, but set in a mudstone matrix.

Triassic

The Triassic succession is divided into four groups, in ascending sequence the Aylesbeare Mudstone, Sherwood Sandstone, Mercia Mudstone and Penarth groups. In the Mappowder Borehole, sited on the northern margin of the Winterborne Kingston Trough, the Triassic sedimentary rocks are 1134 m thick, but thin northwards across the Cranborne–Fordingbridge High to 182 m in the Fifehead Magdalen Borehole. Much of this thinning is due to northward overlap, such that, except in the south, the Mercia Mudstone and Penarth groups are probably the only Triassic deposits on the Cranborne–Fordingbridge High.

Aylesbeare Mudstone Group

The Aylesbeare Mudstone consists dominantly of reddish brown, calcareous, silty and sandy mudstones. Thin beds of orange and reddish brown, fine- to medium-grained, slightly micaceous, calcareous, quartz sandstone also occur. Some sandstones form the bases of fining-upward sequences. In the Mappowder Borehole, the Aylesbeare Mudstone is 183.8 m thick; the group thickens southwards into the Winterborne Kingston Trough, and is about 350 m on the district margin, and over 575 m thick in the Winterborne Kingston Borehole. In the Ryme Intrinseca Borehole, the Aylesbeare Mudstone consists of about 20m of silty mudstone with thin beds of fine-grained, well-sorted sandstone. Seismic and borehole evidence suggests that the northern limit of the group is demarcated by the faults which define the northern limit of the Winterborne Kingston Trough (Figures 4 and 15).

Where fully developed in the Wytch Farm area, the deposits of the Wytch Farm Breccias and Aylesbeare Mudstone Group together form three upward-fining cycles (Bristow et al., 1991) deposited in alluvial fan and playa-sabkha lake continental environments (Smith et al., 1974).

In the Mappowder Borehole, the basal unit consists of about 46 m of reddish brown, firm, blocky, calcareous, slightly micaceous and very slightly sandy mudstone; it produces very high counts on the gamma-ray log, although small-scale, coarsening-upward sequences are evident. At the base of the succeeding unit, 71 m thick, is an orange-brown, friable to moderately well-indurated, fine- to medium-grained quartz sandstone, with some minor feldspar and a calcite cement. The strongly serrated gamma-ray log in the upper part of this unit reflects an interbedded sandstone/mudstone sequence with an overall fining upwards. A third unit, 23 m thick, also commences with fine- to very fine-grained sandstone and fines upwards. The highest unit, 43 m thick, consists dominantly of mudstone, but the serrated nature of the gamma-ray log suggest that the mudstones are interbedded with thin siltstones or clayey sandstones.

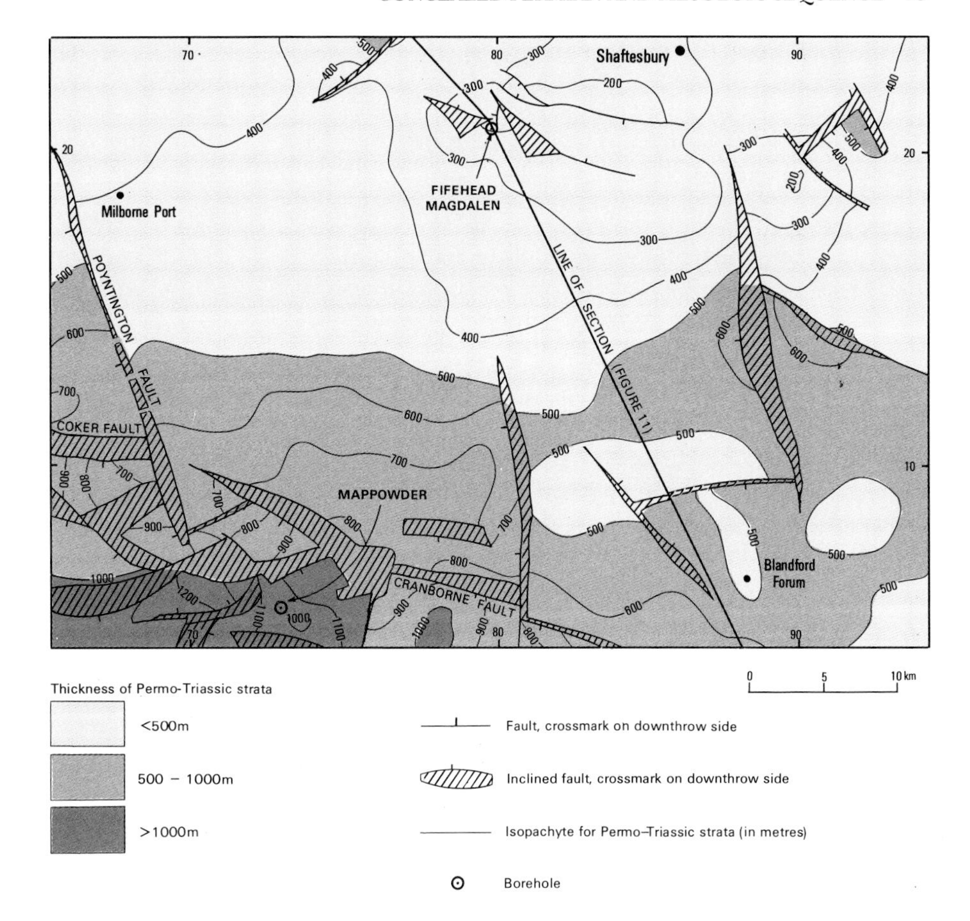

Figure 15 Isopachyte map for the Permian and Triassic strata of the Shaftesbury district

Sherwood Sandstone Group

The Sherwood Sandstone Group extends across the northern boundary fault of the Winterborne Kingston Trough and onlaps on to the southern margin of the Cranborne–Fordingbridge High. As a consequence, it has a more extensive subcrop than that of the Aylesbeare Mudstone. It is thicker in the Mappowder (209 m) and Ryme Intrinseca (207 m) boreholes than in the Winterborne Kingston Borehole (170 m); there are thinner sequences in the Spetisbury and Cranborne boreholes (126 and 115 m respectively).

The Sherwood Sandstone Group is the lower and more productive of the two oil-bearing sandstone reservoirs in the Wytch Farm Oilfield. There, the sandstone rests with angular unconformity on the Aylesbeare Mudstone Group (Colter and Havard, 1981) and passes up into the Mercia Mudstone Group.

At outcrop in Devon and in some boreholes in the Wessex Basin, the Sherwood Sandstone Group can be divided into two formations, the Budleigh Salterton Pebble Beds and the overlying Otter Sandstone. Only the latter has been recognised in the Shaftesbury district. Two divisions of the Otter Sandstone, separated by bedded anhydrite, occur in the Nettlecombe and Ryme Intrinseca boreholes.

In the Winterborne Kingston Borehole, Lott and Strong (1982) showed that the cored sequence of the Otter Sandstone comprises fining-upward, fluvial, clastic sedimentary cycles about 3 m thick, nested within larger cycles, each about 30 m thick. A general fining upwards, accompanied by an upward increase in clay content, is evident. Each cycle commonly has a conglomeratic base, generally a few centimetres thick, but up to 1 m, succeeded by cross-bedded sandstones, which commonly fine up into laminated siltstones or thin mudstones. The clasts within the conglomerates are mostly intraformational and include fragments of mudstone and cornstone; some quartz pebbles occur.

The Otter Sandstone is thought to have formed as a floodplain deposit, the small-scale cycles of which comprise intraformational conglomerates (representing residual, bedload deposits within river channels) overlain by sandstones (formed as accretionary point bars), and then by siltstones and mudstones (formed as temporary lakes and back levée deposits, which were occasionally exposed). Allen and Holloway (1984) suggest that the floodplain may have developed distally from fan deposits, which were banked against the growing Cranborne Fault.

Mercia Mudstone Group

Sedimentation within the Permo-Triassic depocentre continued with the Mercia Mudstone Group and occurred over a wider area; the group underlies the whole of the district. It shows a marked northerly thinning across the Cranborne–Fordingbridge High (Figure 15) from over 600 m at Winterborne Kingston (627.8 m), Mappowder (657 m) and Ryme Intrinseca (c.622 m), to 297 m at Cranborne and to 147 m at Fifehead Magdalen. North of Fifehead Magdalen, the group thickens northwards across a synsedimentary fault into the Mere Basin (Figure 11).

The Mercia Mudstone Group is composed dominantly of reddish brown, silty and sandy mudstones which, at some levels, are calcareous or dolomitic. Minor interbeds of reddish brown, calcareous siltstone, and colourless very fine- to medium-grained sandstone with well-rounded grains, also occur. Beds of halite, totalling 135 to 170 m, were proved near the middle of the sequence at Winterborne Kingston, Mappowder and Ryme Intrinseca. Higher in the sequence, a well-developed anhydrite bed forms a useful geological marker across the district. In the Wytch Farm area, the Mercia Mudstone Group is subdivided into six lithostratigraphical units (A–F, see Bristow et al., 1991, fig. 5). A similar sequence can be recognised in the Winterborne Kingston Borehole, but the divisions are not so clearly defined.

Units A and B, which cannot be distinguished from one another in the Winterborne Kingston and Mappowder boreholes, consist of typical reddish brown mudstones, but include silty and and sandy lenses in their lower part. Their combined thicknesses at Winterborne Kingston and Mappowder are 148 m and 159 m respectively. In the Fifehead Magdalen Borehole, units A and B are probably absent. Unit C commences with a thick halite sequence and passes up into a series of mudstones; thicknesses of the unit are 280 m and 246 m at Winterborne Kingston and Mappowder respectively. At Fifehead Magdalen, the halite sequence is missing and only 50 m of strata intervene between the base of the Mercia Mudstone Group and the base of Unit D. No halite is present in the 14.5 m of Mercia Mudstone Group at Bruton. To the east of the district, no halite is present in the Cranborne Borehole, but the top of Unit C is probably marked by thin beds of anhydrite present in a 50 m interval in the middle of the group.

Unit D has persistent dolomitic and anhydritic sandstone beds at its base; it is absent in the Bruton Borehole. These beds give rise to prominent low gamma-ray and high velocity spikes in the geophysical-log profiles. The beds pass up into a sequence of increasing mud content.

Unit E is typically a mudstone giving high gamma-ray and low velocity values. The mudstones pass gradually upwards into Unit F, which comprises alternating grey to green, more or less calcareous siltstones and mudstones of low gamma-ray and velocity values. This unit constitutes the Blue Anchor Formation, formerly the Tea Green Marl and Grey Marl of western England. This unit is 67.2 m thick at Winterborne Kingston, 46.33 m at Spetisbury and 60.35 m at Mappowder; it thins to 12.19 m at Fifehead Magdalen and 6.6 m at Bruton.

The dominance of mudstones in the Mercia Mudstone Group ensures that the geophysical-log signature is characteristically finely serrated, although the halite and anhydrite beds give a typically low gamma-ray response.

The deposits of the group are thought to have been laid down as playa or inland sabkha deposits, with intermittent marine connections which became more persistent towards the close of the deposition of the group (Allen and Holloway, 1984).

In the Wessex Basin, units C and D fall within the Carnian Stage, and Unit F in the Rhaetian Stage (Warrington et al., 1980). Thus it is likely that units A and B fall within the Ladinian Stage, and Unit E in the Norian Stage.

PENARTH GROUP

The Penarth Group is widespread in the district and comprises the Westbury Formation overlain by the Lilstock Formation. The upper part of the Lilstock Formation, the Langport Member, forms a very prominent reflector on seismic records, and allows the top of the Penarth Group to be plotted with confidence. Beds of the group were probably deposited in marginal marine and lagoonal environments and mark the close of the continental, red-bed sedimentation that dominated the Permo-Triassic sequence. The Penarth Group shows a general northward thinning across the district. It is 28.25 m thick at Winterborne Kingston, 25.3 m at Mappowder and 25.35 m at Spetisbury, thinning to 20 m at Cranborne, 18.6 m at Fifehead Magdalen and 10 m at Bruton.

The **Westbury Formation** typically consists of dark grey mudstones. In the Winterborne Kingston Borehole, the formation commences with a fine- to medium-grained, moderate to poorly sorted, variably argillaceous, feldspathic, glauconitic sandstone. The overlying dark grey, fissile mudstones are variably silty and are calcareous and pyritic (Knox, 1982b). Thin shelly limestones with marine bivalves also occur (Ivimey-Cook, 1982).

At Winterborne Kingston, the 2.66 m-thick **Cotham Member** of the **Lilstock Formation** comprises silty, slightly to highly calcareous mudstones with marine bivalves at its base, which pass up, with increasing calcium carbonate content, into argillaceous, finely sparry or micritic limestones. Thin lenticular silty and sandy layers are also present (Knox, 1982b). The succeeding greenish, very calcareous mudstones contain separated valves of the branchiopod *Euestheria minuta* in tightly packed aggregates, together with plant fragments including *Naiadita*. At the top, a thin, poorly calcareous mudstone yields *Chlamys valoniensis* and *Modiolus* sp. (Ivimey-Cook, 1982; Knox, 1982b).

The **Langport Member** (formerly known as the White Lias) at Winterborne Kingston comprises 12.75 m of pale grey, finely crystalline, slightly argillaceous, sparsely fossiliferous limestones with wisps and thin partings of dark grey mudstone. Styolites are present throughout the limestones and generally cut across the bedding (Knox, 1982b).

Lower Jurassic

LIAS GROUP

The Lower, Middle and Upper Lias may be traced throughout the subsurface of the district, although the boundaries between them are not well defined lithologically and they are difficult to trace in seismic sections.

The Lias ranges in thickness from about 300 to 647 m within the district and adjacent areas, being thickest in the south (578 m at Ryme Intrinsica, 647 m at Winterborne Kingston, 546 m at Spetisbury, 528 m at Bere Regis and 650 m at Mappowder) and thinning to 395 m at Fifehead Magdalen, 270 m at Cranborne and 224 m at Bruton north of the district. However, the thinning is not uniform, and local thickness changes occur within the district.

The Lias was deposited in a marine shelf environment, ranging from outer to inner shelf regions, which were subject to periodic shoaling, giving rise locally to marine barrier sands such as the Bridport Sands (Knox et al., 1982).

The characteristic geophysical-log signatures in boreholes enables the Lias to be correlated regionally (Whittaker et al., 1985) and with the type sections on the Dorset coast (Cope et al., 1980a). The base of the Jurassic System, taken at the lowest occurrence of the ammonite *Psiloceras planorbis*, probably lies about 5 m above the base of the Lias in the district.

Lower Lias

The Lower Lias comprises a series of alternating medium to dark grey, locally silty and calcareous mudstones with interbeds of thin, pale to medium grey, hard, microcrystalline, more-or-less argillaceous limestone. These alternations in lithology give rise to the five lithostratigraphical divisions on the Dorset coast, in ascending sequence: **Blue Lias**, **Shales-with-Beef**, **Black Ven Marls**, **Belemnite Marls** and **Green Ammonite Beds**, which are widespread at both outcrop beyond the district and subcrop beneath it.

Dr I E Penn has examined the geophysical logs of the Fifehead Magdalen and Mappowder boreholes and correlated the Lias sequences with that exposed on the Dorset coast. The member thicknesses deduced are shown in the Geological Sequence (front cover).

Middle Lias

The lower part of the Middle Lias is composed of pale grey, slightly silty, finely micaceous, locally carbonaceous and calcareous mudstones, silts and muddy fine-grained sands, interbedded with thin siltstones which become more prominent upwards; these beds are 45.1 m thick in Fifehead Magdalen Borehole. The upper part of the Middle Lias, 71 m thick in this borehole, consists dominantly of sand and is known as the **Pennard Sands** (Wilson et al., 1958). It is capped by a limestone forming the **Marlstone Rock Bed**, which is about 3 m thick in the Winterborne Kingston Borehole; there, pale grey crinoidal limestone at the base of this unit rests sharply on the underlying greyish green sandy silts.

An intraformational boundary within the Middle Lias is evident in some seismic sections south of the district; it has been intersected in the Mappowder Borehole, where it appears to lie at the base of the Pennard Sands. In general, the sequence boundary is very subtle, with no apparent angular discordance with the underlying beds, so that the contact is probably a non-sequence. South of the district, however, Middle Lias strata are preserved in tilted fault blocks beneath an angular unconformity, above which lie the Pennard Sands seen at Mappowder. This disconformity probably represents a response to local tectonic events.

Upper Lias

The Upper Lias consists of the upper part of the Junction Bed overlain by the Down Cliff Clay and then by the Bridport Sands. The **Down Cliff Clay**, the youngest of the totally concealed formations, is composed of medium grey, soft, calcareous, locally silty and micaceous mudstones, which become increasingly silty and micaceous upwards. Towards the top, pale grey, firm to friable, very fine-grained, moderately sorted, calcareous sandstones occur. The Down Cliff Clay is thickest in the Mappowder and Spetisbury boreholes (120.7 and 118.6 m respectively); it thins southwards to 85.1 m at Winterborne Kingston and 46 m at Bere Regis, and northwards to 44.2 m at Fifehead Magdalen. It appears to be absent in the Cranborne Borehole.

THREE

Lower and Middle Jurassic

Lower and Middle Jurassic[1] strata crop out in the northwest of the district and, except for the Bridport Sands at the base of the exposed sequence, comprise an alternating succession of limestones and mudstones. The formations and members of the exposed Lower and Middle Jurassic are shown in Table 2.

LOWER JURASSIC

The only Lower Jurassic rocks at outcrop in the Shaftesbury district are the Bridport Sands at the top of the Upper Lias (see Figure 19).

BRIDPORT SANDS

Fine-grained silty sands and calcareous sandstones, principally of mid to late Toarcian age, with some of earliest Aalenian age, crop out in an arc through western Dorset and south Somerset. Within this tract, they have been called Upper Lias Sands or Sandstone (Wright, 1856, 1860), Yeovil Sands (Hudleston in Buckman, 1879), Yeovil and Bridport Sands (Woodward, 1888), Midford Sands (Woodward, 1893), Bridport Sands (Woodward, 1893), Yeovil or Briport Sands (Buckman, 1889) and Bridport and Yeovil Sands (Wilson et al., 1958). Arkell (1933), however, argued that the term Yeovil Sands is not synonymous with Bridport Sands, but is equivalent to the Bridport Sands plus the underlying Down Cliff Clay. In this account, the term Bridport Sands is used for the sands that occur between the Down Cliff Clay and the Inferior Oolite.

The upper half of the approximately 60 m thick Bridport Sands is seen at outcrop in the north-western part of the district where it occurs as a series of partially fault-bounded outcrops around Poyntington Hill [654 200] (Figure 19). Characteristic yellow- and orange-weathering, friable, silty, fine-grained sandstones are in rhythmic alternation with more calcareous sandstone doggers; sandy bioclastic limestones crop out locally.

The lithology of the Bridport Sands at outcrop has been studied by Davies (1967, 1969) and Davies et al. (1971), whilst cored material in the Winterborne Kingston Borehole has been described by Knox et al. (1982)

1 To enable direct comparisons to be made with earlier geological literature and the published geological maps, the Middle Jurassic is used in this memoir *sensu* Arkell (1947), and includes only the Aalenian, Bajocian and Bathonian stages. In these, zones and subzones are treated as biozones, and the nominal taxa are recorded in italics.

Table 2 The exposed Lower and Middle Jurassic formations of the Shaftesbury district.

	Stage	Group	Formation	Member
UPPER JURASSIC	Callovian	Great Oolite	Cornbrash	Upper
MIDDLE JURASSIC	Bathonian			Lower
			Forest Marble	
			Frome Clay	
				Wattonensis Beds
			Fuller's Earth	Upper Fuller's Earth
				Fuller's Earth Rock
				Lower Fuller's Earth
	Bajocian		Inferior Oolite	Crackment Limestones
				Rubbly Beds
				Sherborne Building Stone
				Miller's Hill Beds
	Aalenian			Corton Denham Beds
LOWER JURASSIC	Toarcian (pars)	Lias (pars)	Bridport Sands	

and in the Purse Caundle Borehole by Barton et al. (1993). Three principal rock types occur:

1 The main components, in units up to 2 m thick, are olive-grey to greyish green, poorly cemented, micaceous silts and very fine-grained to fine-grained sandstones, with a median grain size of very fine-grained sand. Bioturbation is common, and burrow margins are commonly defined by dark grey silty clay; non-bioturbated beds show faint lamination. Where bioturbated, the sands are poorly sorted; where laminated, they are well sorted. The sands are composed of angular to very angular quartz (65 to 75 per cent), mica (up to 10 per cent), feldspar (1 to 25 per cent) and a variable amount of bioclastic debris, set in a variably clayey, micritic and fine-grained, sparry calcite matrix with sericitic patches. The sands are not calcareous at outcrop.

2 Cemented sandstones, up to 0.5 m thick, occur sporadically within the succession. They consist of abundant shell fragments and scattered fine sand grains set in matrix of fine-grained carbonate cement. Both matrix and bioclastic fragments are predominantly of ferroan calcite, but some shell fragments have resisted replacement. Small-scale cross- and festoon-bedding has been recorded locally; its absence in most sections may be due to bioturbation.

3 Bioclastic limestones generally form a minor part of the sequence and consist primarily of broken, abraded and bored bivalve fragments, with a very small component of chamositic or phosphatised ooliths and fine-grained sand. An exceptional development of these limestones is the 30 m thick Ham Hill Stone north of Crewkerne in the Yeovil district.

Thin beds of chamositic mudstone and sandy mudstone were noted in the Winterborne Kingston Borehole (Knox et al., 1982), where the clay-mineral assemblages are dominated by iron-rich chlorite, with minor amounts of mica and kaolinite. The proportions are reversed in approximately the lower third of the formation.

Thickness data for the Bridport Sands of the region, mainly derived from oil-company exploration boreholes, are shown in Figure 16. In the Shaftesbury district, the formation ranges between about 44 m in the Frith Borehole [707 175] (Whitaker and Edwards, 1926) and about 80 m in the south-east of the district. Northwards, it thins towards the Mendip Hills (62.8 m at Bruton, 19.8 m at Norton Ferris), and the formation is absent on the northern margin of the Glastonbury district. South of the Shaftesbury district, the thickness increases to 102 m at Winterborne Kingston and 122 m at Bere Regis.

The base of the Bridport Sands is taken at the incoming of silt and sand above the mudstones of the Down Cliff Clay. In the Winterborne Kingston Borehole, the junction of the Bridport Sands and Down Cliff Clay is transitional and marked by an upward increase in the number of fine-grained sand laminae. Nevertheless, there is a distinct break between horizontally laminated silty shales and the overlying muddy bioturbated sandstones (Knox et al., 1982).

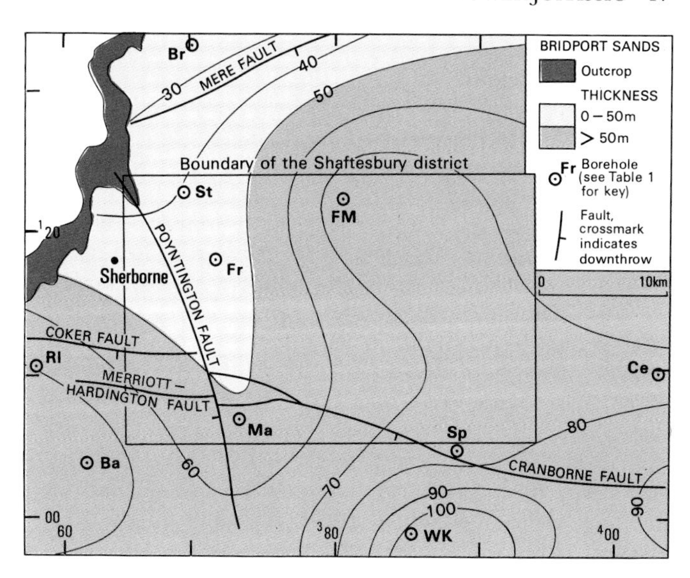

Figure 16 Isopachyte map for the Bridport Sands of the Shaftesbury and surrounding districts (contours at 10 m intervals).

In the Winterborne Kingston Borehole, the basal sands of the Bridport Sands are of late *levesquei* Subzone age and rest on silts and mudstones of the Down Cliff Clay, of *levesquei* and *dispansum* Subzone ages (Ivimey-Cook, 1982). Data from other parts of the sands near Yeovil (Buckman, 1889; Richardson, 1930 —summarised in Wilson et al., 1958) indicate the *levesquei* Subzone at Babylon Hill [c.582 161] and the *moorei* Subzone at Ham Hill [c.482 165] and in the Oborne Borehole [6522 1764]. An *aalensis* Subzone age at the top of the Bridport Sands is proved by *Pleydellia* in the Purse Caundle Borehole (Barton et al., 1993); the same ammonite was found in a dark grey limestone, here regarded as the basal part of the Inferior Oolite, in the Stowell Borehole. Taken together, these data show the top of the Bridport Sands to be close to the Toarcian/Aalenian junction throughout the Shaftesbury district.

Farther west, in the Yeovil district, at Bradford Abbas, North Coker, near Crewkerne, and Chiselborough, *aalensis* Subzone beds have been cut out and the Inferior Oolite rests unconformably on sands and limestones of *moorei* Subzone age (Wilson et al., 1958, pp.71, 92).

Conditions of deposition

The Bridport Sands form part of a coarsening- and shallowing-upward sequence from the Down Cliff Clay, through the Bridport Sands into the Inferior Oolite (Holloway, 1985; Bryant et al., 1988). Individual coarsening-upward cycles have been recognised in the Bridport Sands of the Winterborne Kingston Borehole (Knox et al., 1982).

A wide variety of depositional models has been proposed for the Bridport Sands: as sandbanks in shallow water, with the clays and silts of the Down Cliff Clay deposited in deeper water to either side (Kellaway and Welch, 1948); as a migrating barrier bar, with the postulated bar trending east–west with a fore-bar facies to the south and back-bar facies to the north (Davies, 1967, 1969; Davies et al., 1971); as a storm-dominated deposit in the lower shoreface environment (Colter and Havard, 1981); as low-energy deposits in a shallow restricted sea, with the coarsening-upward cycles representing phases of shoaling (Knox et al., 1982); as a mixture of fair-weather sediments (the relatively clay-rich friable sandstones) and storm deposits (the clay-free, bioclastic cemented sandstones) (Kantorowicz et al., 1987); and on a shoal separating a shallow carbonate shelf to the north, from a silty, deeper basin to the south (Bryant et al., 1988). Large-scale cross-bedding and the low degree of bioturbation in the bioclastic limestones such as the Ham Hill Stone suggested to Davies (1969) rapid emplacement by high-energy currents in a tidal channel, and to Jenkyns and Senior (1991) a fault-induced topographical high where carbonate could accumulate.

On the basis of the heavy minerals, the south-west was thought to be a likely provenance for the sands (Boswell, 1924), although a north-easterly source was thought to be equally possible (Morton, 1982). Measurements of small-scale cross-bedding on the outcrop between the Dorset coast and the Mendips suggest transport dominantly from the north-east, with only a minor south-westerly component (Davies, 1969). Large-scale cross-bedding within the Ham Hill Stone suggests transport dominantly from the west or south-west (Davies, 1969; Cope et al., 1969). Magnetic fabric studies are inconclusive because they indicate deposition by north-easterly or south-westerly directed currents (Hounslow, 1987). Taken together, the data suggest both north-easterly and south-westerly source areas.

Details

An exposure [6539 2017] on the incised trackway called The Ridge, east of Poyntington, showed about 25 m of fine-grained, silty, slightly micaceous, orange-yellow sand interbedded with sandy shelly, slightly micaceous, patchily bioturbated limestones and calcareous sandstones. The presence of mica in the calcareous beds distinguishes them from the Inferior Oolite. Bedding is irregular, but given the limited exposure, the beds appear to be laterally persistent. The following is the measured section:

	Thickness m
Approximate base of the Inferior Oolite	
BRIDPORT SANDS	
Unexposed, mainly sand	0.80
Sandstone, calcareous	0.20
Sand, fine-grained, silty	0.25
Sandstone, calcareous	0.20
Sand, fine-grained, silty	0.85
Sandstone, calcareous, three beds each c.0.3 m thick with intercalated sand	1.50
Sand, fine-grained, silty	4.30
Sandstone, calcareous	c.0.30
Sand, fine-grained, silty	0.70
Sandstone, calcareous	c.0.30
Sand, fine-grained, silty	1.70
Sandstone, calcareous; seven beds with intercalated sand; lowest bed [6543 2011], 0.5 m thick, is a biosparitic sandstone	2.70
Sand, fine-grained, silty, mainly orange-yellow	2.00
Sandstone, calcareous	c.0.20
No exposure, mainly fine-grained silty sand	4.30
Limestone, sandy, micritic with fragmentary large belemnites and bivalves at base	0.25
No exposure, mainly fine-grained, silty sand	2.00
Limestone, sandy	0.20
No exposure, mainly fine-grained, silty sand	3.00

In the Purse Caundle Borehole [7012 1826], 5.29m of olive-grey to greyish green, poorly cemented, micaceous, highly bioturbated sandstone, interwoven with wispy films of dark grey silty clay, were proved (Barton et al., 1993). Interbeds of well-cemented, calcareous sandstone with coarse-grained shell fragments, in units 0.09 to 0.22 m thick, also occur. Fragments of *Pleydellia?* and *P.* aff. *superba* in the top 0.3 m prove the *aalensis* Subzone of the late Toarcian. Other fossils include *Sarcinella socialis, Pseudoglossothyris?, Entolium* cf. *corneolum, Eopecten?, Meleagrinella* sp., *Propeamussium* sp. and *Pseudolimea* sp.

In the Stowell Borehole [6849 2180], the 31.4 m of strata assigned to the Bridport Sands consist of soft dark greenish grey sands with alternations of thin, shelly, sandy limestones (Pringle, 1909, 1910; White, 1923; Ivimey-Cook, 1989; Taylor, 1990). A 1.52 m-thick sandy limestone, regarded by Pringle (1910) as the basal bed of the Inferior Oolite, is here taken (also by White, 1923) as the topmost bed of the Bridport Sands. It yielded *Pleydellia* sp. of *aalensis* Subzone age and a bivalve fauna including *Camptonectes* sp., *Entolium* sp., *Meleagrinella* sp., *Oxytoma* sp., *Placunopsis?, Propeamussium* and a trigonid. The underlying sands yielded a bivalve fauna that included *Camptonectes, Chlamys, Entolium corneolum, Liostrea* sp., *Meleagrinella substriata, Myophorella* cf. *formosa, Oxytoma, Propeamussium pumilum, Tancredia* sp., *Trigonia* sp., *Pleydellia?* and belemnites. About 17.4 m below the top of the Bridport Sands, a *Dumortieria* aff. *externicompta* is indicative of the *moorei* Subzone.

The Oborne Borehole [6522 1764], just west of the district, proved 40.54 m of Bridport Sands, comprising yellow sands with hard 'sandrock doggers' capped by a 0.15 m of very hard crystalline, pale grey shelly limestone which Wilson et al. (1958) equated with the 'Dew Bed' of *moorei* Subzone age described by Buckman (1893, p.485). The limestone yielded a bivalve fauna of *Entolium corneolum, Isoarca* cf. *bajociensis, Meleagrinella* sp., *Plicatula* sp. and *Propeamussium pumilum*, but no ammonite. This limestone may be the uppermost bed of the Bridport Sands, as at Stowell. A grey calcareous sandstone, 17.4 m below the top of the Bridport Sands, contained *Meleagrinella* sp. and a fine-ribbed *Dumortieria* sp. of probable *moorei* Subzone age.

MIDDLE JURASSIC

INFERIOR OOLITE

The Inferior Oolite south of the Mendip Hills consists of a condensed sequence of mud-rich, shallow-water limestones with thin, commonly ferruginous, peloidal limestones, in contrast with the cream-white oolites that dominate the sequence in the Cotswolds. There are frequent and rapid lateral variations in lithology and thickness in the Wessex Basin. Around Sherborne, the limestones are predominantly bioclastic or siliciclastic, commonly nodular, and interbedded with calcareous mudstones. The sequence contains intervals with abundant peloids and intervals that are richly fossiliferous. There are also glauconitic and phosphatic horizons within the succession which, together with regional thickness variations, can be related to the effects of contemporaneous fault movements and variation in the rate of subsidence during sedimentation (Arkell, 1933).

The Inferior Oolite of the Sherborne area has long been known (Buckman, 1893) to be thicker than elsewhere in Dorset and Somerset. The lower part of the succession, the Lower and Middle Inferior Oolite of Arkell (1933, p.193), is markedly attenuated around Yeovil, and is condensed and variable throughout the region, with the lowermost strata commonly absent. The upper part of the succession, the Upper Inferior Oolite of Arkell (1933, p.234), is everywhere substantially thicker, but also thins westwards. Richardson (1916) recorded a progressive northward overstep of the Upper Inferior Oolite across the underlying Middle and Lower Jurassic sequences, until it rests unconformably on Carboniferous Limestone in the Mendip Hills.

It has been pointed out by a number of authors that the area of greatest thickening lies at the eastern apex of a re-entrant in the outcrop (Figure 17). White (1923) attributed the re-entrant to the presence of a south-east-plunging anticline. Arkell (1933, p.237) presumed that the beds exposed farthest down the dip slope were the thickest because they formed in the deepest water. However, the interpretation given in the present account views the thick sequence near Sherborne as the infill of a fault-controlled basin, the geometry of which is described below.

The Inferior Oolite crops out in the north-west of the district around Milborne Port, where the outcrop (Figures 17 and 18) has been offset by the Poyntington Fault. To the east of the fault, dip slopes of limestone are broken by narrow, north-trending fault grabens that have let down small slivers of Lower Fuller's Earth. To the west of the fault, the ground is broken by faults with larger displacements and, for this reason, the Inferior Oolite forms a prominent north-trending ridge above Oborne.

Jurassic faults in the Sherborne area include the Mere Fault (see Whittaker, 1985 for a review) and, 5 km farther south and subparallel to it, the Templecombe Fault (Figure 19). Fuller's Earth and Frome Clay to the north of the latter differ in lithology and thickness from equivalent strata farther south (Taylor, 1990), and comparable changes in the Inferior Oolite occur.

The thickness of the Inferior Oolite near outcrop is known from three boreholes (Figures 3 and 17; Table 1). The BGS Purse Caundle Borehole proved 44.2 m (Barton et al., 1993); the Stowell and Oborne boreholes proved 36.3 and 38.8 m respectively (Ivimey-Cook *in* Taylor, 1990; Pringle, 1909, 1910; Wilson et al., 1958). Some 15 km down dip, 32.6 m were proved in the Fifehead Magdalen Borehole and 32 m in the Mappowder Borehole. Elsewhere in Dorset and Somerset (Figure 17), sequences are typically less than 25 m thick. The Bruton Borehole proved only 6.8 m (Holloway and

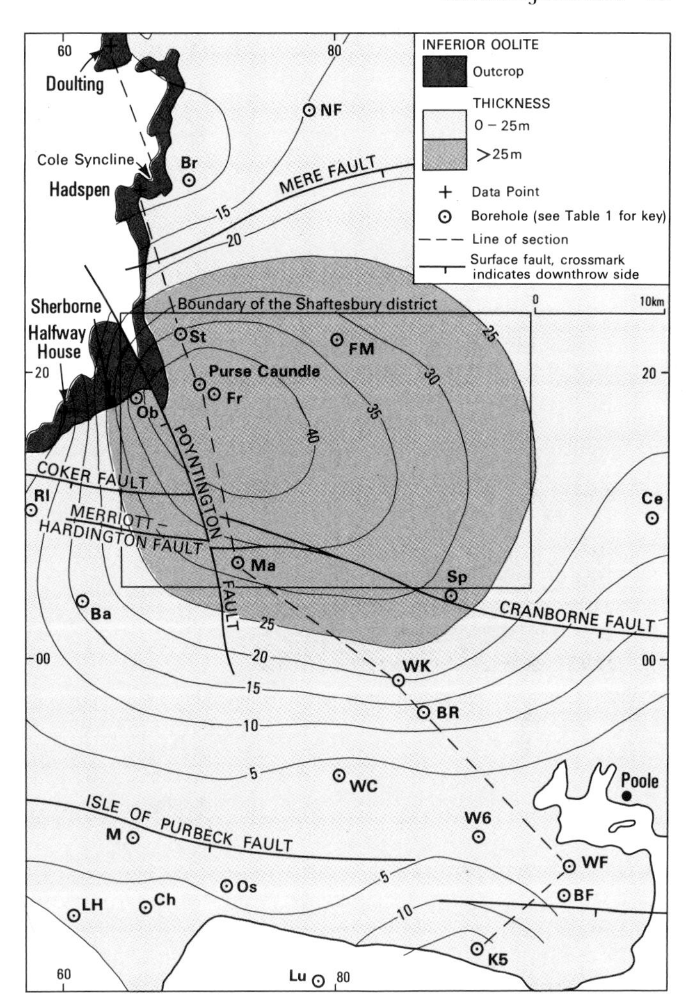

Figure 17 Isopachyte map for the Inferior Oolite. Data points other than boreholes from Wilson et al. (1958) (for correlation along line of section see Figure 20).

Chadwick, 1984); a comparable thin sequence is present in ground between the Mere and Vale of Pewsey faults. West of the district, the formation is attenuated and incomplete as a result of major non-sequences, with only 6.7 m at Halfway House [6015 1640] near Yeovil, and even less at Yeovil Junction [572 140] and in the fragmented succession around Crewkerne. The succession thins progressively toward the south from 20.9 m in the Winterborne Kingston Borehole (Rhys et al., 1982) to approximately 3 m at Wytch Farm (Bristow et al., 1991) and at Burton Bradstock on the Dorset coast (Wilson et al., 1958). The broad outcrop of Inferior Oolite east of Sherborne clearly does reflect the unusually thick sequence in that area.

The Sherborne area is classical ground for the study of Aalenian, Bajocian and early Bathonian faunas, in large part because of the recording of sections and the palaeontological work begun by Buckman (1893) and extended by Richardson (1916, 1932). The macrofaunas used to characterise zonal and subzonal divisions of the thick sequence in the Bath area are present in greatly re-

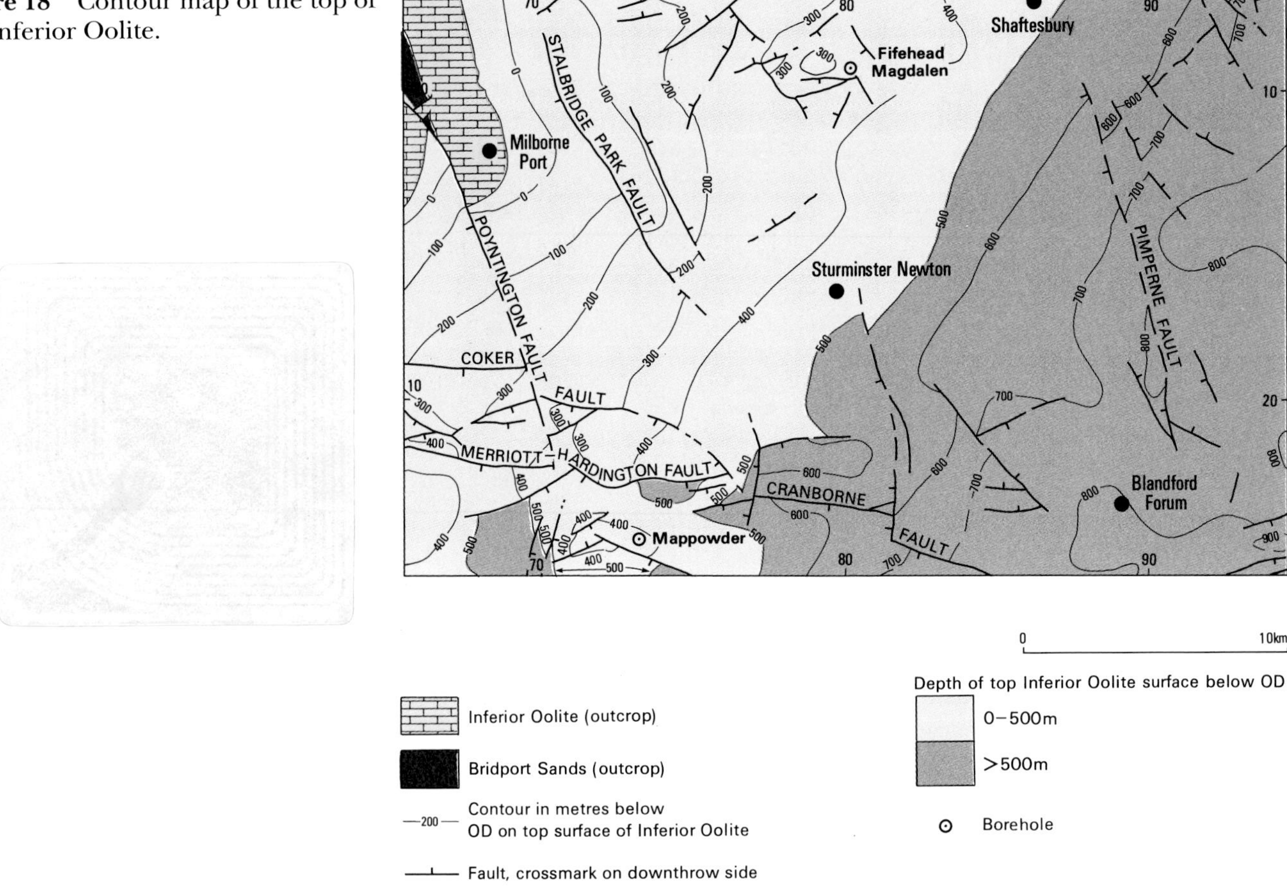

Figure 18 Contour map of the top of the Inferior Oolite.

duced thicknesses of strata in Dorset and Somerset, and successive faunas are often condensed into a few metres, or even a few centimetres, of strata. The modern zonal and subzonal scheme (Table 3; after Parsons in Cope et al., 1980b) owes much to this early work.

Stratigraphy

The Purse Caundle Borehole (Figure 19) proved the thickest sequence of Inferior Oolite in Dorset, and provides a standard lithological succession for the district (Barton et al., 1993). The lithological subdivisions in the borehole, with the exception of the Miller's Hill Beds, follow the nomenclature of Cope et al. (1980b, fig. 3a, column AB12); in the present account, they are considered to be members of the Inferior Oolite Formation and have been named from classical localities described by Buckman (1893), Richardson (1916, 1932), White (1923) and Parsons (1976).

Although the Inferior Oolite typically generates extensive surface brash, significant sections are scarce; the members have been recognised, but they have not been mapped. The basal beds are present in the faulted outliers in the north-west of the district, the middle part of the sequence crops out in fault scarps on either side of the Poyntington Fault, and long dip slopes in the upper strata occur east of the Poyntington Fault.

Corton Denham Beds

The name Corton Denham Beds was assigned (Parsons in Cope et al., 1980b, p.9) to the strata first described by Richardson (1916, 1932) from sections [643 228; 637 212] near the village of Corton Denham (Figure 19). They consist of a basal unit (the Lower Corton Denham Beds of Barton et al., 1993) of pale grey, burrowed and bioturbated limestones (Plates 1G and H) with sandy, calcareous mudstone partings, overlain by an upper unit (the Upper Corton Denham Beds) of olive-grey calcareous mudstones with only thin interbedded limestones (Plate 1F). The uppermost 3 m of the Corton Denham Beds are sparsely glauconitic and contain isolated, small, phosphatic clasts. Extensive burrowing and irregular bed boundaries give the limestones a pseudo-nodular fabric.

The Corton Denham Beds are not well exposed; they occur in faulted outliers on Poyntington Hill and near Corton Denham. The base of the succession is defined by a progressive upward decrease in the amount of matrix sand in sandy limestones at the top of the underlying Bridport Sands.

The thickness of the Corton Denham Beds ranges from 12.8 m in the Purse Caundle Borehole to an estimated 10 m in the Stowell Borehole and 8 m in the Oborne Borehole. Geophysical correlations indicate that the beds are 8.6 m and 8.25 m thick in the Fifehead

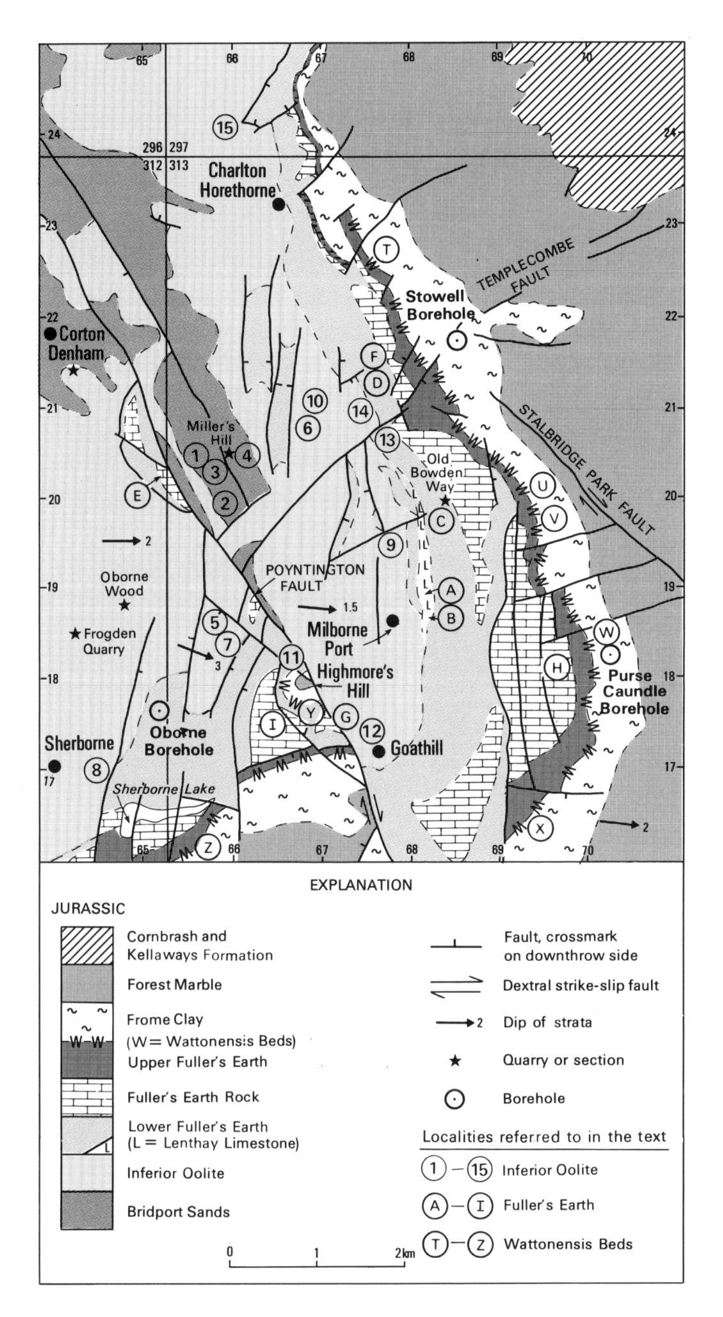

Figure 19 Sketch map of the solid geology of the Milborne Port area showing localities referred to in text.

Magdalen and Mappowder boreholes respectively. Parsons (in Cope et al., 1980b) gave estimates of 8 m and 6 m from surface exposures at Oborne and Corton Denham.

The Toarcian–Aalenian boundary in both the Purse Caundle and Stowell boreholes is at, or close to, the junction of the Bridport Sands and the Inferior Oolite. The Corton Denham Beds are mostly sparsely fossiliferous, although abundant *Homoeorhynchia ringens* c.3 m above the base of the Inferior Oolite indicate the Ringens Bed. According to Richardson (1932), the acme of *H. ringens* is just above the base of the *bradfordensis* Subzone of the *murchisonae* Zone. There is no evidence for beds belonging to the *opalinum* Zone or the early part of the *murchisonae* Zone in the boreholes; both could be represented within the lowest 3 m of beds. Ammonites, possibly indicative of the *concavum* and *discites* zones, occur in the upper part of the Corton Denham Beds in the Purse Caundle Borehole; the Aalenian–Bajocian Stage boundary probably lies between 7 and 8 m above the base of the Corton Denham Beds.

Miller's Hill Beds

The ferruginous and richly fossiliferous beds that include the Oborne Road-stone Bed (Buckman, 1893, p.501), the Cadomensis Beds (Hudleston, 1886, p.193), the *Sphaeroidothyris* Bed (Parsons, 1976, p.126) and the Niortensis Beds (Buckman, 1893, p.507; Arkell, 1933) have been named the Miller's Hill Beds by Barton et al. (1993). The unit consists of glauconitic sands, limestones (Plate 1E) and calcarenites, with clasts of phosphatic limestone and abundant cephalopod fragments; the base of the unit is defined by an erosion surface. Although often described as 'ironshot oolite', the beds are peloidal in part, rather than wholly oolitic. Typical lithologies are intensely bioturbated, glauconitic limestone and sandy limestone with dark greyish green wisps and partings of glauconitic sand, together with thin intervals of phosphatised shell debris and buff-coloured phosphatic pebbles. Burrowed erosional surfaces, across which there are large changes in glauconite abundance, are associated with concentrations of cephalopod fragments and clastic material; they occur at several horizons within the member, and an erosional surface of this type is present near the base. The phosphatic intervals give high natural gamma-ray counts and pronounced peaks on gamma-ray logs. The Miller's Hill Beds form an important gamma-ray log marker.

The member has yielded extensive ammonite faunas which show it to be a condensed sequence representative of several Bajocian zones. Individual subzones of the late Lower Bajocian and early Upper Bajocian are locally absent. At Miller's Hill [6626 2052] (Figure 19), strata with exceptionally well-preserved *romani* Subzone faunas are overlain by a non-sequence which probably spans the rest of the *humphriesianum* Zone and the *subfurcatum* Zone. By contrast, all three subzones of the *humphriesianum* Zone are represented on Poyntington Hill [653 208], 1 km west-north-west. At Oborne Wood [648 188], a highly attenuated sequence less than 2 m thick has yielded fauna indicative of the *laeviuscula* to *baculata* subzones, with the exception of the *sauzei* Zone (Figure 22; Parsons, 1976). Ammonite fragments from the upper part of the Miller's Hill Beds in the Purse Caundle Borehole suggest either a late *subfurcatum* Zone or early *garantiana* Zone age; the overlying *garantiana* Zone strata contain clasts that indicate contemporaneous erosion of beds of the *subfurcatum* Zone (*baculata* Subzone). The principal ferruginous beds in the Oborne Borehole are at least 1.3 m thick and are contained largely within the *subfurcatum* Zone. No comparable strata were recorded from the Stowell Borehole.

Taken together, the faunal data suggest that the Miller's Hill Beds span at least the whole of the *subfurca-*

Table 3 Correlation of late Toarcian, Aalenian, Bajocian and early Bathonian strata in the Shaftesbury and adjacent districts (columns 2–4 after Cope et al., 1980a, b).

Biostratigraphy				Lithostratigraphy			
Stage	Substage	Zone	Subzone	Shaftesbury district (thicknesses from Purse Caundle Borehole)	South Somerset (Corton Denham)	Yeovil district (Yeovil railway cutting)	Dorset coast (Burton Bradstock)
Bathonian	Lower	*Zigzagiceras (Zigzagiceras) zigzag*	*Oppelia (Oxycerites) yeovilensis*	Fuller's Earth	* Fuller's Earth	Fuller's Earth	Fuller's Earth; Scroff *
			M. (Morphoceras) macrescens	* Crackment Limestones 6.2 m	Crackment Limestones ?9 m	Crackment Limestones c.4 m	* Zigzag Bed 0.15 m (Top Beds 2.2 m)
			P. (Parkinsonia) convergens	*			*
Bajocian	Upper	*Parkinsonia parkinsoni*	*P. bomfordi*	? * Rubbly Beds 6.9 m			*Sponge Bed = Burton Limestone 0.76 m
			Strigoceras truellei		* Doulting Stone + 1.0 m	Halfway-House Bed ? m	* Truellei Bed 0.3–0.48 m
		Strenoceras (Garantiana) garantiana	*Parkinsonia acris*	? Sherborne Building Stone 15.6 m † ?	Hadspen Stone + 2.0 m	* Astarte Bed 0.01–0.15 m	* Astarte Bed 0–0.15 m
			St. (Garantiana) tetragona				? ? ?
			St. (G.) subgaranti				
			St. (Pseudogarantiana) dichotoma	? ?			
		Strenoceras subfurcatum	*St. (G.) baculata*	* Miller's Hill Beds 2.7 m			
			Caumontisphinctes polygyralis	*			
			Teloceras banksi	*			†
	Lower	*Stephanoceras humphriesianum*	*T. blagdeni*	*			Red Conglomerate 0.00–0.14 m
			S.humphriesianum	* ?			†
			Dorsetensia romani				
		Emileia (Otoites) sauzei		* †? ?	?		† Red Beds 0.42 m
		Witchellia laeviuscula	*W.laeviuscula*	*			? ?
			Sonninia (Fissilobiceras) ovalis				
		Hyperlioceras discites		† Corton Denham Beds 12.8 m	Corton Denham Beds + 6.0 m		† Snuff-Box Bed 0.1–0.18 m
Aalenian		*Graphoceras concavum*	*Graphoceras formosum* horizon	†			
			G. concavum				
		Ludwigia murchisonae	'*Brasilia gigantea* horizon'				
			Brasilia bradfordensis	† Ringens Bed	Ringens Bed		
			L. murchisonae				
			L. haugi	?	?		
		Leioceras opalinum	*Tmetoceras scissum*	?	?		* Scissum Beds 1.3 m
			L. opalinum	? ?			
Toarcian		*Dumortieria levesquei*	*Pleydellia aalensis*	*		Bridport	* * Bridport Sands
			D. moorei	* Bridport Sands	Bridport Sands	Ham Hill Stone	*
			D. levesquei			Sands	* Down Cliff Clay
			Phlyseogrammoceras dispansum			Junction Bed	* Junction Bed

Inferior Oolite Formation (spans columns Shaftesbury district, South Somerset, Yeovil district from Crackment Limestones to base of Corton Denham Beds / Aalenian)

† indicates Zone proved locally
* indicates Subzone proved locally.

tum, *humphriesianum* and *sauzei* zones, and part of the *laeviuscula* Subzone of the *laeviuscula* Zone. Strata assigned to the Lower Bajocian part of the Miller's Hill Beds nowhere exceed 1 m in thickness. The inferred slow rate of deposition and widespread iron mineralisation now contained in glauconite reflect an important tectonic event.

SHERBORNE BUILDING STONE

The Miller's Hill Beds pass up with rapid and progressive loss of glauconite into a thick sequence of patchily fossiliferous, sandy limestones with thin intervals of calcareous mudstone — the Sherborne Building Stone of Buckman (1893). The Sherborne Building Stone consists of pale grey, sandy-textured limestone (Plates 1C and D) with sparse pale to medium olive-grey, sandy, very calcareous wisps and irregular partings. The siliciclastic component is very fine-grained sand similar to that of the Bridport Sands. The basal strata of the Sherborne Building Stone are extensively bioturbated and relatively shelly, with intervals rich in fragmented terebratulids. Wood fragments, 10 to 50 mm across, are common throughout, and scattered serpulids are abundant, particularly in the upper beds.

The thickness of the Sherborne Building Stone in the Purse Caundle Borehole (15.6 m) is comparable to that in the Oborne Borehole (estimated at 15 m), but substantially thicker than estimates of 3 m from surface exposures north-west of Sherborne (Parsons, in Cope et al., 1980b) and the 3 m recorded in the Winterborne Kingston Borehole (Penn, 1982). The thickening of the Sherborne Building Stone in the district is attributed to an increased input of clastic sediment from fault scarps nearby.

The upper beds of the Sherborne Building Stone contain ammonite fragments that indicate the *garantiana* Zone. Other fossils, typical of a coarse sand environment, are rare; however, epifaunal brachiopods such as *Acanthothiris spinosa* are locally common.

RUBBLY BEDS

The Rubbly Beds (Buckman, 1893, p.497) are cream and pale grey, coarsely shelly, pseudonodular limestones that contain medium grey, anastomosing argillaceous partings, 5 to 30 mm thick (Plate 2). The limestone beds range from 20 to 150 mm in thickness, are intensely bioturbated and contain abundant subhorizontal burrows, 5mm in diameter, infilled with pale calcitic limestone. The lowermost strata consist of pseudonodular limestones with irregular sandy partings and shell fragments that contain phosphatic infillings; they are transitional from the underlying Sherborne Building Stone. The base of the unit is taken at the level where the sand content diminishes markedly.

The thickness of the Rubbly Beds is approximately 4 m in the Oborne Borehole and 6.9 m in the Purse Caundle Borehole. Elsewhere, thickness estimates are 3.3 m in the Fifehead Magdalen Borehole and 3.5 m in the Mappowder Borehole.

Bivalves, including *Catinula bradfordensis*, belemnite fragments and *Parkinsonia* occur in the Rubbly Beds. Most of the fauna is consistent with a *parkinsoni* Zone age. However, *C. bradfordensis* is also found in the basal part of the Rubbly Beds at Bradford Abbas, and is attributed to a late *garantiana* Zone age (Buckman, 1893).

CRACKMENT LIMESTONES

The Crackment Limestones (the Crackment Beds of White, 1923) form the uppermost beds of the Inferior Oolite in the Sherborne area; they consist of a thin lower unit of peloidal limestones (Plate 1B) and an upper unit of argillaceous limestones (Plate 1A) and mudstones. The two units are separated by an erosion surface with a large burrow system. The thickness of the Crackment Limestones in north Dorset ranges from 6.2 m in the Purse Caundle Borehole to an estimated 10.9 m in the Mappowder Borehole.

In the Purse Caundle Borehole, the basal 1.64 m of strata (the Lower Crackment Limestones of Barton et al., 1993) consist of thinly bedded, cream and pale grey, coarsely shelly, nodular, peloidal limestones that contain sparse brown-coated phosphatic clasts and shells; there is an erosional top. The overlying 4.59 m of beds (the Upper Crackment Limestones) consist of greenish grey, argillaceous limestone with partings and thick intervals of olive-grey mudstone, calcareous mudstone and shelly mudstone. The argillaceous limestone and mudstone both show widespread bioturbation; thread-like pyritous burrows are ubiquitous.

An exposure near Charlton Horethorne [6622 2403] (locality 15, Figure 19), proves that the Rubbly Beds there are of *truellei* Subzone (*parkinsoni* Zone) age. Fauna from the peloidal beds at the base of the Crackment Limestones in the Purse Caundle Borehole indicate a *convergens* Subzone (*zigzag* Zone) age, close to the base of the Bathonian Stage. According to Torrens (1967), the uppermost argillaceous beds of the Inferior Oolite near Goathill [6741 1715] (locality 12, Figure 19) contain a *macrescens* Subzone fauna, while the same age is attributed to the basal beds of the Fuller's Earth in the Stowell and Winterborne Kingston boreholes. The lithological change from the Inferior Oolite to the Fuller's Earth appears to have taken place during *macrescens* Subzone times across much of the region.

Regional correlation

The distinctive responses on the gamma-ray geophysical log of the Inferior Oolite allow close correlation of sequences across Dorset and Somerset. The Miller's Hill Beds form a conspicuous log marker in most boreholes in which the Inferior Oolite is more than about 10 m thick. The phosphatic horizon is absent in the Ryme Intrinseca Borehole, where the sequence is thin, and from the Norton Ferris Borehole north of the Mere Fault. It appears to be present near the base of an attenuated sequence in the Bruton Borehole, and can be recognised in the logs of some coastal Dorset boreholes. A smaller gamma-ray peak at the base of the Crackment Limestones in the Purse Caundle Borehole is also evident in other boreholes nearby.

A north–south traverse from the Mendip Hills to the Dorset coast (Figures 20 and 21), shows significant

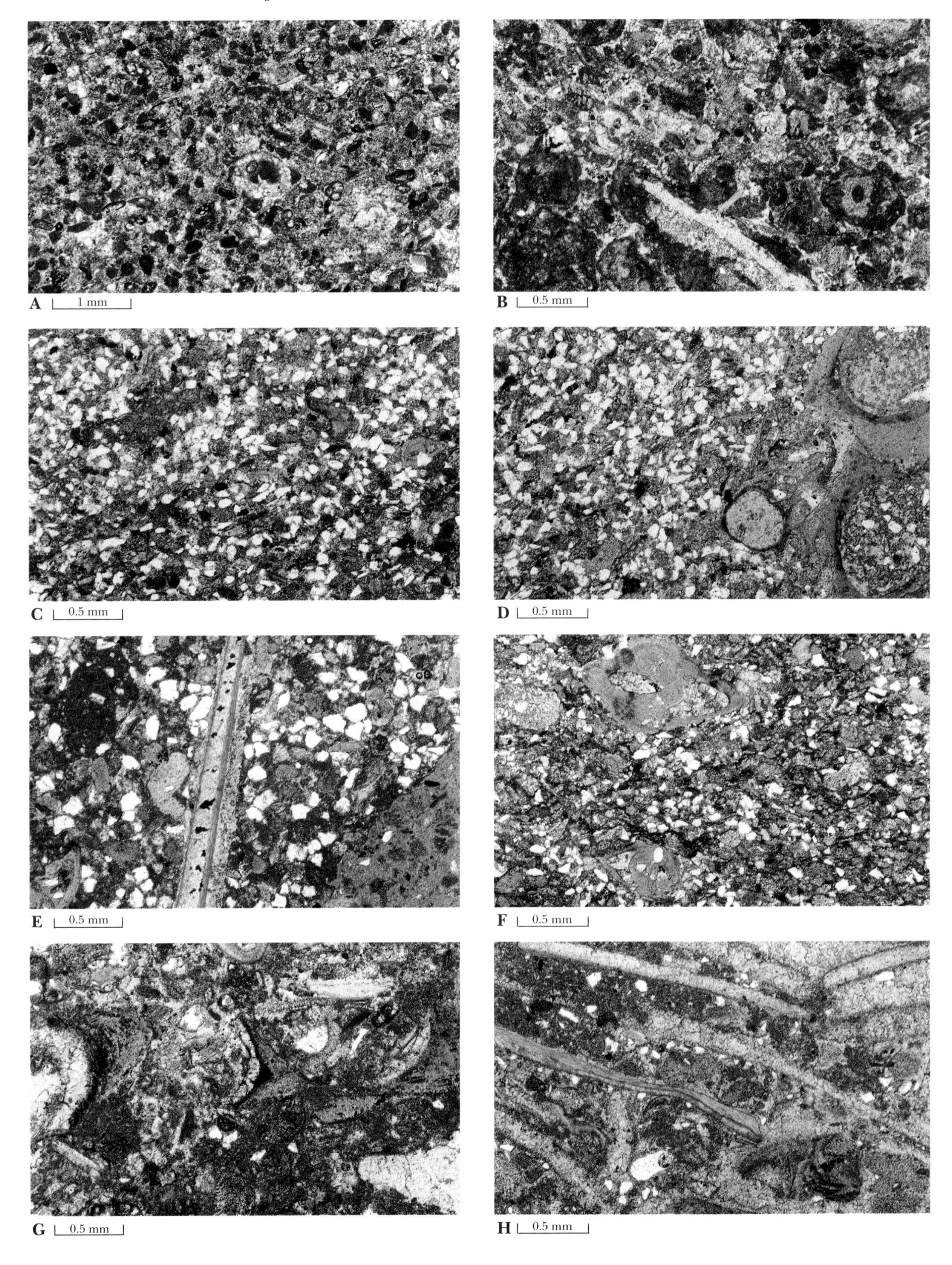
A 1 mm
B 0.5 mm
C 0.5 mm
D 0.5 mm
E 0.5 mm
F 0.5 mm
G 0.5 mm
H 0.5 mm

Plate 1 Photomicrographs of limestones of the Inferior Oolite from the Purse Caundle Borehole (all plane polars, dual carbonate stain).

A. 93.0 m **Crackment Limestones**
Very fine to fine sand grade, micritic peloids (dark brown) and abraded bioclastic fragments (pink), in a slightly ferroan, sparry calcite cement (pale blue).

B. 95.58 m **Crackment Limestones**
Coarse, abraded, non-ferroan calcite bioclasts (pink), finer peloidal grains and large algal-coated grains (dark brown) in a slightly ferroan, sparry calcite cement (pale blue).

C. 105.11 m **Sherborne Building Stone**
Scattered non-ferroan bioclastic (pink) debris debris, together with very fine sand grade dark, peloidal grains and abundant quartz and feldspar grains in an argillaceous and calcareous matrix. Late diagentic ferroan dolomite rhombs (blue) scattered throughout.

D. 107.47 m **Sherborne Building Stone**
Abundant very fine sand grade quartz and feldspar grains, large chambered, non-ferroan calcite serpulid fragment and abraded bioclastic fragments in a slightly argillaceous, calcite matrix. Late diagenetic ferroan dolomite crystals (dark blue) are scattered throughout.

E. 119.05 m **Miller's Hill Beds**
Bioclastic fragments (pink), quartz grains (white) and glauconite (green), in an argillaceous, slightly ferroan sparry calcite cement. Associated with the large bioclastic fragment (centre field), is a later, more strongly ferroan calcite cement (blue). Pale brown to buff phosphatic intraclasts are also common.

F. 122.06 m **Corton Denham Beds**
Calcitic bioclasts (large benthic foraminifera), quartz grains and glauconite (green) in an argillaceous, slightly ferroan, calcite cement. Note the more strongly ferroan calcite cement (blue) occluding the foram chambers.

G. 133.51 m **Corton Denham Beds**
Non-ferroan calcite bioclastic debris (pink) and ferroan spar replaced bioclast (pale blue) in a patchy sparry calcite cement. Other parts of the section show an argillaceous, micritic, occasionally peloidal, matrix. Detrital quartz grains are sparsely present.

H. 135.67 m **Corton Denham Beds**
Coarse bioclastic debris, both non-ferroan (pink) and ferroan spar replaced fragments, in a sparry ferroan calcite cement (pale blue). Patches of brown, ferruginous, peloidal micrite also occur.

changes in thickness and sedimentary facies in the Inferior Oolite. In the Winterborne Kingston Borehole, approximately 10 m of argillaceous nodular limestones, equivalent to the Corton Denham Beds, are overlain by ferruginous and oolitic limestones with phosphatic pebbles (Penn, 1982). Above this condensed and mineralised sequence, beds equivalent to the Sherborne Building Stone are only 3 m thick. The Rubbly Beds are represented by 6 m of argillaceous limestones and mudstones of uppermost *garantiana* to *parkinsoni* Zone age; they pass up into beds corresponding to the Crackment Limestones. The Inferior Oolite thins to the south until, at Burton Bradstock on the coast, there are only 3.3 m of condensed limestones, including limonitic concretions or 'snuff-boxes', interpreted by Gatrall et al. (1972) as limestones of algal origin formed on a submarine swell. The southward condensation across Dorset contrasts with the stratigraphical overstep of the lowermost Inferior Oolite north of Sherborne. The Corton Denham Beds are last seen on Corton Downs [633 234], immediately south of the Mere Fault, where strata of *subfurcatum* and younger age overstep onto Bridport Sands (Arkell, 1933, p.195). The Corton Denham Beds reappear briefly in the Cole Syncline (Richardson, 1916), approximately 5 km north of the Mere Fault, where they comprise 3 m of limestones with corals. These beds were intersected, but not cored, in the Bruton Borehole, where they are estimated to be 2.4 m thick.

Massively bedded limestone up to 3 m thick, with nests of *Acanthothiris spinosa*, equivalent in age to the upper part of the Sherborne Building Stone, occurs between the Mere Fault and the Cole Syncline, where it is known as the Hadspen Stone (Arkell, 1933, p.236). It locally has a conglomeratic base where it unconformably overlies the Bridport Sands. The limestone passes into mudstones and thin limestones north of the Cole Syncline, and is eventually overstepped by the Doulting Stone, a flaggy limestone of *parkinsoni* Zone age. The Doulting Stone is the youngest Inferior Oolite between Bruton and the Mendip Hills, where it rests unconformably on Carboniferous Limestone.

An east–west traverse through the district also shows significant thickness changes in the Inferior Oolite (Figure 22). There is close correlation between the sequences in Purse Caundle and Fifehead Magdalen boreholes, although the thickness decreases toward the latter.

Rapid attenuation of more than 30 m of strata occurs westwards in the 5 km of ground between the Oborne Borehole (Figure 22) and Halfway House [6015 1640] in the adjacent Yeovil district. Most biostratigraphical zones are represented in the condensed sequence at Halfway House, and the strata include limonitic concretions of algal type (the Irony Bed, Buckman, 1893, p.485). At Yeovil Junction, only 1m of Inferior Oolite of largely *garantiana* Zone or younger age separates the Fuller's Earth from the underlying Bridport Sands (Wilson et al., 1958). Between Crewkerne and Beaminster, strata of *levesquei* to *subfurcatum* zone age are commonly absent.

Palaeogeography and provenance

The palaeogeography of the Bajocian–Bathonian interval in Britain has been considered by Martin (1967) and Jones and Sellwood (1989). Their maps show a regional trend from nonmarine sedimentation in the north of England, passing southwards through a belt of shallow-marine, high-energy carbonate sediments into the mud-rich sequences of the Wessex Basin. Muddy sediments are particularly characteristic of the western Wessex Basin and Celtic Sea basins during the Bajocian and Bathonian.

The Inferior Oolite in the Shaftesbury district consists of deposits typical of part of an offshore limestone

Plate 2 Inferior Oolite; junction (arrowed) of the Rubbly Beds (below) and Crackment Limestones (above) near Oborne.

ramp in which the abundance of muddy sediments in the sequence, and the inferred proximity of fault scarps, suggest that sedimentation occurred in a down-ramp, fault-controlled basinal setting. The faunas are diverse and consist of a mixture of suspension-feeding infaunal and epifaunal elements, and free-swimming forms. Periodically, coarse bioclastic and peloidal material was swept into the basin and subsequently developed as the thinly bedded to nodular, bioclastic or peloidal limestone beds. The nodules are interpreted as partly primary in character and partly the result of compactional processes during burial diagenesis (Barton et al., 1993).

The sources, processes and pathways by which the detrital sands and clays reached the Wessex Basin are not clear. The basin margins to the north show little evidence of detrital clastic material being supplied, although the precise location of the shoreline is not clearly established. The detrital sand component, which is similar in grain size, shape and composition to the underlying Bridport Sands, is thought to have been, at least in part, locally derived from the erosion of Bridport Sands exposed in fault scarps. Data from the Bristol Channel Basin, where thick mudstone sequences of Aalenian to Bathonian age characterise the succession, may indicate a westerly source for the detrital clay component in the Wessex Basin (Martin, 1967). Links to the Sherborne sector of the basin through the contiguous Bristol Channel and North Somerset basins seem likely.

A combination of regional lithostratigraphical and isopachyte data for the Inferior Oolite shows that the thick sequence around Sherborne accumulated in a half-graben with a depocentre located approximately 10 km south of the Mere Fault (Figure 21). Contemporaneous removal of strata in the area west of the Poyntington Fault is attributed to erosion of fault scarps formed prior to, and perhaps during, the *garantiana* Zone of the Upper Bajocian. Southward attenuation of the Sherborne sequences, and the formation of a submarine swell in south Dorset, can be accomodated by synextensional rotation along the Mere and related faults.

The fault-controlled basin was starved of sediment during the period of deposition of the Miller's Hill Beds (*laeviuscula* to *baculata* subzones). Sufficient submarine topographical relief was present along the northern and western basin margins during the succeeding *garantiana* Zone to allow both the accumulation of carbonate-rich sand bodies and the erosion of Bridport Sands and basal Inferior Oolite limestones. The abundance of quartz sand grains and the occurrence of large wood fragments, together with the sparse fauna in the Sherborne Building Stone, are consistent with a rapidly deposited

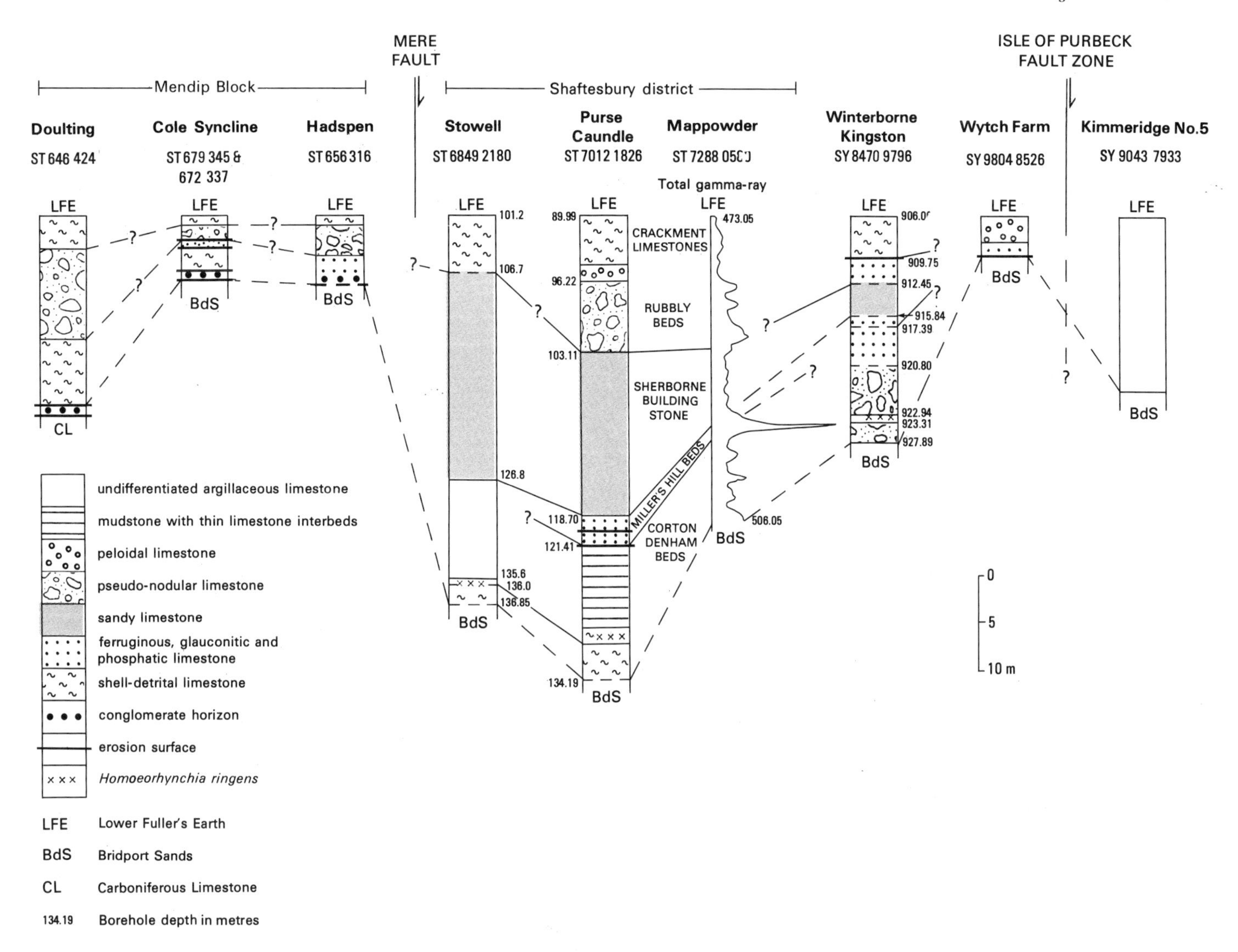

Figure 20 Lithological correlation of the Inferior Oolite along an approximately north–south section, with data from Richardson (1916), Cope et al. (1980b) and Penn (1982) (see Figure 17 for location).

sequence. Continuity of strata younger than late *garantiana* Zone across the Mere Fault suggests that, by that time, topographical relief was less important. Between Oborne and Halfway House, where the succession is discontinuous, fault scarps associated with the Poyntington Fault or related faults may have been more persistent.

Details

Corton Denham Beds

The Corton Denham Beds are present on Poyntington Hill (locality 1, Figure 19) [6553 2018], where they consist of pale grey limestones and pale brown calcisiltites. Field brash yielded *Sphaeroidothyris eudesi, Fontannesia?* (juv.) and *Graphoceras* cf. *magnum* that suggest the *concavum* Zone.

The section along the north side of the track on Miller's Hill [6626 2052], described below, includes the uppermost Corton Denham Beds, which consist of nodular grey limestones with irregular marly and sandy partings. These limestones are overlain by a brownish grey, massive, fine-grained and sparsely glauconitic limestone, 600 mm thick (Bed 2a of Richardson, 1916).

Miller's Hill Beds

Glauconitic limestones of the Miller's Hill Beds form a capping to Southern Hill (locality 2, Figure 19) [6584 1975]. Where excavated (locality 3, Figure 19) [6568 2014] for a reservoir, these beds yielded a rich fauna that includes species of *Chondroceras, Dorsetensia, Stephanoceras* and *Teloceras*, which indicate that all three subzones of the *humphriesianum* Zone are

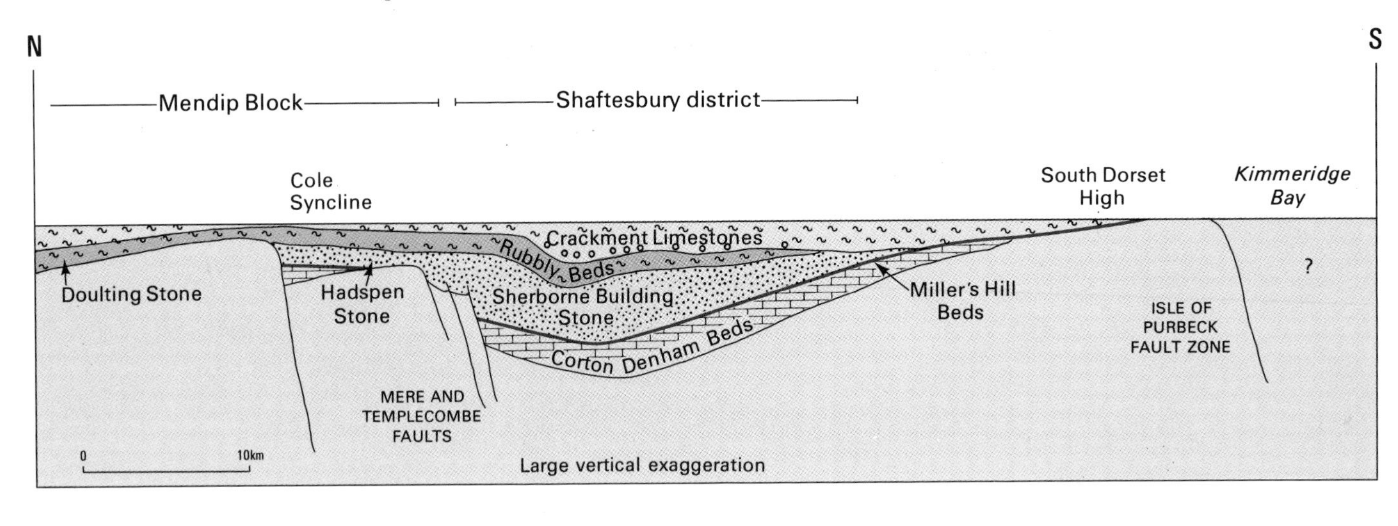

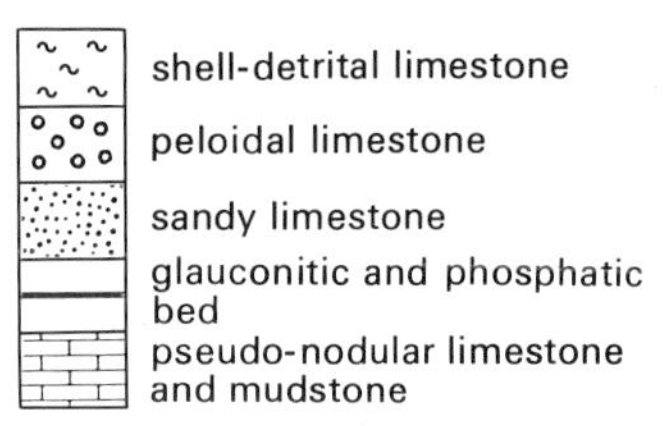

Figure 21 North–south palaeogeographic reconstruction of the Inferior Oolite. Constructed from data given in Figure 20. Top of Crackment Limestones taken as horizontal datum.

present. Field brash in the vicinity includes *Dorsetensia?* and *Normannites* sp. juv. from the *romani* Subzone.

Varied thin-bedded, shelly and ironshot limestones, belonging to the *subfurcatum, humphriesianum, sauzei* and *laeviuscula* zones were formerly exposed in a quarry (locality 5, Figure 19[6562 1856] in Oborne. According to Buckman (1893, p.502), some of the lower limestones are redeposited and form a condensed sequence when compared with the Frogden quarry (Figure 19)[642 185] section less than 1 km farther west. Although the Miller's Hill Beds are no longer exposed in the Oborne quarry, glauconitic limestone with abundant belemnite fragments is present in the lane below the quarry, and fossil brash collected from the base of the Oborne Fault scarp [6558 1860] included *Stephanoceras* sp. from the *humphriesianum* Subzone, a level that equates with the Oborne Road-stone Bed of Parsons (1976).

The Miller's Hill Beds, together with the upper part of the underlying Corton Denham Beds, were best seen in the well-known exposures on Miller's Hill (locality 4, Figure 19; Buckman, 1875, 1893; Richardson, 1916; White, 1923; Kellaway and Wilson, 1941; Parsons, 1976); the sections [6626 2052 to 6628 2054] are now overgrown. Although Buckman (1874) and Richardson (1916) attributed the lowest beds of the 6.5 m thick sequence to the *discites* Zone, Parsons (1976) gives faunal lists for the locality (as Milborne Wick Lane) and shows that the lowest 3.6 m of glauconitic nodular limestones seen there are of *laeviuscula* Subzone age.

Thin (100 mm) beds of highly glauconitic limestone and glauconitic marl of *laeviuscula* Subzone and *?sauzei* Zone age overlie the Corton Denham Beds along the north side of the Miller's Hill track. An irregular erosion surface separates these beds from a 400 mm thick sequence of shelly and patchily glauconitic bioturbated limestones above. These limestones (the *Astarte spissa* Bed, Beds 2 and 3 of Richardson, 1916, p.516, but not the Spissa Bed of Parsons in Cope et al., 1980b) contain abundant belemnite fragments and small ammonites, and, according to Parsons (1976), a basal conglomeratic horizon. The rich and well-preserved fauna belongs to the *romani* Subzone of the *humphriesianum* Zone (Parsons, 1976). A planed surface separates the highly fossiliferous strata from about 4 m of Sherborne Building Stone, which extends along the trackway toward the west. The basal strata of the Sherborne Building Stone are either late *subfurcatum* Zone or early *Strenoceras garantiana* Zone in age. Thus, the non-sequence at the base of the Sherborne Building Stone at Miller's Hill spans much of the *humphriesianum* and *subfurcatum* zones, and possibly into the lowermost part of the *garantiana* Zone.

Sherborne Building Stone

Sparsely bioclastic and massively bedded sandy limestones, approximately 4 m thick, are exposed above the Miller's Hill Beds on the north side of the Miller's Hill track [6628 2054]. The limestones are burrowed and contain common *Acanthothiris spinosa.* A quarry (locality 5, Figure 19)[6562 1856] in Oborne provides one of the few remaining exposures of Sherborne Building Stone in the district; it exposes about 3 m of thick-bedded, orange- or yellow-weathering, sparsely fossiliferous, sandy limestones. Beds are 400 to 800 mm thick, contain vertical burrows and are interlayered with thin smears of ferruginous sandy marl; the limestones weather rubbly. The section corresponds with part of Buckman's (1893, p.502) section XVI, in which he described a more complete sequence, 6m thick, of sandy limestone, the 'Sherborne Building Stone Equivalent' overlying the Miller's Hill Beds. Nearby brash included *Leptosphinctes* sp. and *Parkinsonia* sp. of *parkinsoni* Zone age or, possibly, of *garantiana* Zone.

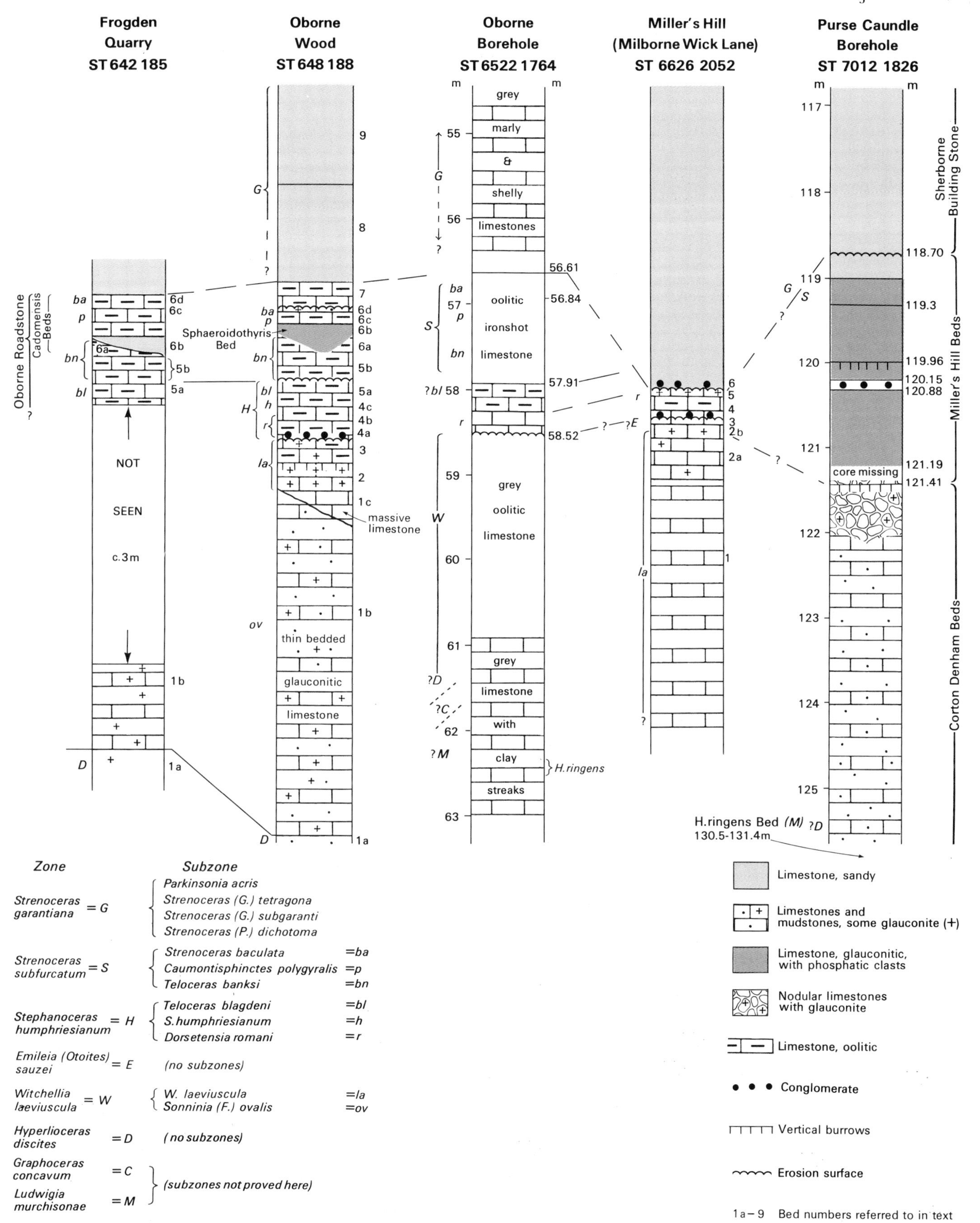

Figure 22 Correlation of condensed sequences in the Inferior Oolite of the Milborne Port and adjacent areas (partly after Parsons (1976), with bed numbers).

An exposure (locality 6, Figure 19) [6702 2081] in the Sherborne Building Stone occurs below the fort at Milborne Wick:

	Thickness m
Topsoil and cryotubated limestone rubble	0.7
Limestone, pale grey, fine-grained and irregularly lenticular	0.2–0.4
Sand, buff, fine-grained, weakly cemented, with small limestone nodules	0.3
Limestone, irregularly bedded	0.3
Sand, as above	0.1–0.15
Limestone, as above	0.1–0.2
Sand, weakly cemented, with a c.0.1 m-thick limestone bed; vertical and horizontal 5 mm diameter burrows in upper part	c.0.3
Limestone, pale grey, sandy in part	1.0

Some tubular horizontal and vertical burrows are seen in the upper thin limestones.

RUBBLY BEDS

A railway cutting 300 m east of the Oborne quarry (locality 7, Figure 19) [6592 1852] exposes the following section (Plate 2):

	Thickness m
?CRACKMENT LIMESTONES (?*zigzag* Zone)	
Limestone, thinly bedded (50 to 100 mm scale), sandy weathering with thin marl partings	1.50
Limestone, greyish blue, with sparse ferruginous peloids and *Procerites* aff. *schloenbachi*	0.20
Marly limestones and marls; basal marl is ferruginous, 0.12 m thick and transitional to	0.35–0.45
?RUBBLY BEDS	
Limestone, massive grey and greyish white weathering, unbedded to rubbly; *Parkinsonia* cf. *pachypleura* fragment 0.25 m below top of unit	2.60

A temporary section (locality 8, Figure 19) [6475 1708] on Castle Farm, 250 m north-north-west of Sherborne Old Castle, exposed 2 m of rubbly shelly limestone with fragments of fossil wood, *Pholadomya lirata, Pleuromya* sp., *Parkinsonia* aff. *pachypleura, Parkinsonia* sp., *Procerites* sp. and *Belemnopsis* sp. This fauna suggests an horizon at about the junction of the *parkinsoni* and *zigzag* zones.

The upper part of the Rubbly Beds is exposed below Crackment Limestones in the type-section (locality 11, Figure 19) [6677 1830] on the A30 near Crackment Hill (Richardson in White, 1923, p.17). There, approximately 3.75 m of grey, rubbly, bioclastic, seminodular limestones with marl partings yielded *Pholadomya* sp., *Pleuromya* sp. and *Belemnopsis bessina*.

A strike section near the southern end of the quarried exposure (locality 15, Figure 19) [6622 2403 to 6622 2407] beside the Charlton Horethorne to Blackford road, just north of the district showed:

	Thickness m
Soil and rubbly limestone head	0.50
?RUBBLY BEDS	
Limestone, buff- to cream-weathering, peloidal or oolitic, with irregular lenticular bedding; *Sphaeroidothyris* cf. *sphaeroidalis* and *Strigoceras truellei* from 0.3 m above the base	0.50
Marl parting, ochreous to brown-weathering, with small lenticular limestone nodules and *Pholadomya lirata* and *Parkinsonia* sp.	0.10
Limestone, moderately peloidal or oolitic, irregularly bedded to seminodular, with a soft, creamy buff-weathering matrix; *Acanthothiris spinosa Sphaeroidothyris?, Cenoceras* cf. *inornatum* and *Procerites* sp. from 0.4 m below the marl, and *Sphaeroidothyris sphaeroidalis* 0.5 m below the marl. Loose material included *Sphaeroidothyris* cf. *globosphaeroidalis, Homomya?* and *Pleuromya* cf. *calceiformis*	1.35

At the base of a fault scarp 1 km north of Milborne Port (locality 9, Figure 19) [6764 1954], 2 m of rubbly, sandy, sparsely fossiliferous limestones with *Acanthothiris spinosa, Pholodomya lirata* and *Parkinsonia?,* are tentatively assigned to the *parkinsoni* Zone. Similar sequences in the Rubbly Beds crop out along the base of the fault scarp as far south as the Anglican church [6763 1853] in Milborne Port.

Some 5 m of Rubbly Beds occurs on the valley side (locality 10, Figure 19) [6711 2099] above the road to Milborne Wick, briefly mentioned by Richardson (1916):

	Thickness m
Soil and limestone rubble	c.0.80
Limestone, medium grey, seminodular and rubbly, with ochreous to brown, irregular marl intercalations; limestone nodules contain vertical burrows and shell debris that includes ostreids, *Procerites* sp. and *Belemnopsis* sp. (Plate 3)	3.00
Marl, brown	0.02
Limestone, seminodular, cohesive, with *Cenoceras* cf. *inornatum*	0.35
Marl, brown	0.02–0.05
Limestone, seminodular, cohesive	c.0.20
Marl	0.20
Limestone, seminodular, cohesive	0.25
Marl	0.03
Limestone, seminodular	0.20

Faunas obtained from the upper, rubbly limestones in nearby extensions to this section include *Parkinsonia* cf. *convergens* from 2 m above the uppermost marl and indicative of the *zigzag* Zone, *convergens* Subzone, [6711 2101]; *Kallirhynchia* cf. *expansa, Pholadomya* sp., *Garantiana* cf. *garantiana* and *Planisphinctes*? of possible *garantiana* Zone age [6713 2102]; *Pleurotomaria* sp., *Pholadomya lirata, Parkinsonia* cf. *pachypleura* and *Parkinsonia* sp., of possible *zigzag* Zone age [6715 2104]; *Garantiana* cf. *garantiana* and *Parkinsonia* sp., possibly from the top of the *garantiana* Zone [6718 2109], and *Sphaeroidothyris* sp. juv., *Stiphrothyris?* and *Parkinsonia* aff. *frederici* from the *parkinsoni* Zone [6720 2108]. At this locality, the Rubbly Beds appear to span the full range of the member, but preservation of the fauna is poor. Ammonites are crushed, distorted or fragmented; belemnites are usually fragmented or broken. It is likely that much of the damage is penecontemporaneous.

CRACKMENT LIMESTONES

The basal 2.1 m of massive limestones with marl partings of the overlying Crackment Limestones are still seen on the south side of the cutting on the A30 near Crackment Hill (Richardson in White, 1923), while an adjacent quarry (now filled) showed an additional 3.85 m of alternating flaggy limestones and marls.

Rubbly Beds and Crackment Limestones from the southern outcrop have been described in a section (locality 12, Figure

19) [6741 1715] near Goathill by Torrens (1967), but are now partly overgrown or obscured. According to Torrens, a fault separates 2.36 m of Rubbly Beds with a fauna that includes many *Parkinsonia*, and which he places in the middle or lower part of the *parkinsoni* Zone, from 1.65 m of regularly bedded limestones with thick marl partings, of the Crackment Limestones. An ammonite (*Morphoceras macrescens*) from the latter beds shows them to belong to the middle part of the *zigzag* Zone (*macrescens* Subzone).

The exposure (locality 13, Figure 19) [6752 2090] in the road cutting at the disused Milborne Port railway station (Richardson, 1916) reveals 3.5 m of shelly and recrystallised limestones, with an irregular seminodular layering. They are overlain by flaggy limestones, with abundant brown-weathering ooliths or peloids in a sparsely bioclastic, micritic matrix, of the Crackment Limestones.

North-north-east of the old station, a trench (locality 14, Figure 19) [6751 2100 to 6753 2103] showed 400 mm of fossiliferous and oolitic limestones from which the following fauna was obtained: *Kallirhynchia* sp., *Amphitochilia duplicatus*, *Ampullospira?*, *Modiolus imbricatus*, *Myoconcha crassa*, *Protocardia lycetti*, *Protocardia stricklandi*, *Tancredia* sp., *Trigonia* sp. and *Parkinsonia* sp., indicative of a *parkinsonia/zigzag* Zone age. The limestones resemble the Crackment Limestones in the cutting at Milborne Port station.

GREAT OOLITE GROUP

The Great Oolite Group was studied in the Bath area by William Smith in the early nineteenth century. He described thick oolitic limestones above a mudstone unit that contains commercial fuller's earth (montmorillonite-rich clay). His observations still form the basis of the detailed subdivision of the Great Oolite Group in the area north of the Mendip Hills where four principal units, the Fuller's Earth, Great Oolite, Forest Marble and Cornbrash formations, are recognised. Marked facies changes occur southwards within the Great Oolite Group between Bath and Shaftesbury; in particular, the oolitic limestones of the Great Oolite Formation are replaced by the Frome Clay in the Frome district. Thus the Bathonian interval in the Wessex Basin is a mudstone-dominated sequence (Figure 23) (Penn et al., 1979). In the present district, the Great Oolite Group comprises the Fuller's Earth, Frome Clay, Forest Marble and Cornbrash formations in ascending sequence (Figure 23).

Where the lateral transition from limestone to mudstone occurs, north of Frome, there is little change in thickness. Farther south, across the Wessex Basin, the mudstone successions of the Fuller's Earth and Frome Clay are greatly expanded (Figure 23). Lithological and thickness changes at the level of the Forest Marble and Cornbrash are much less significant. The facies boundary between the oolitic limestones of the Great Oolite Formation and the mudstones of the Frome Clay occurs at depth along a line trending south-east from the Mendip Hills to the Hampshire coast.

Fuller's Earth

The Fuller's Earth was divided by Woodward (1894) into argillaceous lower and upper members separated by a limestone interval, the Fuller's Earth Rock. These divisions are applicable in the Shaftesbury district. Strata now included in the Frome Clay (Penn and Wyatt, 1979) were formerly included with the Upper Fuller's Earth (White, 1923). A thin argillaceous limestone unit, the Wattonensis Beds, locally rich in brachiopods, marks the base of the Frome Clay. The Purse Caundle Borehole (Figure 23) provides a reference lithological succession for the Shaftesbury district (Barton et al., 1993).

The Fuller's Earth shows a progressive thickness increase between the Mendip Hills and the Dorset coast (Figure 23). In the Shaftesbury district, the formation thickens south-westwards from less than 60 m to more than 110 m (Figure 24). Close to the outcrop, the BGS Purse Caundle Borehole (Barton et al., 1993) and the Stowell Borehole (Pringle, 1909, 1910) proved 77.5 and 70.9 m of Fuller's Earth respectively. Thin sequences are

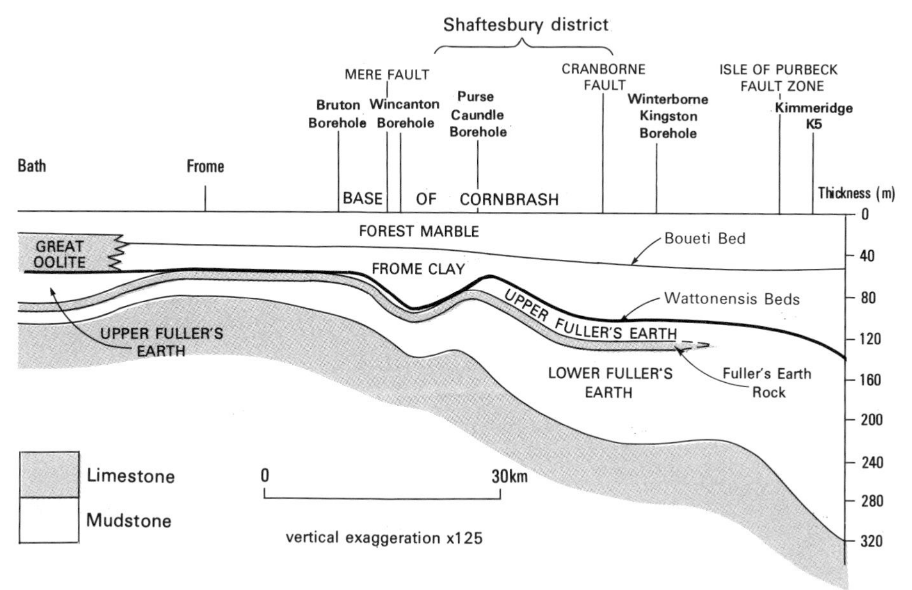

Figure 23 Diagrammatic cross-section through the Great Oolite Group between Bath and the south coast (data north of Frome taken from Penn et al., 1979).

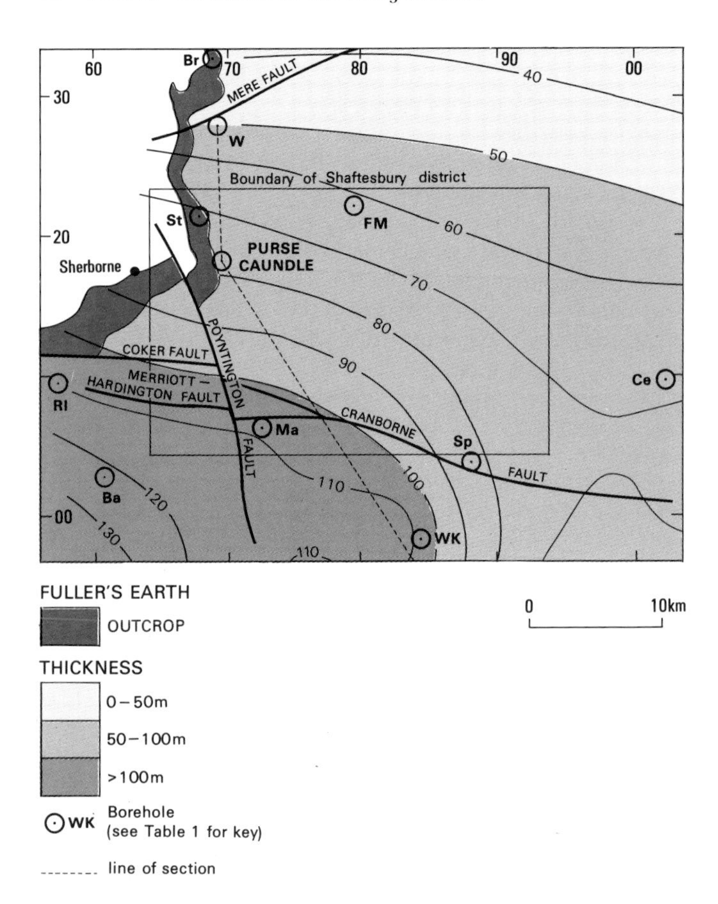

Figure 24 Isopachyte map for the Fuller's Earth of the Shaftesbury and surrounding district (contours at 10 m intervals).

typical of the area north of the Mere Fault; less than 50 m of Fuller's Earth is present in the Norton Ferris and Shrewton boreholes, and in the Bruton Borehole (Holloway and Chadwick, 1984).

Ammonites from the base of the Fuller's Earth in the Purse Caundle (Barton et al., 1993) and Winterborne Kingston (Penn, 1982) boreholes suggest a mid-*zigzag* Zone age (?*macrescens* Subzone). This date is slightly earlier than the *yeovilensis* Subzone age suggested for the basal strata in the Bath–Frome area (Penn and Wyatt, 1979).

LOWER FULLER'S EARTH

A combination of lithology and fauna allow a subdivision of the Lower Fuller's Earth in the Purse Caundle Borehole into five named units (Figure 25). Whilst each unit can be recognised at outcrop, exposures are insufficient to map their distribution.

Lenthay Beds The basal 6 to 8 m of Lower Fuller's Earth in the district consist of medium olive-grey, calcareous, silty mudstones with a single interval, less than 1 m thick, of argillaceous, creamy grey, bioclastic limestone with abundant *Sphaeroidothyris lenthayensis* (the **Lenthay Limestone**). The same limestone has been described in temporary sections west of Sherborne as the *Sphaeroidothyris* bed (Fowler, 1957) and the Lenthay Bed (Torrens, 1968; Cope et al., 1980b). The Lenthay Limestone is traceable in the area east and north-east of Milborne Port (Barton, 1989; Figure 19), above a lenticular unit of up to 2 m of calcareous silty clays which forms the basal bed of the Fuller's Earth.

In the Purse Caundle Borehole, the Lenthay Beds are 6.96 m thick. There, the Lenthay Limestone is 0.32 m thick and close to the base of the unit; it lacks the diagnostic brachiopod. The limestone has a coarse shelly base that passes up into extensively bioturbated micrite. A resistivity log of the borehole shows a well-defined peak at this level. Similar peaks occur close to the base of the Fuller's Earth in the resistivity logs of both the Winterborne Kingston and Fifehead Magdalen boreholes. These data suggest that, though thin, the Lenthay Limestone can be traced across much of north Dorset.

The fragmentary ammonite fauna in the Lenthay Beds of the Purse Caundle Borehole includes *Parkinsonia*?, *Oxycerites* cf. *yeovilensis* and *Zigzagiceras*?, suggesting a *zigzag* Zone age, with possible *macrescens* Subzone beds, overlain by beds of *yeovilensis* Subzone age. Ammonites indicative of a *yeovilensis* Subzone age were reported by Torrens (in Cope et al., 1980b) from the 'Lenthay Bed' west of Sherborne.

Knorri Beds (= *knorri* Clays and *Ostrea knorri* Clays of Richardson, 1916, from the Doulting–Milborne Port district) The Knorri Beds in the Purse Caundle Borehole consist of 10.73 m of medium olive-grey, silty, predominantly fissile, weakly calcareous mudstones with thin beds of mudstone containing abundant *Catinula knorri* (Plate 3). The range of the oyster is discontinuous, with the greatest abundances approximately 0.5 to 1.5 m and 5 to 8 m above the base of the unit. Other fauna includes *Sphaeroidothyris* and bivalves including *Entolium* and *Pinna*. Ammonite fragments include *Oecotraustes (O.)* cf. *bomfordi*, indicative of a *yeovilensis* Subzone age; this is older than the *tenuiplicatus* Zone age assigned to the Knorri Beds by Torrens (in Cope et al., 1980b). A *yeovilensis* Subzone age is also suggested by the presence of *Zigzagiceras?* in the Knorri Beds, approximately 10 m above the base of the Fuller's Earth, in the Stowell Borehole.

Hanover Wood Beds (Barton et al., 1993; named from Hanover Wood [682 172], Milborne Port) The middle part of the Lower Fuller's Earth succession in the district consists of soft, dark olive-grey, fissile mudstones. The beds are extensively bioturbated and contain abundant minute pyritised burrows and larger, bedding-parallel burrow traces. The Hanover Wood Beds in the Purse Caundle Borehole are 19.12 m thick and contain shelly intervals and thin, non-fissile calcareous beds. Shell debris, dominated by *Bositra buchii*, commonly pyritic, is scattered throughout. The range of *Oecotraustes (O.)* cf. *bomfordi* extends to within approximately 2 m of the top of the Hanover Wood Beds; if this taxon is restricted to the *yeovilensis* Subzone (Hahn, 1968), it suggests that the junction of the *zigzag* and *tenuiplicatus* zones is within the member, at c.35 m above the base of the Fuller's Earth (Table 4). However, Penn (1982) placed this zonal boundary within the Knorri Beds.

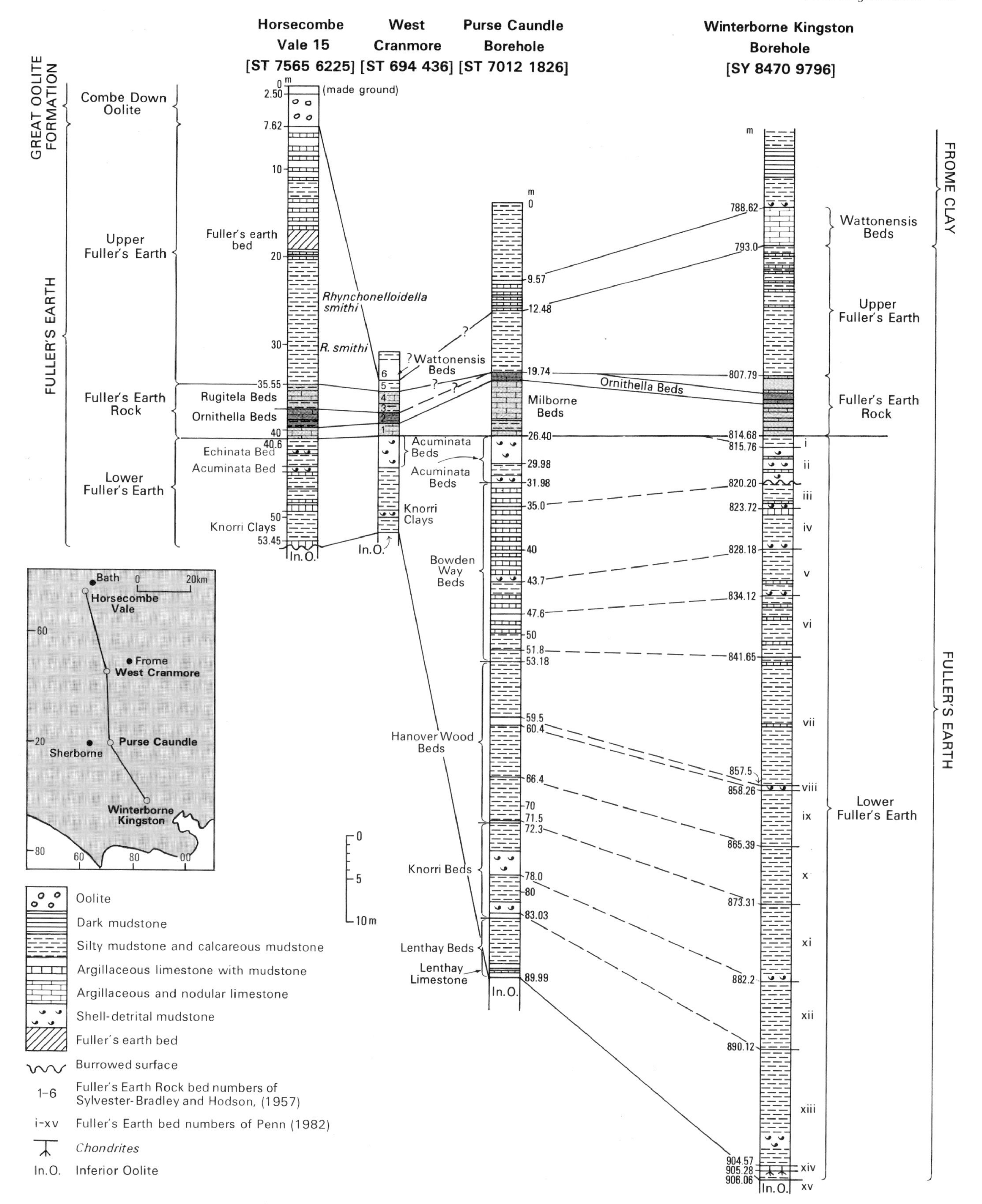

Figure 25 Correlation of the Fuller's Earth in Somerset and Dorset (with data from Barton et al., 1993; Penn and Wyatt, 1979; Penn, 1982; Torrens, 1969; Cope et al., 1980b).

Table 4
Correlation of the Great Oolite Group in the district.

Stage	Substage	Zone	Subzone	Sherborne area (after Cope et al., 1980b)	Shaftesbury district (this work)	
Callovian		Herveyi	Kamptus	Kellaways Clay	Kellaways Formation	
			Terebratus	Upper Cornbrash (c. 0.2–7.5 m)	Upper Cornbrash 2.5–9 m	Great Oolite Group
			Keppleri			
Bathonian	Upper	*Clydoniceras (Clydoniceras) discus*	*C. (C.) discus*	* Lower Cornbrash (c. 3.7 m)	Lower Cornbrash (3–5 m)	
			C. (C.) hollandi	Forest Marble 39 m	Forest Marble 37–55 m	
		Oxycerites orbis		Boueti Bed Upper Fuller's Earth Clay 30 m	Frome Clay: Boueti Bed Frome Clay 45–65 m	
		Procerites hodsoni		† Wattonensis Beds + 5 m Middle F.E.C. 9 m † Ornithella Beds 0.6 m	Frome Clay: Wattonensis Beds 0–3 m Upper Fuller's Earth (2–21 m) Ornithella Beds 0.6–3.4 m	
	Middle	*Morrisiceras (Morrisiceras) morrisi*		Fuller's Earth Rock: † Linguifera Bed 0.6–1 m	Fuller's Earth Rock: Milborne Beds 6.1–7.3 m	
		Tulites (Tulites) subcontractus		Fuller's Earth Rock: † Thornford Beds +7.5 m		
		Procerites progracilis		Lower Fuller's Earth Clay: † Acuminata Beds ?c.1.5 m	Lower Fuller's Earth: Acuminata Beds 5.6 m	
	Lower	*Asphinctites tenuiplicatus*		Lower Fuller's Earth Clay: Clay	Lower Fuller's Earth: Bowden Way Beds 16–21 m	
		Zigzagiceras (Zigzagiceras) zigzag	*Oppelia (Oxycerites) yeovilensis*	* Lenthay Bed 0.3 m Clay 2 m	Hanover Wood Beds 19.1 m Knorri Beds 10.7 m Lenthay Beds 6–8 m	
			M. (Morphoceras) macrescens	* Crackment Limestones	Lenthay Limestone 0.3 m Clay 0–2 m	
			P. (Parkinsonia) convergens	*	Inferior Oolite	

Bowden Way Beds (Barton, 1989; type-section [6837 2015 to 6870 2005] locality C, Figure 19, Old Bowden Way, Milborne Port) The Bowden Way Beds consist of 16 to 21 m of medium olive-grey, slightly silty, calcareous mudstones that contain numerous thin intervals of micritic argillaceous limestone, particularly in the upper part of the unit. The Bowden Way Beds and the Acuminata Beds have been mapped together, undivided, from the southern boundary of the district as far north as the Laycock railway cutting [6785 2134] (locality D, Figure 19); north of the cutting, the units are less easily recognised and have not been mapped.

Limestones within the Bowden Way Beds are pale olive-grey, sparsely shelly and very slightly bioturbated, and have transitional boundaries with calcareous mudstones. Twenty limestone beds in the range 0.10 to 0.34 m thick were recognised in the Purse Caundle Borehole; about half this number crop out in the section along Old Bowden Way. Mudstone intervals are faunally dominated by comminuted and pyritised *Bositra buchii*, except in a 2 m thick, sparsely bioclastic bed near the top of the unit. Beds with abundant intact *Meleagrinella echinata* occur locally, together with *Kallirhynchia*, *Modiolus* and *Pholadomya* cf. *socialis*.

Acuminata Beds (= *Ostrea acuminata* Beds of Welch and Crookall, 1935; Arkell and Donovan, 1952; based on sections in Somerset and the southern Cotswolds) The uppermost beds of the Lower Fuller's Earth consist of olive-grey, silty, shell-detrital mudstones with abundant *Praeexogyra acuminata*. Thin argillaceous limestones that have gradational boundaries with calcareous mudstones, and which also contain *P. acuminata*, are present at the top of these strata; they are 5.58 m thick in the Purse Caundle Borehole. The oyster forms a sparse lumachelle, together with *Kallirhynchia*, *Camptonectes*, *Oxytoma* and *Meleagrinella*, and is most abundant in the uppermost 3.5 m of beds. Fragments of *Oecotraustes?*

occur in the Purse Caundle Borehole, but are not zonally diagnostic.

The uppermost beds of the Lower Fuller's Earth in the Stowell Borehole include a 1.5 m thick oyster-rich bed with abundant *P. acuminata.* The Acuminata Beds correlate with *acuminata*-rich mudstones, approximately 4.5 m thick, in the Winterborne Kingston Borehole (unit ii of Penn, 1982).

FULLER'S EARTH ROCK

The Fuller's Earth Rock in the district is a 5 to 11 m-thick sequence of bioturbated argillaceous limestones and calcareous mudstones, with some bioclastic limestones. The junction with the Lower Fuller's Earth is a rapid upward transition from medium olive-grey, calcareous, silty mudstone into pale olive-grey, very shelly, muddy limestone containing *Radulopecten vagans* and crustacean fragments. Torrens (in Cope et al., 1980b) puts the base of the Fuller's Earth Rock at about the base of the *subcontractus* Zone (Table 4).

The Fuller's Earth Rock in the Purse Caundle Borehole is 6.66 m thick and can be divided into a lower 6.1 m thick unit, the Milborne Beds, and an upper 0.56 m thick unit, the Ornithella Beds. Kellaway and Wilson (1941) suggested an average thickness of 10.7 m for the Yeovil and Sherborne area, a value comparable with that in the Stowell Borehole. However, a thickness in the range 5 to 7 m appears to be more typical of the present district (proved in the Purse Caundle, Fifehead Magdalen, Mappowder and Winterborne Kingston boreholes). The Fuller's Earth Rock is thin or absent from western and southern Dorset (Wilson et al., 1958; Penn, 1982).

The argillaceous limestones in the lower part of the Fuller's Earth Rock contain a relatively sparse fauna. The common *Morrisiceras morrisi* (Plate 3) collected as field brash near Purse Caundle village, and a fragment of *Morrisiceras?* from near the top of the Milborne Beds in the Purse Caundle Borehole, indicate a *morrisi* Zone age. Elsewhere in the district, older horizons have yielded *Tulites* spp. of *subcontractus* Zone age (Table 4).

Strata in the middle of the Fuller's Earth Rock in the Purse Caundle Borehole are argillaceous shelly limestones that show extensive bioturbation and anastomosed intervals of dark grey, shelly, silty mudstone. The fauna from this interval includes *Kallirhynchia, Anisocardia fullonica, Camptonectes, Ceratomya?, Homomya gibbosa, Inoperna plicatus,* ostreid fragments, *Pholadomya lirata* and belemnite fragments. The uppermost beds are richly fossiliferous, pale olive-grey, bioturbated, argillaceous limestones.

An abundant ornithellid fauna within the top 0.56 m of strata in the Purse Caundle Borehole suggests that these are the Ornithella Beds, described from the Fuller's Earth Rock of the Frome area by Sylvester-Bradley and Hodson (1957). The fauna contains some ornithellids with centrepetal infills, and includes *Kallirhynchia platiloba, Ornithella bathonica, O.* sp., *Wattonithyris fullonica, W.* sp., *Camptonectes laminatus, Chlamys?, Entolium corneolum, Modiolus imbricatus* and *M.* sp. The overlying Rugitela Beds, described from the Frome area by the same authors, were not proved in the borehole. They are also missing in the Stowell Borehole, where 3.4 m of earthy limestones (Ornithella Beds) overlie 7.3 m of sandy limestones (Milborne Beds).

Torrens (in Cope et al., 1980b) reported ammonites of basal *hodsoni* Zone age from a small pit [6707 1756] in the Fuller's Earth Rock adjacent to the Poyntington Fault (locality G, Figure 19). Strata from the same locality yielded abundant dwarf ornithellids (McKerrow, 1953), comparable with those described by Muir-Wood (1936) from Troll Quarry [594 127], 5 km south-west of Sherborne. In the Winterborne Kingston Borehole, the characteristic faunas of the Ornithella Beds occur in a unit 0.9 m thick (unit ii of Penn, 1982); its dignostic ornithellids are missing from the overlying unit, which is lithologically comparable with the Rugitela Beds near Bath (Penn et al., 1979).

UPPER FULLER'S EARTH

The Upper Fuller's Earth is a pale to medium olive-grey, calcareous silty mudstone sequence that includes thin, highly calcareous mudstone or argillaceous limestone beds, together with some shelly mudstone intervals. The member is absent in the district north of Charlton Horethorne, where the Frome Clay rests on Fuller's Earth Rock (Taylor, 1990). The Upper Fuller's Earth is 2.1 m thick in the Fifehead Magdalen Borehole and 7.26 m thick in the Purse Caundle Borehole, and attains a maximum thickness of 21 m in the Mappowder Borehole in the south of the district.

The lowermost strata of the Upper Fuller's Earth in the Purse Caundle Borehole consist of bioturbated and burrowed, shell-detrital mudstones and thin beds (0.08 to 0.25 m thick) of argillaceous limestone or calcareous silty mudstone. The basal mudstones are burrowed in part, although they retain an indistinct horizontal lamination.

The upper beds are moderately fissile, silty, shell-detrital mudstones that contain a uniform scatter of fine-grained shell debris, together with intact bivalves and larger shell fragments. Sparse small flakes of muscovite occur and pyritised *Chondrites*-type burrows are abundant; one thin bed has a waxy texture. Uniform olive-grey, soft, silty mudstones with sparse *Bositra* debris at the top of the member pass up, by progressive increase in shell material, into shelly mudstones of the Wattonensis Beds.

Oysters are common near the base of the member and include *Catinula matisconensis, Praeexogyra hebridica subrugulosa* and *P. acuminata.* Other fauna include *Camptonectes laminatus, Ornithella bathonica, Modiolus* and *Pinna*; fossils decline in abundance upwards. The calcareous interval is bioturbated and shelly, with abundant rhynchonellid fragments that include *Rhynchonelloidella wattonensis* and *Kallirhynchia,* while beds above contain the bivalves *Bositra buchii, Entolium corneolum, Modiolus imbricatus, Pinna subcancellata,* the ammonite *Oecotraustes* sp. and crinoid fragments. No zonally diagnostic ammonite was recovered. Both the underlying Fuller's Earth Rock (Torrens in Cope et al., 1980b) and the overlying Frome Clay have yielded *hodsoni* Zone ammonites.

1a
1b
1c
2a
2b
3
4
5a
5b
6a
6b
6c
7a
7a
7b
7c
8
9a
9b
10a
10b
10c
11
12
13a
13b
14a
14b
15
16a
16b
16c
17

Plate 3 Fossils from the Middle and Lower Jurassic.

1a–c. *Obovothyris magnobovata* S S Buckman; Lower Cornbrash; Long Burton, Sherborne. GSM 62732.
2a–c. *Goniorhynchia boueti* (Davidson); base of Forest Marble, Boueti Bed; Milborne Port. CM 207.
3. *Pholadomya lirata* (J Sowerby); Frome Clay, Wattonensis Beds; Milborne Port. CM 173
4. *Modiolus anatinus* (Wm Smith); Fuller's Earth Rock: Stowell. CL 440
5a,b. *Morrisiceras morrisi* (Oppel); Fuller's Earth Rock, *morrisi* Zone; Gospel Ash, Purse Caundle. CM 371
6a–c. *Rhynchonelloidella smithi* (Davidson); Fuller's Earth Rock; West Hill, Sherborne. GSM 62789
7a–c. *Wattonithyris nunneyensis* (S S Buckman); Fuller's Earth Rock; Stowell Borehole between 36–37.5 m. Pg 4565.
8. *Praeexogyra acuminata* (J Sowerby); Lower Fuller's Earth, Acuminata Beds; Laycock railway cutting, Stowell. CM 476.
9a,b. *Catinula knorri* (Voltz); Lower Fuller's Earth, Knorri Beds; Milborne Wick. CL 371.
10a–c. *Sphaeroidothyris lenthayensis* (Richardson & Walker); Lower Fuller's Earth; Milborne Wick. CM 260.
11. *Procerites incognitus* Arkell; Crackment Limestones, *convergens* Bed, *zigzag* Zone; Ven House, Milborne Port. GK 276, × 0.5, Holotype.
12. *Belemnoposis* sp.; Upper Inferior Oolite, Rubbly Beds; Milborne Wick. CL 266.
13a,b. *Acanthothiris spinosa* (Linnaeus); Upper Inferior Oolite; Charlton Horethorne. CL 321.
14a,b. *Dorsetensia subtecta* S S Buckman; Inferior Oolite. *humphriesianum* Zone; Frogden Quarry, Oborne. GSM 32010, Holotype.
15. *Normannites formosus* (S S Buckman); Inferior Oolite, *humphriesianum* Zone; Frogden Quarry, Oborne. GSM 32010, Holotype.
16a–c. *Homeorhynchia ringens* (von Buch); Lower Inferior Oolite, Ringens Bed; Purse Caundle Borehole at 130.50 m. BKA 5133.
17. *Pleydellia aalensis* (Zieten); Bridport Sands, *aalensis* Subzone; Haselbury Mill. GSM 69983.

The lithology and fauna of the Upper Fuller's Earth in the district correlate well with the 14.8 m thick sequence in the Winterborne Kingston Borehole; both sequences contain *Bositra, Modiolus* and *Pinna subcancellata.* There is no indication of the sedimentary and faunal cycles of the type described by Penn and Wyatt (1979) from the 24 m-thick sequence in the Upper Fuller's Earth in the Bath–Frome area.

The palaeogeography of the Fuller's Earth is discussed on p.41.

Details

LOWER FULLER'S EARTH

The Lenthay Limestone in a temporary section (locality A, Figure 19) [6806 1896] just south of Milborne Port cemetery consists of approximately 0.3 m of rubbly limestone crowded with *Sphaeroidothyris lenthayensis,* together with *Acanthothiris* sp. juv. and *Pholadomya lirata,* above alternating mottled grey clays and silts 1 to 2 m thick. Extensive brash also occurs on a 200 m-wide dip slope of Lenthay Limestone in ground north of the cemetery where *Sphaeroidothyris* is locally abundant. Similar brash occurs on the eastern slopes of Highmore's Hill, adjacent to the Poyntington Fault [e.g. 6699 1802].

The Lenthay Beds are thought to occur within the transition from Crackment Limestones to Lower Fuller's Earth in a temporary section (locality B, Figure 19) [6808 1823] near Ven, described by Kellaway (1938), in which clays, with occasional hard calcareous mudstone or shale beds (Lower Fuller's Earth), rest on beds here reclassified as follows:

	Thickness m
Soil	0.30
LENTHAY BEDS	
Limestone, thin-bedded and pink	0.45
Clay, green and yellow with nodules of soft brown limestone in part	0.45
Marl, calcareous, buff	0.30
Clay, mottled	0.23
Limestone, impersistent, hard brown	0.05–0.13
Clay, green and yellow	0.30
Limestone with thin clay partings: yellow with hard blue centre	0.45–0.61
LENTHAY LIMESTONE	
Limestone, hard, blue with clay partings, *Sphaeroidothyris* sp., *Acanthothiris* sp., fossil wood	0.91
section gap	
CRACKMENT LIMESTONES	
Limestone, blue, weathering yellow with *Oppelia* sp.	3.03–3.33
section gap	
Ochreous sandy limestones with sandy clay partings yielding abundant *Procerites, Gonolkites* etc.	0.30–0.61

Up to 2 m of fine-grained, slightly sandy, flaggy and buff-weathering limestone, the Lenthay Limestone, occurs in a ditch [6789 2010] within the northernmost outcrop. The limestone is separated from the Inferior Oolite by about 0.9 m of buff-weathering clay and calcareous silt. The base of the limestone yielded *Sphaeroidothyris lenthayensis* (Plate 3).

Although the Knorri Beds are unexposed, thin calcareous silts overlain by pale brown silty clays with abundant oysters that include *Catinula knorri* can be augered south of Ven [6820 1814], and also south of Goathill [6771 1695]. *Catinula* cf. *ampulla, C. knorri, C. matisconensis* and ostreids were found near Spurles Farm [6888 1958], low down in the formation.

There is no permanent exposure of Hanover Wood Beds, although sparsely fossiliferous, mottled grey-brown, weakly calcareous silts and clays can be seen in tree roots within Hanover Wood below the scarp of Fuller's Earth Rock. Medium to dark olive-grey clays or slightly silty clays crop out [6624 1818; 6627 1836] on either side of the A30 road between Oborne and Vartenham Hill. Foraminifera obtained from an augered sample [6620 1821] include *Citharina clathrata, Lenticulina muensteri, L. exgalatea, L. tricarinella, Planularia beierana, P. eugenii, Epistomina stelligera* and *Dentalina* sp.; ostracods included *Nophrecythere rimosa* (poor preservation), *N. bessinensis* and *Paracypris* sp. The ostracod fauna contains the index of the *N. rimosa* ostracod Zone, which ranges from the highest *parkinsoni* Zone to approximately the top of the *zigzag* Zone (Sheppard, 1981).

The Bowden Way Beds exposed along Old Bowden Way (locality C, Figure 19) [6837 2015 to 6870 2005] consist of at least 11.6 m of thinly bedded, sparsely bioclastic, argillaceous limestones interbedded with calcareous silts and silty clay. One of the limestones [6837 2014] yielded *Kallirhynchia* sp. Each limestone is about 200 mm thick, and at least ten occur within the sequence, principally at three levels.

Oyster-rich marls and clays, c.4 m thick, occur above the Bowden Way Beds in the same locality, and can be augered

[6840 2005] immediately below the Fuller's Earth Rock next to an old lime kiln. The beds yield abundant *Praeexogyra acuminata*, together with *Meleagrinella*. The Acuminata Beds are also exposed [6833 1591] close to an old quarry in Fuller's Earth Rock near Tripp's Farm, where a 3 m-thick clay–limestone succession in the quarry floor [6836 1588] includes at least 0.5 m of oyster-rich marl. The same oyster-rich interval can be augered in other localities, for example near Pinford [6618 1735], south of Gospel Ash Farm [6916 1884], and east of Poyntington village [6522 2011 and 6534 2014], where the beds are repeated by faulting.

At the bottom of the Laycock railway cutting near Milborne Wick (locality D, Figure 19) [6785 2134], 2 to 3 m of soft, thinly bedded, yellow- to buff-weathering, bluish grey, calcareous siltstones interbedded with mudstones are exposed. Loose fossils include *Rhynchonelloidella smithi crassa*, *Pholadomya lirata* and *Praeexogyra acuminata* (Plate 3) from the top of these beds, and fragments of sandy limestone with *Procerites* sp. and *Belemnopsis* sp. (see also below).

Fuller's Earth Rock

The Ridge trackway east of Poyntington has a succession of small exposures on the track surface (locality E, Figure 19) [6529 2010 to 6536 2016], of which three are described. The first [6529 2010], shows c.2 m of fine-grained, silty, greyish buff weathering, soft to friable, slightly micaceous limestone, in irregularly nodular beds 100 to 300 mm thick; shell fragments, up to 3 mm across, and some carbonaceous traces occur. The second [6532 2011], exposes about 1 m of buff, pseudonodular limestone in a soft marly matrix; these are probably the Milborne Beds. The irregular limestone pseudonodules, 10 to 200 mm in diameter, are separated by up to 10 mm of anastomosing marl partings. The exposure yielded *Rhynchonelloidella wattonensis*, *Wattonithyris* sp., *Amberleya* sp., *Anisocardia minima*, *Inoperna plicatus* and tulitid fragments; the brachiopods occur mainly in the matrix. The third exposure [6535 2015 to 6536 2016], shows fine-grained, silty, buff- to ochreous-weathering, rubbly, nodular limestone. The limestone is hard and medium grey when fresh. Fossils include worm tubes, terebratuloids, large bivalves, including *Ceratomya?* and *Pholadomya lirata,* and *Belemnopsis* sp.

Brash from the Ornithella Beds about 150 m north-west of the track yielded numerous brachiopods, including *Rhynchonelloidella tutcheri*, *R. smithi*, *R. smithi crassa*, *Ornithella bathonica bathiensis*, *O.* cf. *constricta*, *'Terebratula' richardsoni* and *Wattonithyris fullonica*, and the bivalves *Inoperna plicatus* and *Modiolus anatinus* (Plate 3).

A quarry [6745 2240], 1 km south-east of Charlton Horethorne, exposes 1.5 m of fine-grained, slightly bioclastic, buff-weathering limestone, flaggy at the top, but more massive and poorly bedded below, with marly partings 0.1 to 0.3 m thick. The fauna includes *Montlivaltia* sp., *Rugitela?*, *Sphaeroidothyris* sp. juv., *Entolium* sp., *Pholadomya moschi*, *Protocardia* sp. and *Quenstedtia* sp.; there are some vertical tubular burrows.

A disused quarry [6753 2185] near Starve Acre, 1 km north-east of Milborne Wick, exposes 0.1 to 0.2 m of topsoil and rubble, overlying 0.3 to 0.4 m of shelly, massive limestone with irregular joints and partings. The fauna includes *Wattonithyris* cf. *nunneyensis*, *Rhychonelloidella smithi*, *Pleuromya uniformis* and fragmentary small belemnites. Loose material nearby yielded *Montlivaltia* sp., *Kallirhynchia* sp., *Ornithella cordiformis*, *Rhynchonelloidella smithi*, *R. wattonensis*, *Rugitela?*, *Wattonithyris* cf. *nunnyensis*, *Inoperna plicatus* and *Modiolus anatinus*.

A roadside quarry (locality F, Figure 19) [6772 2152] 1 km south-west of Stowell exposes:

	Thickness m
Topsoil and rubble	c.0.04
Ornithella Beds	
Limestone, shelly, fine-grained, hard, buff-weathering, greyish mottled, bioturbated; fauna includes *Ornithella* cf. *cinctaeformis*, *Wattonithyris fullonica*, *Pleurotomaria* sp. and *Modiolus anatinus*	0.24
Marly clay, ochreous- to buff-weathered; *Ornithella* cf. *cinctaeformis* (commonly crushed), indet. rhynchonellids, *Belemnopsis* sp.	0.14
Limestone, rubbly, friable with marly partings and seams. Ochreous inclusions in the lower 0.15 m and marly seam at base, 20 mm thick. Bivalves and brachiopods in marly layers. Top surface of the bed very uneven, almost nodular in appearance. Ornithellids?, *Rugitela?*, *Rhynchonoidella smithi crassa*, *Gresslya peregrina*	0.56
Milborne Beds	
Limestone, slightly shelly, fine-grained, buff-weathering; more massive and hard when fresh. Close, very irregular jointing; thin marly parting at base c.0.01 m. The fauna includes *Anisocardia minima* and *Limatula gibbosa*	0.25–0.30
Limestone, massive, as above, thin ochreous parting 10 mm thick with *Modiolus anatinus* and *Pleuromya uniformis* at base	0.30
Limestone, massive, as above, ochreous parting with *Rugitela*? at base	0.25
Limestone, massive, as above. *Tulites mustela* at top indicates the *subcontractus* Zone	0.60

The jointing is irregularly spaced and 100 to 300 mm apart, and the beds are inclined at 18° toward N 070°.

The Laycock railway cutting (locality D, Figure 19) [6785 2134] still exposes 3 m of Lower Fuller's Earth, overlain by 3 to 4 m of thickly bedded, sparsely fossiliferous limestone. The section as described by Woodward (1894), and here reclassified and metricated, is as follows:

	Thickness m
Fullers's Earth Rock	
Grey and brown earthy limestones, rubbly on top and very fossiliferous	2.43–2.74
Thicker beds of buff earthy limestone, shelly in places, the shells weathering out on joint surface	2.74–3.05
Lower Fuller's Earth	
Clays [*Acuminata* Beds] (fallen material) Dark bluish grey marls, with indurated bands of light bluish grey earthy limestone. Casts of *Myacites* in natural position, and *Pholadomya*	4.57
Clays, not well exposed	

A good section of the Ornithella Beds is in Goathill Quarry (locality G, Figure 19) [6707 1756], 650 m north-west of Goathill Church (Plate 4). The beds are located in the hangingwall of the Poyntington Fault and dip 15° west. The limestones are extensively veined by calcite and limonite.

	Thickness m
Soil	
ORNITHELLA BEDS	
Limestone, argillaceous, rubbly and broken	0.60
Limestone, argillaceous, massive	0.65
Marl	0.05
Shell-detrital limestone, rubbly, with thin and irregular marl partings; fauna dominated by brachiopods including *Rugitela?*, *Rhynchonelloidella smithi*, *R. wattonensis*, *Stiphrothyris* aff. *birdlipensis* and *Ptyctothyris* sp.; other fossils present include *Amberleya* sp., *Anisocardia* cf. *gibbosa*, *A. minima?*, *Pleuromya* sp. and *Pseudolimea duplicata*	0.70
Limestone, argillaceous, massive, with irregular and biotubated base	0.30
Limestone, rubbly, sparsely shelly, also with irregular base	0.25
Limestone, massive, shelly, crowded with small or dwarf *Ornithella*	0.65
MILBORNE BEDS	
Limestone, massive, thick-bedded (0.5 m-scale)	0.90

McKerrow (1953) also recorded the ornithellid-rich beds in this quarry. Torrens (in Cope et al., 1980b) reported ammonites, including *Choffatia* and *Procerites*, indicative of a basal *hodsoni* Zone age.

Fossils from the *subcontractus* Zone up to the level of the Ornithella Beds are well represented in brash material collected from dip slopes of Fuller's Earth Rock. In ground between Spurles Farm and Gospel Ash Farm [e.g. 6905 1920 and 6905 1932], as well as farther south near Purse Caundle [6952 1850], the ammonites *Tulites cadus*, *T.* cf. *mustela* and *Procerites* sp. occur. However, *Ornithella bathonica*, *O.pupa*, *Kallirhynchia* cf. *platiloba*, *Pleurotomaria* sp. *Gresslya* cf. *peregrina*, *Mactromya impressa*, *Modiolus imbricatus*, *Trigonia elongata* and *Belemnopsis* sp., indicative of the Ornithella Beds, were also collected from the same ground. Evidently, the fossils represent a mixture of *subcontractus* to *hodsoni* zone age.

Brash of Fuller's Earth Rock near Purse Caundle (locality H, Figure 19) [6966 1840] includes numerous specimens of the *morrisi* Zone ammonites *Morrisiceras morrisi*, *M. sphaera* and *M.* sp.

Fossils from brash west of the Poyntington Fault, north of Pinford [6654 1770], include *Ornithella bathonica*, *O. haydonensis*, *O. pupa*, *Ptyctothyris arkelli*, *Rhynchonelloidella smithi* (Plate 3), *R. wattonensis*, *Wattonithyris fullonica*, *Pholadomya lirata* and *Pleuromya calceiformis*. Fowler (1957) recorded *Ornithella haydonensis*, *O. cordiformis*, *O. boxensis*, *O. pupa*, *O. bathonica* and *Rhynchonelloidella wattonensis* from a disused quarry [probably 663 189] in Fuller's Earth Rock at the base of Vartenham Hill.

Plate 4 Ornithella Beds (Fuller's Earth Rock) at Goathill Quarry.

UPPER FULLER'S EARTH

Brash from the Upper Fuller's Earth consists in large part of broken oyster shells and unidentified bivalves. The clays and silts immediately below the Wattonensis Beds yield more shell fragments than elsewhere; they include *Modiolus lonsdalei* just north of Gospel Ash Farm [6934 1931] and *Liostrea* sp. near Purse Caundle [6997 1800].

A representative auger hole [6729 2298] within the Upper Fuller's Earth 1 km north-east of Waterloo Crescent showed:

	Thickness m
Topsoil	0.2
HEAD Stony clay, buff-brown	0.3
UPPER FULLER'S EARTH	
Clay, calcareous, silty, very firm, buff- and grey-mottled, with race; crinoid ossicle and shell fragments at base	0.5
Clay, becoming very silty or fine-grained sandy, buff- and grey-mottled by 1.2 m, crumbly and very firm	0.5
Clay predominantly pale grey, shell fragments at 1.7 m	0.3
Clay, very silty, light grey- to buff-mottled, continues with alternations of silty clay and clayey silt	0.55
Clay, calcareous, silty, medium to light grey, brownish buff-mottled, very firm; crinoid and echinoid debris	to 0.10

A stream exposure and auger holes between Charlton Horethorne and Stowell [6757 2255] showed the following sequence below the Wattonensis Beds:

	Thickness m
UPPER FULLER'S EARTH	
Mudstone, silty, calcareous, very firm	1.6
Siltstone or mudstone, moderately calcareous, moderately lithified, pale grey, buff-weathered (this unit produces a step in the stream bed)	0.2
Clay, silty, blocky, medium grey; bivalves	1.0
Clay, silty, firm	2.3
Siltstone, hard	0.1
Clay, as above	0.1

Exposures in the faulted wedge of strata at the eastern end of Sherborne Lake, show part of the sequence between the Bowden Way Beds and Frome Clay. A sample augered [6595 1671] midway between the Fuller's Earth Rock and the Wattonensis Beds yielded the foraminifera *Lenticulina muensteri, L. subalata, L. tricarinata, Citharina clathrata, Planularia beierana, Saracenaria cornucopiae, Trochammina* cf. *squamata* (crushed) and *Lagena* sp., together with the ostracoda *Oligocythereis fullonica, Glyptocythere oscillum, Progonocythere polonica, Nophrecythere rimosa, Micropneumatocythere* cf. *brendae* (juveniles) and *Eocytheridea* cf. sp. *A* sensu Bate 1978. The concurrent range of *P. polonica* and *G. oscillum* occurs only in a narrow band in the middle part of the Upper Fuller's Earth (at the base of Zone 5 sensu Bate, 1978). The fauna can be dated to the highest part of the *P. polonica* ostracod Zone sensu Sheppard (1981); by inference the middle part of the *hodsoni* Zone.

Frome Clay

The Frome Clay of the district was formerly included with the Fuller's Earth (White, 1923) but, following Penn et al. (1979), the beds between the Upper Fuller's Earth and the Forest Marble have been renamed the Frome Clay. A thin sequence of argillaceous limestones and mudstones, the Wattonensis Beds, forms the basal unit of the Frome Clay in Dorset and southern Somerset. Its base is defined by a marked increase in shell-detritus in calcareous mudstone.

The Frome Clay varies in thickness from 45 to 65 m in the district, and shows a progressive increase towards the south. The Fifehead Magdalen Borehole proved 50.6 m, the Mappowder Borehole 61.27 m and the Spetisbury Borehole 68.13 m. South of Spetisbury, the thickness decreases, with 60.3 m proved in the Winterborne Kingston Borehole, 57.6 m in the Bere Regis Borehole and 48.78 m in the Waddock Cross Borehole. North of the district, the formation varies from 36 to 45 m thick between the Mere Fault and Frome; 33.23 m was proved in the Norton Ferris Borehole.

The Frome Clay crops out below the scarp of the Forest Marble in two parts of the district: in Sherborne Park, west of the Poyntington Fault, and between Caundle Marsh and the area north of Stowell. The Wattonensis Beds have been traced from their eastern occurrence in the Yeovil district, across the district and into the Wincanton area. The Wattonensis Beds are 2.91 m thick in the Purse Caundle Borehole, 1.5 to 4 m in the Sherborne area (Kellaway and Wilson, 1941; Torrens in Cope et al., 1980b) and 4.4 m in the Winterborne Kingston Borehole (Penn, 1982).

The formation in the Shaftesbury area contains rare ammonites indicating the *hodsoni* zone of the Upper Bathonian. The succeeding *orbis* [*aspidoides*] Zone is proved in the Winterborne Kingston Borehole. Poorly preserved ammonites from the basal strata of the Frome Clay in the Purse Caundle Borehole include *Oecotraustes (Paroecotraustes) maubeugei*, a species placed by Hahn (1968) in the '*P. retrocostatum*' Zone, which was later renamed the *Procerites hodsoni* Zone by Torrens (in Cope et al., 1980b). Faunas of *hodsoni* Zone age occur in both the Upper Fuller's Earth and Frome Clay of the district; there is no evidence for the stratigraphical break at the base of the Frome Clay described by Penn and Wyatt (1979) north of the Mendip Hills.

The Frome Clay consists predominantly of mottled buff and olive-grey calcareous clays. The basal 12.48 m of beds were proved in the Purse Caundle Borehole; the middle and upper strata are known from shallow auger holes.

The Wattonensis Beds at the base consist of medium olive-grey, bioturbated, shell-detrital, silty mudstones interbedded with five thin, pale olive-grey, argillaceous shelly limestones, ranging from 80 to 120 mm in thickness, and a single limestone bed 50 mm in thickness. Each limestone has a sharp, irregular margin, and most have *Chondrites*-burrowed tops. Mudstone beds are approximately 0.3 m thick and contain pyritised bivalve and rhynchonellid debris, and wood fragments.

The fauna of the lowermost 2 m of the Wattonensis Beds is sparse; it includes *Kallirhynchia, Rhynchonelloidella wattonensis, Bositra buchii, Entolium?, Gervillella, Modiolus?*, ostreids, *Pholadomya?* and *Pinna?*. The fossils listed by Kellaway and Wilson (1941, p.160) as characteristic of the Wattonensis Beds occur principally in the uppermost 1m at Purse Caundle, and in the overlying silty mudstones. They include *Acanthothiris powerstockensis, Rhynchonelloidella smithi, Rugitela?* and *Camptonectes sp.* The mudstones overlying the Wattonensis Beds also yielded specimens of *Rugitela bullata* and *Ornithella pupa. Rugitela* occurs in surface brash above the Wattonensis Beds in a number of localities in the Milborne Port district (Barton, 1989).

Frome Clay above the Wattonensis Beds in the Purse Caundle Borehole consists of 9.5 m of medium olive-grey, calcareous, silty mudstones with variable abundances of shell debris. The beds are fissile in part and contain traces of bioturbation and abundant *Chondrites*-type burrows, some pyritised, together with a faint lamination due to the preferred planar orientation of shell debris. The fauna which indicates a *hodsoni* Zone age, includes *Bositra buchii, Camptonectes laminatus, Ceratomya concentrica, Cucullaea, Entolium corneolum, Falcimytilus sublaevis, Isognomon isognomonoides, Meleagrinella, Modiolus imbricatus, Pholadomya lirata* (Plate 3), *Praeexogyra hebridica, Trigonia, Choffatia?, Oecotraustes (Paroecotraustes) maubeugei, Oppelia sp.* and *Belemnopsis.*

The Frome Clay in the Purse Caundle Borehole does not show the distinctive, 5.4 m thick, black shales and mudstones unit, recorded above the Wattonensis Beds in the Winterborne Kingston Borehole (Penn, 1982). These beds yielded scattered *Praeexogyra hebridica*; a solitary *Oecotraustes* (*Paroecotraustes*) sp. was also recovered from close above the base. In the Winterborne Kingston Borehole, Penn (1982) attributed all the Frome Clay, above the Wattonensis Beds, an *orbis* [*aspidoides*] Zone age as *Oppelia* (*Oxycerites*) *aspidoides* was identified from 10.8 m up. The presence in the black shales of *Praeexogyra hebridica* was also correlated with a lumachelle of that oyster above the Wattonensis Beds on the Dorset coast (Torrens, 1969, p.A33). The junction of the *hodsoni* and *orbis* zones may be within a few metres above the top of the Wattonensis Beds in this district.

At outcrop, the upper 35 m of Frome Clay not proved in the Purse Caundle Borehole consists of mottled, pale yellowish brown, gypsiferous, silty clays that yield a calcareous microfauna which ranges from *hodsoni* to *orbis* Zone in age. The highest c.4 m of beds are pale grey, highly calcareous clays with a burrowed upper surface. These beds correlate with the uppermost Frome Clay in the Winterborne Kingston Borehole on foraminiferal evidence.

Palaeogeography and provenance of the Fuller's Earth and Frome Clay

Thickness variations in the Fuller's Earth and Frome Clay sequences suggest that the rates of subsidence in the Wessex Basin south of the Mere Fault increased steadily during the Bathonian and led to the development of a basinal mudstone facies. There is limited evidence for sedimentary cyclicity within the mudstones, but no evidence of faunal cyclicity. The rhythmic repetition of mudstones and argillaceous limestones in the Bowden Way Beds may be due to fluctuations in sea level.

Clay-mineral determinations from the Purse Caundle Borehole (Barton et al., 1993) show that the kaolinite-to-illite ratio is low throughout the Fuller's Earth and Frome Clay. Smectite is abundant in the Inferior Oolite, but declines gradually into the Lower Fuller's Earth, increases again in the Upper Fuller's Earth and falls abruptly in the Frome Clay. Similar trends were recognised in the Winterborne Kingston Borehole by Knox (1982), who related the abundance of smectite (of probable volcanic derivation) in the Upper Fuller's Earth to periods of decreased fluvial input, which allowed the concentration of smectite, rather than to increased volcanic activity. He supported earlier suggestions (see Hallam, 1975 for summary) that high kaolinite-to-illite ratios are characteristic of nearshore environments, while low ratios indicate offshore environments.

The sediment source of the mudstones has not been identified. Martin (1967) suggested that the Middle Jurassic clays were winnowed from argillaceous sequences along the western margins of the Wessex Basin. Derivation from the west is supported by data from the Celtic Sea and Western Approaches, where thick mudstone sequences of Bathonian age characterise the succession (Shannon, 1991). Links to the Wessex Basin through the Bristol Channel and North Somerset basins, as well as through the English Channel Basin, seem likely. However, derivation from the shelf located between central England and the southern North Sea is also possible.

Details

WATTONENSIS BEDS

A stream gully (locality T, Figure 19) [6760 2249 to 6757 2255], north-west of Stowell, exposed approximately 2.95 m of Wattonensis Beds:

	Thickness m
FROME CLAY	
Clay	1.0 to 2.0
WATTONENSIS BEDS	
Limestone, silty, soft, buff, with *Rhynchonelloidella wattonensis, Wattonithyris nunneyensis, Amberleya fowleri* and *Catinula* sp.	c.0.15
Clay, silty, shelly, soft	0.20
Limestone, shelly,harder	0.15
Clay	c.0.25
Limestone, silty, crumbling, buff; rhynchonellid and small oysters	0.20
Clay	0.05
Limestone, shelly, fine-grained, hard, buff-weathering	0.15
Clay	0.30

The section continues c.5 m downstream (less than 0.1 m gap):

Limestone, irregular top and base	c.0.15
Clay	0.05–0.10
Limestone, marly, soft, grey to buff	0.15
Limestone, fine-grained, slightly shelly, buff-weathering, blue-hearted, very hard, with poorly preserved bivalves at base	0.15

	Thickness m
Clay, pale grey, ochreous and buff-mottled, with silty bands; becoming medium grey, brown-mottled, silty, firm and calcareous; augered interval	1.00

Section continues (less than 0.1 m gap):

Limestone, silty, and very calcareous siltstone, soft, crumbling and friable, but hard in parts. *Rhynchonelloidella smithi, R. smithi crassa, R. wattonensis,* indeterminate terebratuloids, *Catinula matisconensis, Gresslya* sp. *Modiolus anatinus, Pholadomya lirata, Plagiostoma* sp. and *Protocardia* sp.	0.10

A gully near Henstridge Bowden (locality U, Figure 19) [6921 2011] exposes three beds, 10 to 200 mm thick, of soft, decomposed, muddy, silty limestone, with a few harder remnants of medium to pale grey limestone, separated by pale grey, ochreous-mottled clay. A sparse fauna from the lower bed includes *Kallirhynchia* sp., *Rhynchonelloidella smithi* and *Cucullaea* cf. *minchinhamtonensis.* Larger blocks of limestone on the banks downstream indicate the presence of other limestone beds; *Acanthothiris powerstockensis, Rhynchonelloidella smithi, R. wattonensis, Rugitela?, Liostrea* sp. and *Pholadomya* sp. were recovered from them.

Wattonensis Beds are exposed in a ditch 800 m north-north-east of Gospel Ash Farm (locality V, Figure 19) [6937 1975], where at least eight muddy limestone layers, interbedded with calcareous silts, crop out in a 3 m thick section. Individual limestones do not exceed 250 mm in thickness; they have irregular margins. Fossils from this locality, and from brash along strike, include *Acanthothiris* cf. *powerstockensis, Rhynchonelloidella wattonensis, R. smithi, Wattonithyris* sp., *Lopha* sp. and *Trigonia elongata.*

The Wattonensis Beds form a small scarp east of Gospel Ash Farm [6960 1897], on which *Rhynchonelloidella smithi crassa, Rugitela?, 'Terebratula' richardsoni?* juv., *Tubithyris* cf. *whatleyensis, Wattonithyris fullonica, Anisocardia truncata, Catinula* cf. *knorri* and *Pholadomya lirata* were found.

The Wattonensis Beds occur in a ditch (locality W, Figure 19) [6992 1835] north of the A30 near Purse Caundle. Argillaceous limestone rich in brachiopods is exposed along 150 m of the ditch, which defines the eastern margin of the outcrop. The fauna includes *Acanthothiris, Kallirhynchia?, Rhynchonelloidella smithi, R. wattonensis, Tubithyris* sp., *Liostrea, Amberleya fowleri* and *Lopha costata.*

South of Purse Caundle, the Wattonensis Beds form a small scarp 300 m south-east of Manor Farm, on which sparse *R. smithi* and oyster fragments occur as brash. More extensive brash occurs farther south on Manor Farm, where the Wattonensis Beds are let down between two faults. Beneath the limestone, beds in the western outcrop (locality X, Figure 19) [6910 1660] consist of alternating pale clay and calcareous silt at least 2.5 m thick, which yielded *R. smithi, R. wattonensis* and *Trigonia.*

West of the Poyntington Fault, the Wattonensis Beds occur on the lower slopes of Highmore's Hill [668 181]; they yielded *Acanthothiris powerstockensis, Rhynchonelloidella globosa, R. smithi, R. wattonensis, Rugitela powerstockensis, Tubithyris powerstockensis, Wattonithyris pseudomaxillata, W. fullonica, W. nunneyensis* and *W. wattonensis,* together with the bivalve *Trigonia elongata* (Fowler, 1944); *Trigonia* is also common in brash east of the Poyntington Fault. The thin limestones are best seen on the south-west slopes of the hill, where they form a small scarp below Crackmore Wood (locality Y, Figure 19) [6661 1777].

An outcrop in the Wattonensis Beds south of Sherborne Lake (locality Z, Figure 19) [6542 1599] shows argillaceous limestone brash with *R. smithi* resting on 1.7 m of alternating mottled clay and limestone. Farther west, mapping has proved the continuity of the Wattonensis Beds between Sherborne Park [6600 1667] and the Dorchester Road (A352) south of Sherborne [642 155], although the beds are displaced by a fault in the intervening ground.

Shelly limestone with crushed rhynchonellids in a stream [6818 1395] near Tut Hill Farm may be the Wattonensis Beds.

Frome Clay above the Wattonensis Beds

Grey calcareous clay augered [6660 1784] immediately above the Wattonensis Beds on the southern slopes of Highmore's Hill yielded the foraminifera *Spirillina infima, Haplophragmoides* cf. *canui, Paalzowella feifeli* and *Massilina dorsetensis*? The last has previously been recorded only from the Forest Marble.

Fault slivers of Frome Clay occur between Haydon and Bishop's Caundle, on either side of the Poyntington Fault. Clay just above the Wattonensis Beds [6829 1395] yielded the foraminifera *Lenticulina exgaleata, Ammobaculites* cf. *fontinensis, Ophthalmidium carinatum* and *Dentalina pseudocommunis,* and the ostracods *Glyptocythere oscillum, Lophocythere fulgurata, L.* sp. cf. *batei, Oligocythereis fullonica, Pontocyprella* cf. *harrisiana* and *Micropneumatocythere subconcentrica,* indicative of the topmost *hodsoni* or *orbis* zones.

Beds just above the Wattonensis Beds in Sherborne Park [6698 1712] yielded the foraminifera *Lenticulina muensteri* (abundant), *L. exgaleatea, L. subalata, L. varians, Trochammina* sp., *Haplophragmoides* cf. *canui* and *Cornuspira liasina,* and the ostracods *Progonocythere polonica, Micropneumatocythere quadrata, M. postrotunda, Lophocythere* cf. *scabroides, Terquemula acutiplicata* and *Paracypris* sp. The ostracods are indicative of the *hodsoni* Zone. Ostracods from the uppermost Frome Clay in Sherborne Park [6603 1612] include *Micropneumatocythere brendae* and *Glyptocythere oscillum,* indicative of the *blakeana* ostracod zone (uppermost *hodsoni* to the upper part of the *orbis* ammonite zones).

Pale grey, calcareous silts and clays, about 4 m thick, form the highest part of the Frome Clay, e.g. above Purse Caundle [7031 1865], north-west of Redhouse Farm [6910 2120] and 250m north of Hanglands Lane [6747 2334]; they form a small subsidiary feature beneath the Forest Marble.

Forest Marble

The Forest Marble is a widespread formation which takes its name from the old Forest of Wytchwood in Oxfordshire, where limestone which could be polished was once quarried. The name was first adopted for geological purposes by William Smith in 1799.

The formation crops out in an arc in the north-west of the district (Figure 26). East of Haydon [670 158], the outcrop is offset by some 3.5 km to the south-south-east by the Poyntington Fault. Topographically, the western edge of the formation locally forms a prominent scarp where limestones and sandstones occur near the base, while its long dip slope, together with that of the Cornbrash, forms the western boundary of the Blackmoor Vale.

The Forest Marble is a variable sequence, 40 to 45 m thick at outcrop, of olive-grey, brownish-grey-weathering, calcareous mudstones, interbedded with flaggy, coarse, shell-detrital, variably oolitic limestones and cal-

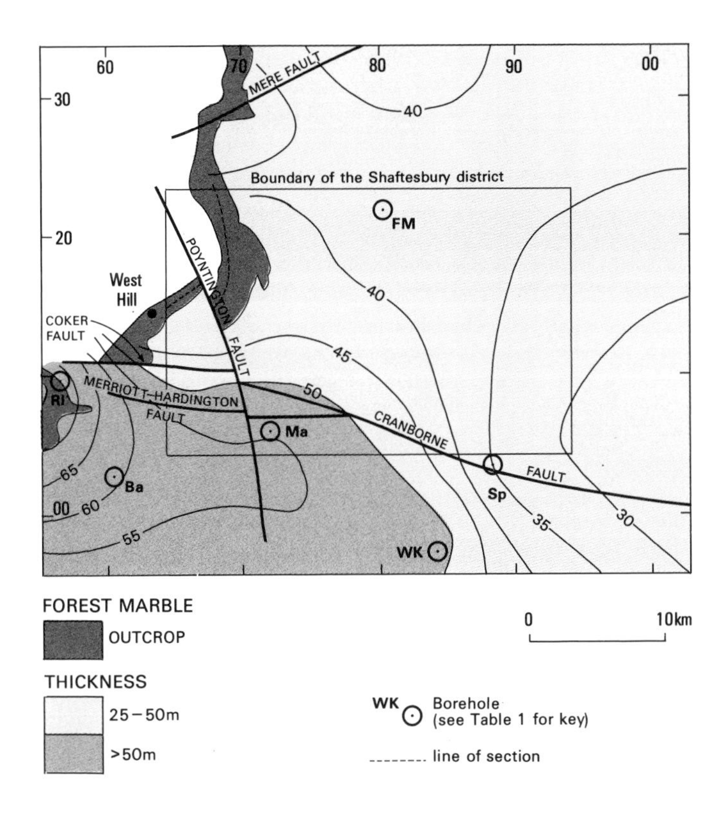

Figure 26 Isopachyte map for the Forest Marble (contours at 5 m intervals).

careous sandstones (Figure 27). At the base, the Boueti Bed, a thin shell-fragmental muddy limestone, up to 0.25 m thick, in which there can be an abundance of *Goniorhynchia boueti* (Plate 3, 2a–c), forms a persistent marker bed that extends from the Dorset coast to West Cranmore [682 417], near the northern edge of the Wincanton district (Cope et al., 1980b). The limestones and sandstones pass laterally and vertically into each other and form lenticular units extending tens to hundreds of metres laterally, although their three-dimensional form is largely unknown. Some lenticular units of reddish orange, uncemented and noncalcareous, clay-laminated sands also occur, mainly within the mudstones; they also occur within and at the top of some of the thicker limestone units. They resemble the Hinton Sands in the Forest Marble near Bath (Woodward, 1894; Green and Donovan, 1969) and in the Yeovil area (Kellaway and Wilson, 1941; Fowler, 1957).

The thickness of the formation at or near outcrop remains relatively constant at between 40 and 45 m (Figure 26); almost 40 m were recorded by Woodward (1895) at West Hill [642 147] near Sherborne, and there are less reliable figures of 40.2 m, 45.7 m and 40.5 m from old boreholes at Copse House Farm [710 180], Henstridge [7243 1989] and near Templecombe [7100 2279] respectively (Richardson, 1928). Geophysical logs of the Fifehead Magdalen and Mappowder boreholes suggest 37 m and 55 m respectively. The overall thickness appears to increase to the south-west (Figure 26).

The Forest Marble is assigned to the late *Oxycerites orbis* Zone and the *Clydoniceras hollandi* Subzone of the *C. discus* Zone of the late Bathonian (Torrens in Cope et al., 1980b), although Penn (1982) regarded the formation as falling wholly within the *discus* Zone. The rarity of ammonites from the Forest Marble of the district makes further biostratigraphical refinement difficult. Cifelli (1959, 1960) placed the Forest Marble within the uppermost of four Bathonian foraminiferid faunules; the species of foraminifera in the Forest Marble, however, differ only slightly from those in the underlying Frome Clay (his 'Upper Fuller's Earth').

The Boueti Bed at the base of the formation may be a condensed deposit; in the Seabarn Farm and Winterborne Kingston boreholes, it rests on a burrowed surface, and in the basal part of the bed near Redhouse Farm [6895 2158], Henstridge Bowden, a bored limestone nodule was found. In addition to *Goniorhynchia boueti*, the fauna of the Boueti Bed includes serpulids, *Montlivaltia* cf. *trochoides, Thecocyathus?, Avonothyris langto-*

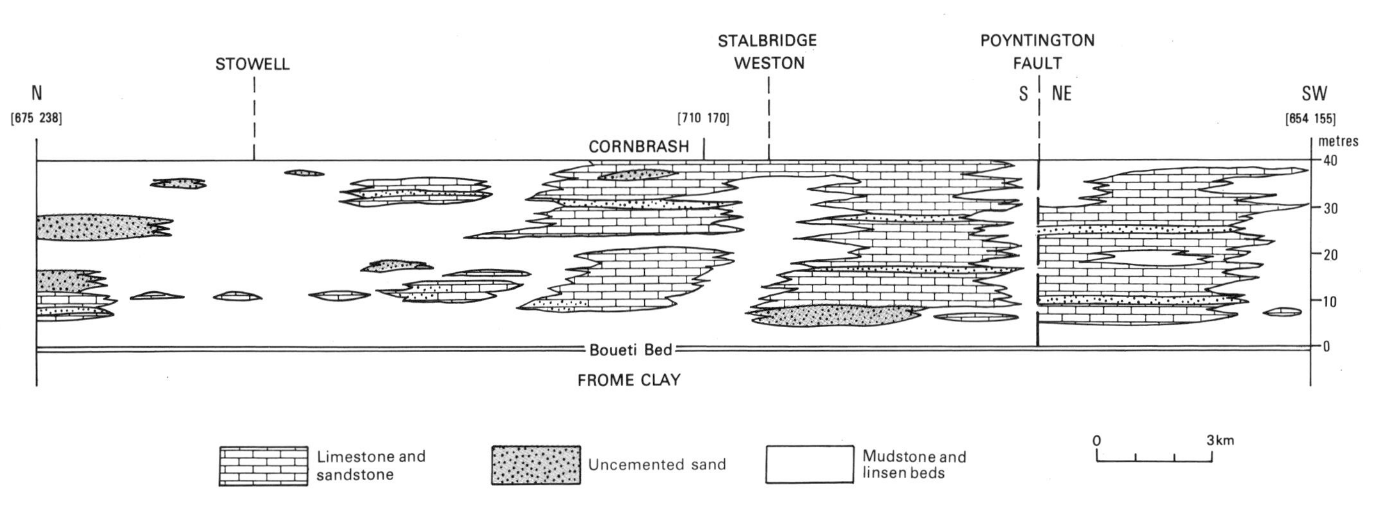

Figure 27 Schematic section along the strike of the Forest Marble of the district (see Figure 26 for line of section).

nensis, Cryptorhynchia?, Digonella digonoides, Kallirhynchia sp., *Obovothyris* sp., *Ptyctothyris arkelli, 'Bathonella' scotica, Chartrionella?, Anisocardia truncata, Camptonectes (C.) laminatus, Catinula* sp., *Gervillella* sp., *Liostrea?, Lithophaga* sp., *Lucina* sp. and *Placunopsis socialis.* The only ammonite recorded from the Boueti Bed of the district is *Clydoniceras (Delecticeras)* sp. from Honeycomb Wood [?635 142], near Sherborne (Arkell, 1958, p.244). In thin section, the Boueti Bed is seen to be a dark grey, muddy micrite with dispersed bioclastic debris. A few angular to subrounded quartz grains, less than 0.2 mm in diameter, also occur.

In the southern part of the district, around Bishop's Caundle, the Forest Marble consists dominantly of mudstones with only minor beds of sandstone and limestone. In the central tract between Stourton Caundle and Henstridge, flaggy sandstones and limestones are dominant. From Henstridge northwards, olive-grey mudstones comprise much of the Forest Marble succession, but they are rarely exposed and are consequently poorly known. At Templecombe, railway cuttings [7000 2230 to 7028 2235] and an old well [7100 2279] (Richardson, 1928) show the lower half of the succession to be mudstone-dominant, with only two horizons, each less than1 m thick, of flaggy sandstone and limestone. The sequences in the Fifehead Magdalen and Spetisbury boreholes appear to consist mostly of sandstone and limestone, with minor amounts of mudstone; the Mappowder Borehole proved a similar sequence, but with a predominance of mudstone in the lowest third of the sequence. Where limestones and thin sandstone beds occur near the base of the formation, they give rise to a prominent scarp.

Comminuted plant debris is common throughout the formation; larger fragments occur in the sandstones and limestones, and may include logs in the latter. Trace fossils include plaited trails of *Gyrochorte comosa,* which occur commonly on the surface of linsen and thinly bedded sandstones, especially near the base of the formation, *Teichichnus, Skolithos* and, more rarely, *Arenicolites,* and faecal pellets. Generally, the molluscan fauna above the Boueti Bed is sparse and of limited diversity. The uppermost 6 m of Forest Marble proved in the Combe Throop Borehole yielded a more diverse bivalve fauna, particularly in the uppermost 0.7 m (see p.48).

The Digona Bed, with abundant brachiopods of the genera *Digonella* and *Avonothyris,* occurs between 9 and 18 m above the base of the formation in south Dorset (Arkell, 1947), and has been recorded as far north as the Winterborne Kingston Borehole (Penn, 1982). In the Shaftesbury district, a possible occurrence, but without the characteristic fossil, is a creamy, grey-weathering limestone with nuculoids and high-spired gastropods, 12 m above the base of the Forest Marble in the railway cutting [6991 2225] west-south-west of Templecombe. The recent survey showed that the reputed occurrence of the Digona Bed at Caundle Break [7050 1825] (Sylvester Bradley, 1957) is the Boueti Bed.

In thin section, the limestones contain much shell material, ooliths, 0.15 to 0.25 mm across, accreted on quartz and organic grains, and sparse fine-grained, angular to subrounded quartz grains, set in a matrix of sparry ferroan calcite. The sandstones have grains of fine-grained, angular to subrounded quartz, with scarce metaquartzite and feldspars, shell debris and ooliths, set in a matrix of ferroan calcite which commonly amounts to 25 to 35 per cent of the whole rock. Some of the quartz grains have thin oolitic coatings.

Lenses of cross-bedded calcareous sandstones, 2 to 25 mm thick and up to 20 cm across, occur throughout the mudstones. The term linsen has been applied to lenticular sandstones of this type (De Raaf et al., 1977). The ratio of linsen to clay is variable, from a few per cent to 80 per cent sandstone for the larger linsen; the linsen can pass into flaggy sandstone.

Palaeogeography

The sedimentary structures, including opposed cross-bedding and linsen, suggest shallow-water deposition (De Raaf et al., 1977). The lenticular limestone and sandstone units are thought to have formed as shoals on the predominantly muddy sea floor. Vigorous, but probably intermittent and/or localised current action is indicated by argillaceous clasts in the base of limestone beds. Prod and groove casts also occur. It seems likely that the limestones formed during storm-related events, which introduced bioclastic detritus from outside the area. Holloway (1983) suggested that the large wood fragments were stranded on emergent shoals, because deposition by waterlogging should result in a more uniform distribution through the succession. The absence of strong bioturbation and the abundance of land-derived plant debris have been suggested by Hallam (1970) to indicate reduced salinity. This conclusion is supported by the the more brackish elements in the fauna recovered from the uppermost Forest Marble in the Combe Throop Borehole. The trace fossils indicate a relatively shallow littoral to sublittoral depositional environment (Ekdale et al., 1984). Measurements of current-related sedimentary structures from a number of localities indicate a general west to west-north-westerly derivation (Figure 28).

Details

BOUETI BED

Sherborne Park

Shelly, pale-coloured, calcareous silts with fragments of *Goniorhynchia boueti* occur in Sherborne Park [6666 1650; 6591 1589] and south-west of Milborne Port, on Highmore's Hill [e.g. 6675 1790] (see also Fowler, 1944).

Frith to Toomer Hill

The Boueti Bed, a 0.2 m-thick calcareous silty clay with common *G. boueti,* lies above a small scarp south of Frith Farm [e.g. 7027 1714; 7038 1739] and west of Toomer Hill [e.g. 6982 1984]. At the second locality, *Anisocardia truncata, Avonothyris?, Digonella* cf. *digonoides, Ptyctothyris arkelli* and *Tubithyris* cf. *whatleyensis* were found.

Specimens of *Digonella digonoides* found by Sylvester Bradley (1957b) at Caundle Break [c.7050 1825], thought to represent the Digona Bed, actually fall on the mapped outcrop of the Boueti Bed.

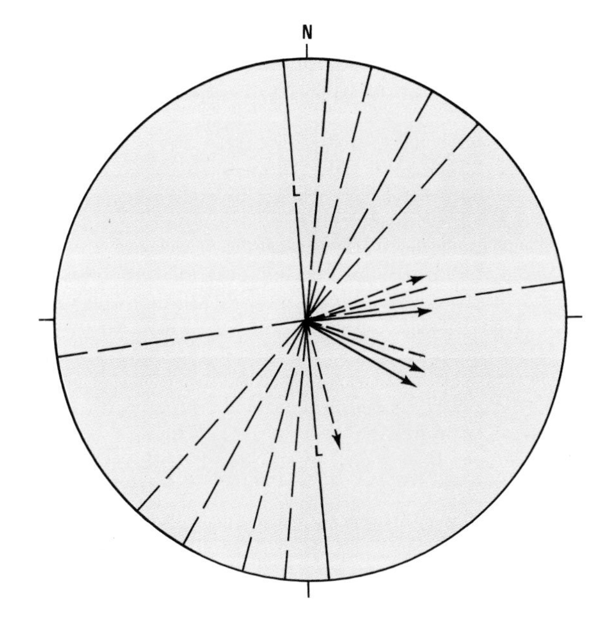

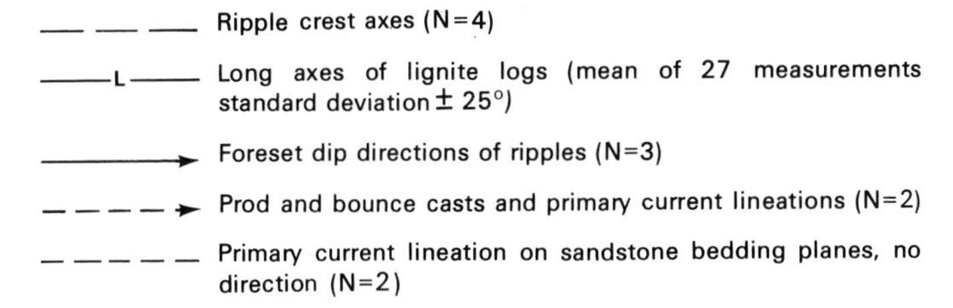

Figure 28 Orientation of current-related sedimentary structures in the Forest Marble.

Henstridge Bowden to Wilkin Throop

An exposure [6972 2014] near Manor Farm, Henstridge Bowden, shows 0.6 m of soft, muddy, medium grey, very shelly limestone with contorted serpulid tubes up to 5 mm in diameter, *G. boueti*, *Digonella digonoides* and many small *Liostrea* sp., on silty clay.

G. boueti and soft, argillaceous, shelly limestone are ploughed up below the Forest Marble scarp at Henstridge Bowden [6912 2078; 6903 2083] and near Redhouse Farm [6911 2127]. At the first locality, the fauna also included *Kallirhynchia* sp., *Digonella* cf. *digonoides* and *Obovothyris* sp.

Soft, muddy, creamy buff-weathering limestone with *Avonothyris* sp., *Digonella digonoides, Obovothyris* sp., *G. boueti*, '*Bathonella' scotica* and *Lucina* sp. is exposed [6892 2162 to 6898 2148] north of Redhouse Farm. A much-bored nodule of micritic limestone [6895 2158], probably from the base of the bed, contained coated pellets, serpulid and coral fragments, *Avonothyris*?, rhychonellid fragments, *Liostrea* sp., *Lithophaga*? and *Placunopsis socialis*. The Boueti Bed is overlain by about 5 m of clay, and then by sandstone and limestone which form the Forest Marble feature.

Debris from a rabbit hole [6889 2189] included *Montlivaltia* cf. *trochoides, Thecocyathus?, Digonella* cf. *digonoides, G. boueti* and *Anisocardia truncata*.

Typical limestone debris with *G. boueti*, overlain by shelly, sandy clay with *Digonella* sp. and *Lucina* sp., was found [6980 2191] east of West Wood. Farther east, a ditch [6990 2198] exposed 0.3 m of argillaceous, brownish grey limestone, which passsed up into a buff, shelly, sandy clay with serpulids, *Avonothyris* sp., *Digonella digonoides, G. boueti, Obovothyris* sp., *Chartrionella?, Camptonectes (C.) laminatus, Catinula* sp., *Gervillella* sp. and *Liostrea*?

A pit [6925 2220] in the Boueti Bed near North Side Wood exposed:

	Thickness m
Topsoil and clayey subsoil	0.30
Boueti Bed	
Clay, silty, shelly, ochreous-buff with abundant small *Liostrea* sp.	0.25
Limestone, shelly, soft, greyish brown, with *G. boueti*	0.20
Limestone, silty, or calcareous silt, partly nodular	0.20
Frome Clay	
Silt, buff, calcareous (augered)	1.30

In a gully [6791 2311] west of Wilkin Throop, about 0.2 m of greyish buff-weathering, argillaceous, burrowed limestone with serpulids, *Avonothyris langtonensis, G. boueti, Camptonectes (C.) laminatus* and *Protocardia?* is exposed.

Beds above the Boueti Bed

Outcrops west of the Poyntington Fault

North-east of North Wootton, field brash [around 6608 1485] comprises slabs, up to 4 cm thick, of very shelly limestone with lignite debris. A patchy, pale greenish grey, fine-grained limestone matrix surrounds some shells.

Thin sandstone and very shelly limestone, with a spring [6533 1370] at the base, forms the scarp slopes on the west side of the valley to the south of North Wootton.

In Sherborne Park [6615 1629], a 2 m-thick sandstone, with 20 to 50 mm-scale lenticular beds, occurs in the lower part of the succession. Similar sandstone forms brash along the escarpment edge west of the Poyntington Fault (though not on Highmore's Hill [668 181] where bioclastic limestone occurs) and east of Haydon [around 675 155].

North of Haydon, well-bedded (50 to 100 mm-scale) clastic limestones, above thick, greenish grey clay and silt, cap a small hill [669 162]. Eastwards, higher beds show an upward transition from coarse-grained and in part sparsely oolitic, cross-bedded, clastic limestones, through more massive limestones with uniformly small bivalves, to the sandy, sparsely shelly limestones of the Cornbrash.

Bishop's Caundle to Stalbridge Weston

North of Bishop's Caundle [around 690 140], sandy to extremely sandy clay with small lenses of fine-grained sandstone and impersisent very shelly limestone, forms an extensive dip slope, inclined 2° SE.

North of Holtwood, exposures in grey sandy clay with thin (usually less than 1 cm) lenses of fine-grained sandstone are common [e.g. 6973 1434; 6982 1423]; at the first, the beds rest on 3 cm of very shelly sandy limestone, on over 0.2 m of sandy grey clay. Upstream, a section [6955 1441] was seen in yellowish brown, very sandy clay, with shelly sandy limestone lenses (less than than 10 mm thick) and a nodular cemented limestone layer 1 m from the base, with oysters below; it passes downwards in the basal 0.3 m into slightly sandy clay.

Flaggy, coarse-grained, shelly, patchily oolitic, bioclastic limestone is well developed between Bishop's Caundle and Stourton Caundle; extensive brash occurs on dip slopes north-east [701 147] and east of Holtwood Farm [700 140]. Fossils are abraded and include corals, brachiopods, bivalves and crinoids.

Sandstones are extensive on Woodrow Farm, and good exposures that include symmetrical ripple-marked and trace-fossil-burrowed pavements occur in the stream to the south [e.g. downstream of 6970 1509]. Ripple-mark crests are orientated approximately east–west. Thin discontinuous partings, up to 2 m thick, of foxy red, silty sand, interbedded with grey clay on a 10 mm scale, occur in the lower part of the Forest Marble [695 156; 690 150].

Massive limestone flags up to 5 m thick are extensively developed around Woodrow [699 155] and Cockhill farms [703 163], but elsewhere, the limestone is interbedded with clay and calcareous sandstone.

South-west of Stalbridge Weston [715 160], greenish grey clay near the top of the Forest Marble is interlayered with mottled silt and orange sand, and is overlain by sandstone flags with clay and sand partings, and a 2 m-thick shelly limestone.

Stalbridge Weston–Stalbridge

Sequences dominated by thin calcareous sandstone flags form a conspicuous scarp north of Frith Farm, but pass southwards into sand- and clay-rich sequences between Haddon Lodge [7022 1626] and Stalbridge Weston.

West of Stalbridge Weston, fine- and medium-grained sand, up to 5 m thick, occupies hummocky, ?landslipped ground. Farther west, the sand is increasingly interlayered with clay; regular centimetre-scale alternations of sand and clay occur north of Cockhill Farm [e.g. 7026 1665]. Along the ridge between Cockhill and Manor farms, dark silty clay is overlain by 0.2 m of very pale calcareous sand and silt, overlain by 0.1 m of calcareous sandstone with shelly limestone [e.g. 7103 1648]. Calcareous sandstone occurs below brown silt and sand north of Cockhill Farm [7046 1694].

Mottled orange-weathered, grey clayey silt and clayey sand occur in the valleys north-west [e.g. 7171 1724; 7050 1709] and south-west [around 7180 1642] of Stalbridge Weston. At the latter locality, thin partings of calcareous sandstone also occur.

Limestones, overlying sand, crop out west of Stalbridge. A section [7106 1944] near Quarry Farm shows 4 m of thinly bedded, coarsely shelly, patchily oolitic limestone. Similar beds, 6 m thick, were formerly exposed in the Stalbridge Quarry [7158 1851]; they were used to build the walls surrounding Stalbridge Park. Abraded brachiopods, bivalves, corals and crinoids occur.

Core from two boreholes [7146 1846; 715 1854] north-east of Copse House, close to the Stalbridge Quarry, consists mainly of shelly coarse-grained limestone with some calcareous mudstone intervals. Fossils include wood, serpulids, *Anisocardia, Chlamys, Corbula?, Meleagrinella* sp., *Placunopsis socialis, Praeexogyrya hebridica, Protocardia* sp. and echinoid fragments.

An extensive body of fine- to medium-grained orange sand, locally more than 2 m thick, occurs near the top of the formation north-west of Stalbridge Park [around 720 190]. In its southern part, clay is interbedded with the sand, but this decreases northwards.

Toomer Hill–Henstridge Bowden

The quarry [7031 1903] adjacent to the A30 road shows 6 to 7 m of cross-bedded sandstone with narrow seams and partings of orange sand and grey silty clay. Beds range from 10 to 200 mm thick (20 to 50 mm thick, on average) and are flaggy or lenticular, with some symmetrical ripple-like structures; thicker beds are planar laminated. Horizontal burrow trails resembling *Gyrochorte comosa* occur. Thin sections show moderately to well-sorted, angular silt or very fine-grained (0.05 to 0.1 mm) quartz, cemented by a single phase of ferroan calcite. Ooliths and shell fragments account for variable proportions, but average 5 per cent, while quartz grains form 70 to 80 per cent of the less calcareous rock types. The dip is about 8° to N036°.

Exposures [6944 2054; 6926 2070; 6927 2067 to 6929 2063] and quarries [6922 2070; 6963 2035; 6970 2027; 6932 2056] near Bowden Lane and Henstridge Bowden show essentially similar sections in up to 2 m of bioclastic, sparsely oolitic limestone, with ochreous clay clasts, variably bioclastic and oolitic sandstones and a thin flaser-bedded unit (Taylor, 1991). The oolith content is very variable, being more abundant in the thinner, finer-grained beds. Lignitic plant fragments, up to 10 mm across, also occur. Some beds are lenticular, and opposed cross-bedding directions occur in adjacent beds. Symmetrical ripple marks trending N015° occur on some bedding surfaces. *Gyrochorte* trails and prod marks occur on the bed bases.

About 1.5 m of flaser- and planar-bedded, flaggy bioclastic limestones with clayey partings and clasts, together with linsen, are exposed in a quarry [6975 2060] 250 m north-east of Bowden Lane. Abraded shell fragments, dominated by the bivalves *Camptonectes (C.) laminatus, Praeexogyra hebridica* and *P. h. elongata*, sparse ooliths and carbonised plant fragments are present. Worm burrows and scour casts occur on the bases of beds underlain by the thicker clay units. Sinuous asymmetrical ripples occur on some beds. The dip is about 6° to N013°.

Exposures in two quarries [6913 2108; 6915 2105] at Redhouse Farm are basically similar and show up to 1 m of flaggy, shelly, patchily oolitic sandstone, shelly limestone, linsen beds and flaser beds. The linsen beds lenses are ripple cross-bedded, up to 0.01 m thick and 0.1 m wide, and are set in a clay matrix. Horizontal sandstone-filled burrows, and small trails and prods occur on the bases of some linsen. Small, symmetrical, non-orientated ripples, shell fragments, and small complete bivalves in a convex-up position occur on the tops of beds. The bases of beds show *Gyrochorte* trails, prod and load casts. Thicker sandstone beds at floor level, up to 0.15 m thick, contain many clay clasts. More massive shelly limestones in beds up to 0.2 m thick, with shelly, marly, ochreous partings also occur in the quarry floor. Dips measure 9° to N045° and 5° to N104°.

In the gully [6909 2187 to 6906 2195] north of West Wood, about 2 m of flags of shelly bioclastic limestone, up to 0.25 m thick, overlie pale grey, blocky and conchoidal-jointed, silty, calcareous mudstones, with a thin, 10 mm thick, argillaceous limestone and a 20 to 25 mm thick, pyritous, clayey limestone with shell fragments on the bedding surface.

Stowell to Windmill Hill

East of Stowell, the Boueti Bed is overlain by about 12 m of mudstone, capped by a 1 m-thick limestone [6998 2230].

Up to 1.7 m of silty, slightly clayey, fine-grained, foxy red, orange and buff-mottled sand occur north-east of Henstridge Bowden [6960 2140], south-west of Horsington [6964 2342; 6929 2321], at Wilkin Throop [6860 2314] and near Sandhills [6811 2362].

The upward passage from sand to limestone along the crest of the Forest Marble scarp near Windmill Hill [6775 2335 to 6740 2370] is transitional; a section [6712 2409] in a quarry just north of the sheet boundary is recorded by Taylor (1991).

Inwood–Templecombe

A trench [7000 2056; 7017 2022] near Inwood exposed thin, flaggy, coarsely bioclastic, oolitic limestones, with some well-preserved bivalves and a rhynchonellid, interbedded with clay. The limestone thins eastwards along the trench [7041 2006 to 7055 2000] to about 0.5 m.

Some 700 m north-north-west of Inwood, a lenticular body of flaggy sandstone, limestone and sand has been extensively quarried [7075 2076] and exposes:

	Thickness m
Sandstone, calcareous, flaggy, with linsen and thin clay partings. The sandstones contain clay clasts and burrows. Ripple cross-bedding has a hummocky form without orientation	0.78
Sandstone, calcareous, flaggy, ripple cross-bedded	0.14
Clay	0.23
Sandstone, fine-grained, calcareous	0.14
Linsen beds	0.18

The dip is about 2.5° to N172°.

South of Yenston, a degraded exposure [7157 2086] at the top of the Forest Marble shows about 2 m of greenish grey, calcareous mudstones with small ostreids and crushed rhynchonellids (see also Page, 1988). Opposite, a hard, bluish grey, bioclastic limestone about 0.2 m thick, and about 3 m below the Cornbrash, projects from a grassed bank of mudstone.

Quarries at Windmill Hill [7080 2103] now expose only 0.75 m of irregularly flaggy, sparsely lignitic, coarsely bioclastic limestone consisting mainly of poorly aligned ostreid fragments with a few intact valves, set in a micritic matrix. The apparent dip is about 5°E. Manuscript notes by F H A Engleheart, believed to date from about 1930, record 5.5 m of limestone, overlain by 4 m of sandstones, linsen beds, thin limestones and clays in this area.

West, south-south-west and north-east of Windmill Hill, a lenticular, fine-grained, ochre-coloured sand can be traced in ditches [7050 2226 and 7045 2206], by mole hills and auger holes [7005 2162; 7041 2212; 7021 2197].

Between Yenston and Templecombe, much flaggy, bioclastic limestone and some sandstone debris occurs [7059 2129 to 7070 2153]. A stream [7065 2166 to 7075 2167] exposes dark bluish grey, variably oolitic, patchily lignitic, flaggy, bioclastic limestones and ripple cross-bedded, calcareous sandstones, underlain at the western end by linsen beds with 50 to 75 per cent sandstone linsen in a clay matrix. The limestone and sandstone beds, 0.02 to 0.2 m thick, include clayey partings and some thicker clayey beds with linsen. About 45 m downstream, two thin calcareous sandstone beds, separated by 65 mm of clay, occur. The base of the lower bed, about 15 mm thick, is a sparry limestone, of which the lower surface carries a dense scatter of irregular protuberences, about 5 to 10 mm across, into the underlying clay; these overprint traces of prod and bounce casts. These structures superficially resemble those described by Douglas (1951), but they are more probably small load casts. The upper 10–15 mm of the bed is a calcareous sandstone enclosing small ripple-like lenses of limestone, with lignitic debris in the ripple laminae. The upper surface carries a few small faint trails of *Gyrochorte* and isolated vertical and horizontal burrows 2 to 3 mm in diameter which project above the surface. The upper bed, 10 mm thick, has well-developed prod, bounce and brush casts on the base, and traces of primary current lineation and some aligned shells on the upper surface (data incorporated in Figure 28). These two beds, separated by a few millimetres, have current flow directions about 90° apart. The dip is 5° to N091°.

West of Templecombe, along the side of the valley south of the railway, ditches and gullies expose several sections (Taylor, 1991). One [7007 2192 to 7007 2195] shows the following section:

	Thickness m
Linsen beds, disturbed seen	c.2.0
Clay, calcareous, grey	1.0
Sandstone, calcareous, flaser- and linsen-bedded	0.4
Clay with calcareous sandstone linsen	0.6
Sandstone, calcareous, flaser- and linsen-bedded	0.3
Clay with calcareous sandstone linsen	1.0
Limestone, bioclastic, sparsely and variably oolitic, in four main beds with clay partings up to about 0.1 m thick	1.0
Mudstone, silty, weakly calcareous, slightly fissile, micaceous	seen 1.5

The sandstone linsen and beds carry numerous trails of *Gyrochorte* and penetrating sandstone-filled burrows up to 10 mm in diameter. Symmetrical sinuous ripples on the surface of the upper sandstone bed trend N030°; the foresets dip south-east (Figure 28). The limestone has planar-bedded subunits, a few tens of millimetres thick, separated by thin ochre-coloured clay partings with wisps incorporated into the overlying limestone. Ooliths are more common in the finer-grained, less shelly units and tend to be concentrated at the base. Carbonised and lignitic plant fragments up to 10 mm in length are common. The upper bed shows truncated cross-bedding foresets which give a planar-bedded structure in strike section. The foresets dip 18° to N124°. The dip is 6° to N108°.

An oyster lumachelle, about 0.25 m thick, within buff clay [7084 2202], and a concentration of serpulids, a crushed rhynchonellid and ostreids in clayey soil brash [7116 2211] were found near Templecombe.

The mainly overgrown cuttings for the disused railway sidings [6980 2218 to 7030 2235] west-south-west of Templecombe provide a representative section through the Forest Marble (Figure 29)(for more detail see Taylor, 1991). The creamy to pale grey-weathering, fairly hard, very slightly sandy, moderately argillaceous, micritic, 0.25 to 0.4 m thick limestone, about 12 m above the base, is possibly the Digona Bed. The top 20 to 30 mm of the bed is intensely burrowed and shelly, and contains common, small, 4 to 6 mm long bivalves and small high-spired gastropods. The upper surface of the overlying bioclastic limestone is covered by small ostreids and abundant small, 3 to 4 mm, high-spired gastropods [6998 2228; 7010 2229]; this limestone dies out westwards [at 6991 2225].

Combe Throop Borehole

The Combe Throop Borehole [7260 2350] proved 6.41 m of Forest Marble, comprising 5.68 m of olive-grey calcareous mudstone with patchy lamination, overlain by 0.73 m of alternating muddy calcareous sandstones and calcareous mudstones with a varied fauna. A lignite fragment 0.15 m below the base of the Cornbrash carried borings filled with grey limestone, indicating the presence of an erosional event prior to the deposition of the Cornbrash.

The 5.68 m mudstone unit at the base has millimetre-scale silty and fine-grained sand laminae and wisps, associated with scattered thin sandstone linsen. Finely comminuted plant debris is common in the silty and sandy laminae, but is also present in the intervening mudstone. Some of the laminae are irregularly disrupted. Traces of sand-filled burrows also occur. Intact shells and shell debris occur thinly scattered throughout the core and include: serpulid tubes, *Corbula* sp., *Costigervillia crassicosta, Eocallista antiopa, Eomiodon* sp., *Gervillia ovata, Isocyprina* sp., ostreids, *Placunopsis socialis, Protocardia stricklandi* and

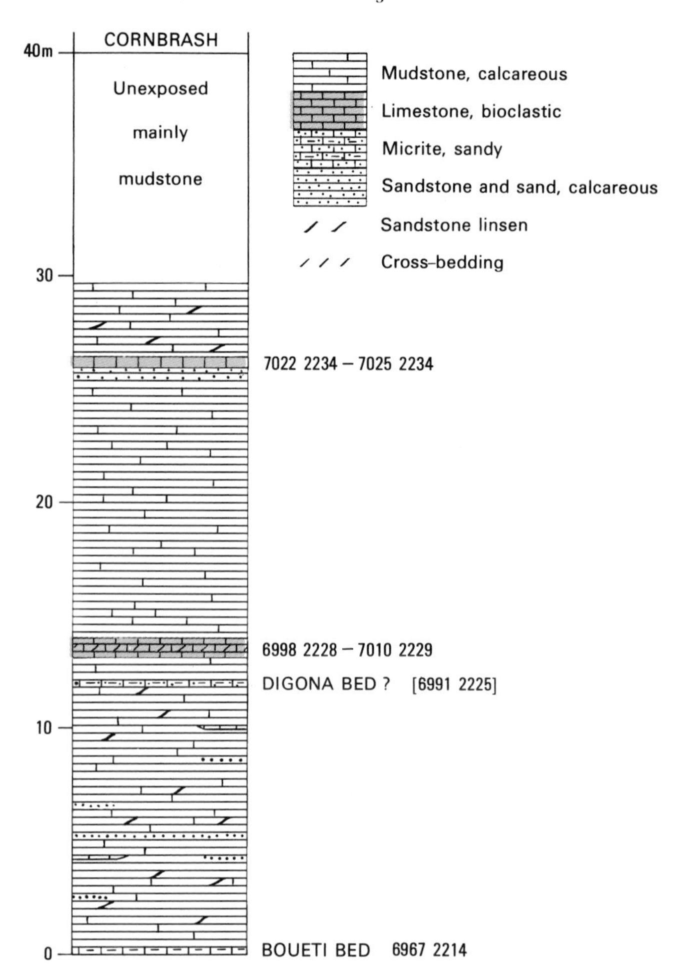

Figure 29 Section through the Forest Marble in the Templecombe railway cutting.

an echinoid spine. The upper 0.73 m of sandstones and mudstones has a more varied fauna with, in addition to the species listed above (but lacking *Corbula* sp., *Eocallista antiopa* and *Gervillella ovata)*, a rhynchonellid fragment, astartids indet., *Camptonectes* sp., *Isocyprina depressiuscula, Lithophaga* excavations in wood, *Modiolus* cf. *lonsdalei, Praeexogyra hebridica, Plicatula* sp., *Pseudolimea* sp., *Pteroperna?* and fish fragments including *Eomesodon*. Both faunas are typical of the late Bathonian.

Cornbrash

The Cornbrash of the district consists of up to 12 m of bioclastic, micritic, slightly oolitic, partly sandy limestones. The limestones crop out in the north-west where they are broken by faults into three principal outcrops: around Alweston [660 140], Stourton Caundle [715 150] and between Stalbridge [735 175] and Horsington [700 238]. The formation dips gently southwards or south-eastwards and is proved to depths of 600 m below OD in boreholes to the east of the district.

The base of the Cornbrash is taken at the incoming of hard, thickly bedded, bioclastic, micritic limestones, but these give rise only to a featureless feather edge on the dip-slope of the Forest Marble.

The formation shows an overall southerly thickening from 3.43 m at Gillingham, just north of the district, to 6 to 7 m along much of the northern outcrop in the district (6.4 m in the Fifehead Magdalen Borehole; 6.8 m at Combe Throop), then to 8.19 m at Holwell [693 120], 9.5 m at Holnest, 12.19 m at Mappowder and 14.06 m at Winterborne Kingston (Figure 30).

Traditionally, the Cornbrash is divided into a lower and an upper unit, each with characteristic ammonite and brachiopod faunas (Douglas and Arkell, 1928). The boundary between them, a sedimentary break in most of the United Kingdom, marks the Bathonian–Callovian stage boundary. The Lower Cornbrash ammonite fauna includes smooth, laterally compressed *Clydoniceras* indicative of the Upper Bathonian *discus* Zone; the Upper Cornbrash includes species of the strongly ribbed and inflated genus *Macrocephalites,* indicative of the Lower Callovian Herveyi (formerly Macrocephalus) Zone (Page, 1988).

Lower Cornbrash

The Lower Cornbrash consists of up to 4 m of hard, bioclastic, micritic, locally slightly oolitic, very slightly sandy, bioturbated and irregularly bedded limestone in beds up to 0.8 m thick. A regional threefold division of the Lower Cornbrash was recognised by Douglas and Arkell (1928): at the base, limestones up to 2 m thick, rich in the brachiopod *Cererithyris intermedia* (the '*intermedia* beds' or 'zone'); a median, cross-bedded, coarse-grained

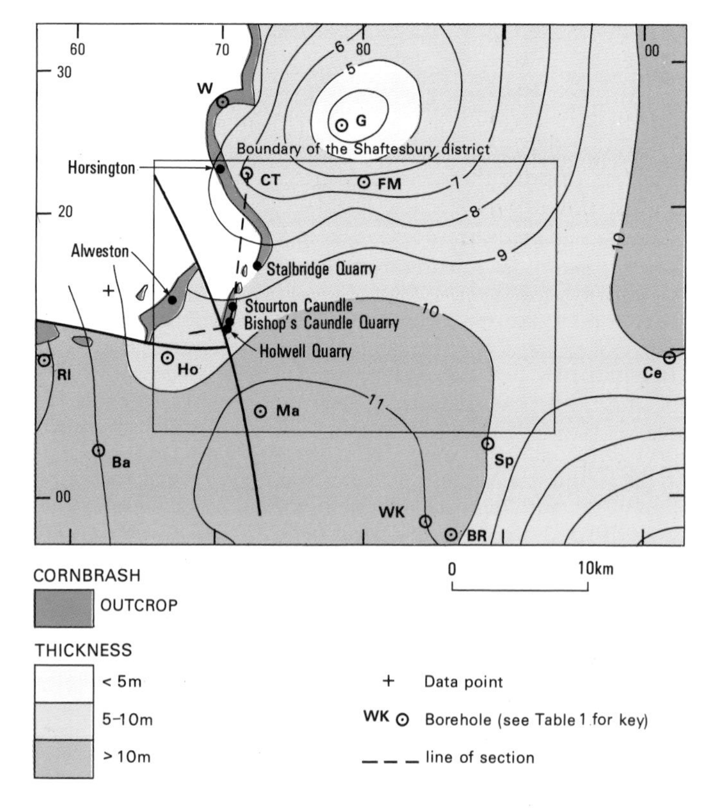

Figure 30 Isopachyte map for the Cornbrash (contours at 1 m intervals).

shell-sand limestone and calcarenite (Corston Beds); locally, at the top, 0.3 m of micritic, locally argillaceous, shelly limestone with abundant molluscs, including *Astarte hilpertonensis* and *Trigonia* spp. (the *Astarte-Trigonia* Bed). Within the district, rubbly, highly fossiliferous limestones (the '*obovata* limestones', or 'zone', of Douglas and Arkell (1928)) are the lateral equivalent of the Corston Beds. The most southerly known occurrence of the Corston Beds is at South Cheriton, just beyond the northern margin of the district.

The Combe Throop Borehole [7260 2350] (Figure 31) proved a Lower Cornbrash sequence consisting of 3.74 m of hard, bioclastic, micritic, in part peloidal, very slightly sandy, irregularly bedded, bioturbated limestone with a mixed muddy to sandy, and ferroan calcite matrix.

The Lower Cornbrash in the borehole can be divided on gross lithology into six units (Units 1 to 6, Figure 31). The basal unit, 0.93 m thick, is a pale green, muddy, bioturbated limestone that rests with an irregular base on the Forest Marble; it forms a distinctive hard bed at outcrop. Unit 2 consists of pale grey, ?peloidal limestone, 0.5 m thick, with partings of pale grey, shelly mudstone; the fauna includes serpulids, *Camptonectes* sp., *Exogyra* sp., *Meleagrinella echinata, Placunopsis socialis* and *Pleuromya* sp. Unit 3, 1.06 m thick, comprises extensively bioturbated, bioclastic limestone, with wisps and patches of mudstone. The fourth unit consists of 9 cm of calcareous sandy shelly mudstone with *Entolium corneolum, Meleagrinella echinata,* ostreids and echinoderm fragments. The mudstone passes up into ?peloidal, shelly (bryozoa and ostreids), highly bioturbated limestone, 0.41 m thick, that is extensively bioturbated and contains muddy wisps and patches (Unit 5). The uppermost unit (Unit 6) consists of pale grey, bioclastic limestone, 0.72 m thick, which is muddy, shelly and bioturbated in the lowest 0.29 m, and ?peloidal in the interval 84.66 to 84.82 m (Figure 31). The fauna of Unit 6 includes serpulids, *Obovothyris obovata, Kallirhynchia* sp., *Meleagrinella echinata,* ostreids and fish fragments, including *Eomesodon. Obovothyis obovata* in this unit links these strata with the flaggy and rubbly limestones at Stalbridge [7320 1867] and Holwell [6933 1198] (Douglas and Arkell, 1928) (Figure 31).

The thickness of the Lower Cornbrash at Combe Throop is similar to that at Holwell. The base was not exposed in quarries at Bishop's Caundle [6990 1320] and Stalbridge [7320 1867], but the thicknesses there are probably comparable (Figure 31).

In the Winterborne Kingston Borehole, the Lower Cornbrash is only 0.73 m thick and has a well-developed pebble bed at its base, with bored and phosphatised clasts, and angular fragments derived from the underlying formation. The attenuation at Winterborne Kingston is thought to be due to the absence of the basal beds of the Cornbrash (Penn, 1982).

Upper Cornbrash

At outcrop in the district, the Upper Cornbrash comprises between 2.5 and 4.5 m of sandy, moderately to sparsely bioclastic and micritic limestone, in irregular beds 0.2 to 0.4 m thick, separated by sandy calcareous partings. The limestones are pseudonodular in part, commonly shelly, and with some clay intraclasts. Fine-grained quartz sand constitutes up to 20 per cent of the Upper Cornbrash and, together with ooliths, pelletal calcite clasts and shell fragments (all with extensive calcite overgrowths), is cemented in large part by sparry calcite.

The Upper Cornbrash in the Combe Throop Borehole consists of 3.13 m of sandy limestones, calcareous muddy silts and fine-grained sandstones. The junction with the Lower Cornbrash appears to be erosional. This sequence is broadly comparable with that of the Tytherton No. 2 [9558 7282] and No. 3 [9440 7445] boreholes near Chippenham (Cave and Cox, 1975), although the Tytherton sequences are thinner, with only 1.56 m of Upper Cornbrash (Figure 31).

In the Combe Throop Borehole, the Upper Cornbrash is divided into five lithological units (Units 7 to 11, Figure 31). A 0.48 m thick basal unit (Unit 7) of muddy sandstone with serpulids and *Obovothyris* sp., is succeeded by 1.25 m of sandy, slightly argillaceous, sparsely shelly, limestone (Unit 8) with *Nanogyra nana* and echinoderm fragments. Unit 9 comprises 0.48 m of pseudonodular, pale grey, micritic limestone with no recognisable fossil, and is overlain by 0.32 m of calcareous muddy sandstone with abundant serpulids and *Obovothyris* sp. (Unit 10). The uppermost unit (Unit 11), 0.6 m thick, consists of sandy limestone that passes up into a sparsely shelly, micritic limestone.

Thicker Upper Cornbrash occurs south-west of the district at Melbury Osmond (= Gallica Bridge) [5777 0835], where 7.3 m of strata were recorded (Page, 1988), and at depth in the Mappowder (9.19 m), Spetisbury (5.18 m) and Winterborne Kingston (13.3 m) boreholes.

Chidlaw and Campbell (1988) suggest that the Upper Cornbrash was deposited during a period of regression.

Details

North Wootton

An exposure [6553 1470] at North Wootton shows 1m of marly limestone with subhorizontal *Thallasinoides* burrows, interbedded with rubbly micritic limestone layers, 0.05 to 0.1 m thick.

A dip slope between Longburton and Alweston is covered in a brash of shelly micrite. Locally, e.g. west of Alweston [around 6585 1410], the limestone weathers to a reddish brown, stone-free clay (?terra rossa). Exposures of 1 to 2 m of buff, fairly fissile, shelly micrite occur along a lane [6558 1295 to 6562 1282]. Some of the layers are hard and blue hearted; others are soft and decalcified.

Caundle Marsh to Bishop's Caundle

The Cornbrash forms a narrow, probably steeply dipping, faulted outcrop [6821 1344 to 6829 1258] south of Hawkins's Farm. Micritic limestone, 0.2 m thick, occurs south [6828 1286] of an old quarry, and alongside a roadside excavation [6820 1328].

East of Bishop's Caundle, the Cornbrash forms a dip slope descending south-eastwards from east of the church [698 133].

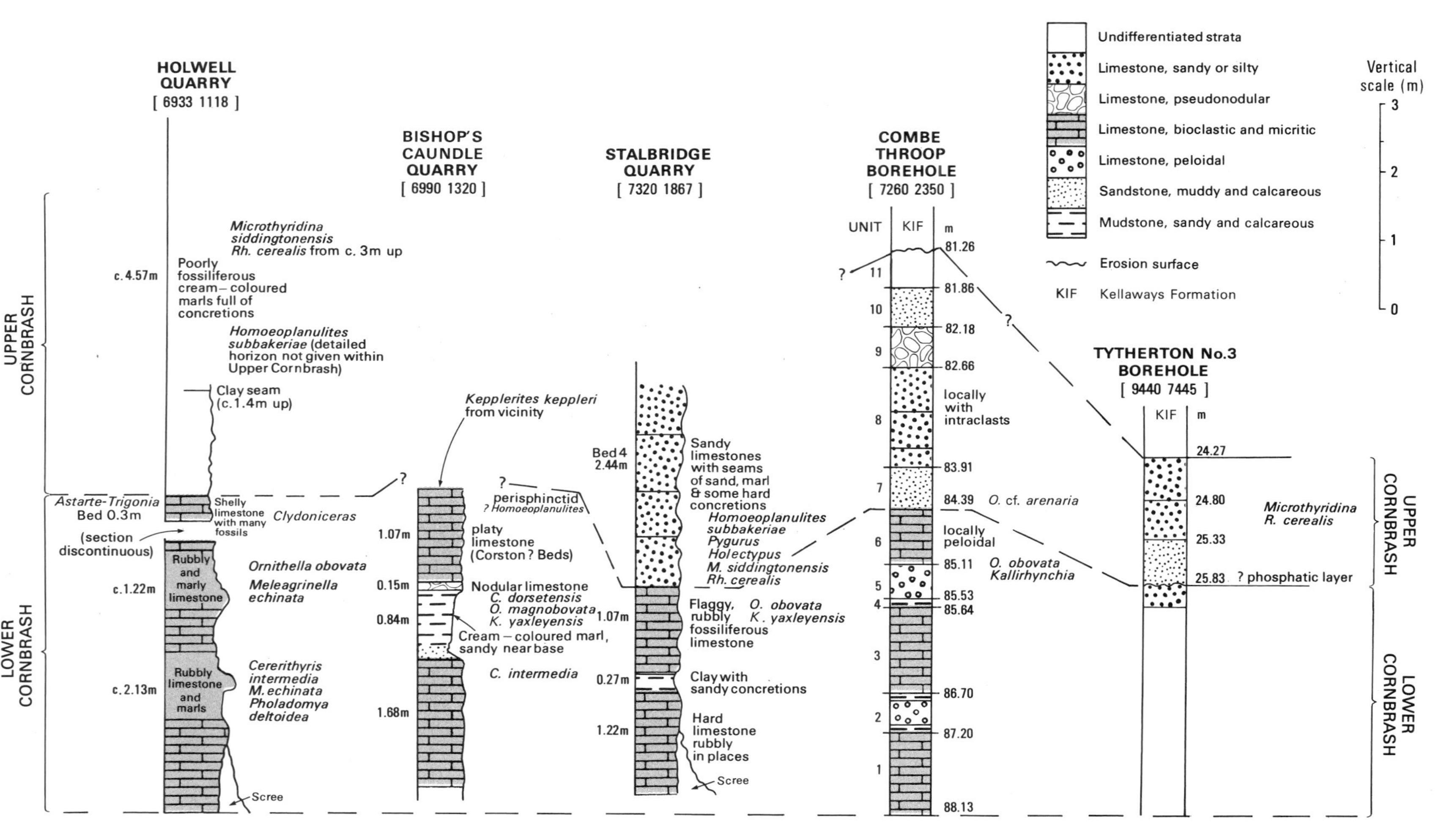

Figure 31 Comparative sections in the Cornbrash of the Shaftesbury and adjacent districts (see Figure 30 for location) (Data from Douglas and Arkell (1928), Arkell (1958), Cave and Cox (1975) and Page (1988)). Units in the Combe Throop Borehole are described in the text.

An exposure of shelly micritic limestone occurs in the foundations of Manor Farm [6978 1328]. South-east of the farm, Douglas and Arkell (1928) described the following section (here metricated) from the now-filled quarry [6990 1320]:

	Thickness m
Lower Cornbrash	
Limestone, platy, unfossiliferous, weathering brown (probably Corston Beds)	1.07
Limestone, nodular with '*Pseudomonotis echinata, Pholadomya* and *Pinna reticulata*'	0.15
Marl, cream-coloured, thin-bedded with beds of clay, sandy towards the base; '*Cererithyris dorsetensis, Ornithella magnobovata Kallirhynchia yaxleyensis Modiola bipartita, Goniomya* sp., *Pholadomya lyrata, Pecten vagans, Nucleolites clunicularis, Holectypus depressus*'	0.84
Limestone, hard, blue-hearted, compact, in massive beds with abundant '*Cererithyris intermedia, Homomya gibbosa, Pholadomya deltoidea*'	1.68

Holwell to Stalbridge

The Holwell quarry [6933 1198] formerly exposed the following section (Douglas and Arkell, 1928) (here metricated):

	Thickness m
Upper Cornbrash	
Marls, pale cream-coloured, poorly fossiliferous, full of concretions. '*Microthyris siddingtonensis*' and '*Rhynchonelloidea cerealis*' about 3 m from the base. '*Homoeoplanulites subbakeriae*', probably from a hard band in the basal 1.83 m, in a fallen block. A seam of blue clay, 1.37 m from the base, contains many bivalve casts: '*Pholadomya, Ceratomya, Pleuromya, Anisocardia* etc.'	4.57
Lower Cornbrash	
Astarte-Trigonia Bed	
Limestone, grey-centred, very shelly; only top then seen, but quarried blocks lie on the floor of the quarry; '*Clydoniceras sp., Astarte hilpertonensis, Trigonia rolandi, T. angulata, Anabacia complanata, Entolium demissum, Pecten vagans, Pseudomonotis echinata*'	0.30
Faulted break in section	
Limestones, rubbly and marly, yielding '*Ornithella obovata* and *Pseudomonotis echinata*'	1.22
Limestones and marls badly obscured by scree (= Blakes' Bed 3 with '*Avicula echinata, Terebratula intermedia* and *Pholadomya deltoidea*')	c.2.13
Forest Marble	
Limestone, blue, shelly, and oyster-bearing marls	seen to 1.22

North-west of the quarry, a second quarry [6919 1215] shows about 1.5 m of rubbly micritic limestone; a second exposure [6924 1217] shows 1m of micritic limestone and very sandy clay on 0.1 m of very shelly micritic limestone, overlying marly, very shelly limestone.

Douglas and Arkell (1928) described sections in the Lower Cornbrash in a quarry [714 146] at Stourton Caundle. The succession is similar, but less complete than that at Bishop's Caundle. The '*intermedia* Zone', about 2.13 m thick, contains abundant separated valves of *Cererithyris intermedia* throughout. Douglas and Arkell suggested that this was due to rapid deposition of material transported by strong currents. At the top of the '*intermedia* Zone', there are abundant casts of *Pholadomya* and *Homomya*. This bed is overlain by cream-coloured or grey, thinly bedded marl, at least 0.6 m thick, which contrasts with the rubbly limestone ('*obovata* Zone') that occupies the same horizon at Stalbridge. The brachiopods *Cererithyris dorsetensis* and *Obovothyris magnobovata* occur in the marl bed at Stourton Caundle, but are rare in the rubbly beds at Stalbridge. One specimen of *Clydoniceras* was found below the soil, confirming a *discus* Subzone age.

In a former quarry [possibly 7358 1593] at Poolestown, a limestone with *Cererithyris intermedia* at the base (Arkell, 1958, p.226) falls within the lower part of the Lower Cornbrash. The ammonites *Homeoplanulites homeomorpha, Macrocephalites* sp. and *Kepplerites (K.) keppleri* recorded from this area (Page, 1988), probably came from another pit [possibly 7370 1621], as the latter ammonite is the index fossil of the Herveyi Zone, Keppleri Subzone (Page, 1989), at the base of the Upper Cornbrash.

Douglas and Arkell (1928) gave the following section (here metricated) in the now overgrown Stalbridge Quarry [7320 1867]:

	Thickness m
Upper Cornbrash	
Sandy limestone, with seams of sand and marl; largely made up of hard concretions of varied shapes and sizes. '*Microthyris siddingtonensis* and *Rhynchonelloidea cerealis* abundant at the base, also *Homoeoplanulites subbakeriae, Pygurus michelini* and *Holectypus depressus*'	2.42
Lower Cornbrash	
Flaggy and rubbly fossiliferous limestone, with abundant '*Ornithella obovata, O.grandobovata, Kallirhynchia yaxleyensis, Acrosalenia hemicidaroides*' and lamellibranch-casts; conspicuous band of '*Pholadomya deltoidea*' about 150 mm from base	1.07
Clay with sandy concretions	0.25
Hard limestone, rubbly in places, obscured by scree; '*Pseudomonotis echinata*'	1.22

Only the uppermost 1.5 m of the Upper Cornbrash are still exposed.

The overgrown quarry [724 167] at Stalbridge Weston yielded '*Meleagrinella echinata, Pholadomya lirata, Modiolus plicatus, Pleuromya uniformis, Gervillia* sp. and a fragment of *Choffatia (Homeoplanulites) homoeomorphus*' (from the uppermost 0.6 m of the Lower Cornbrash) (Arkell, 1958).

Henstridge to Horsington

Creamy-weathering, biomicritic, sparsely shelly, burrowed limestone of the Upper Cornbrash is exposed in a silage pit [7202 2052] north of Henstridge Ash. The sinuous burrows are 5 to 7 mm wide; some have grey or ochreous clay filling.

Temporary exposures [7160 2083 to 7165 2075] in the Lower Cornbrash along the A357 south of Yenston showed buff-weathered, bioclastic, sparsely oolitic limestones with marly interbeds containing a few poorly preserved bivalves. The lowest limestone, 0.5 m thick, is probably close to the base of the Cornbrash. It is bioturbated, with ochre-coloured marly clay beneath and a clay-laminated marly bed above. At the southern end of the trench, and about 2 m higher in the succession, the limestones are more uniform, less bioclastic, sparry micrites of the Upper Cornbrash.

On the west side of the A357 south of Yenston, a degraded exposure [7157 2086] formerly showed (Page, 1988):

	Thickness m
LOWER CORNBRASH	
?*obovata* Biozone	
Limestone, thinly bedded, bioclastic and shelly; lower part consists of grey, shelly calcarenite with a marl seam and a band of common bivalves near the base. Near the top, it becomes a thinly bedded, grey, shelly oolitic limestone. Fauna includes *Meleagrinella echinata, Homomya* sp., *Pleuromya* sp., *Liostrea* and a loose fragment of *Clydoniceras* cf. *discus*	1.60
?*intermedia* Biozone	
Limestone, buff-weathering, bioclastic and calcarenitic, in two beds separated by a marl seam. *Clydoniceras discus* collected from loose block	0.65
Limestone, grey, shelly, bioclastic; shelly patches with *M. echinata* near irregular top	0.25
Limestone nodules, micritic, greenish grey, in brown marl with shell grit; fauna of *Lingula* sp. and fragmentary bivalves	0.25
FOREST MARBLE	
Clay and mudstone, greenish grey, some layers of small bivalves	2.40
Limestone, bluish grey, bioclastic with abundant oyster and wood fragments, in three beds with marl seams between	0.40
Clay, slightly silty, greenish grey	0.40

At Yenston, a roadside exposure [7181 2116] shows about 1 m of flaggy-weathering, sparry micritic limestone of the Upper Cornbrash.

The Cornbrash is exposed at several locations around Templecombe village (Taylor, 1991); only the more important sites are described here.

The disused railway cutting [7157 2199 to 7143 2222] at Templecombe provides intermittant exposure along the strike of the Upper Cornbrash. At [7155 2203], the beds weather to irregular, thin, buff, flaggy, even-grained, sparry, slightly sandy and bioclastic limestones; they may be peloidal or contain sparse recrystallised ooliths. From this point to the end of the cutting [7143 2222], small exposures are of hard, more massive, irregularly blocky-jointed, light grey, even-textured, sparry and finely bioclastic limestone, with some burrowing.

At Templecombe Station, a cutting [7075 2250 to 7080 2251] shows a dip section with a massive limestone bed, about 0.8 m thick, at the base of the Cornbrash, projecting from the grassy bank, with an apparent dip of 5° E. Nearby [7071 2247], 1.25 m of massive, creamy buff, very hard, very bioclastic, patchily burrowed and bioturbated limestone of the Lower Cornbrash is exposed.

A temporary exposure [7088 2254] of Upper Cornbrash east of the station, showed c.3 m of buff, even-textured, finely bioclastic, sparry, sandy limestones in irregular beds up to 0.3 m thick, separated by bioturbated, sandy, marly, ochreous-weathering partings about 0.1 m thick. Casts of bivalves occur in the partings. Page (1988) saw the junction of the Lower and Upper Cornbrash:

	Thickness m
UPPER CORNBRASH	
Limestone, soft, sandy with sandy beds between containing small concretions. '*M. siddingtonensis*' common	0.45
Limestone, hard, buff, slightly sandy, in three beds with sandy marl seams between	0.70
Calcarenite, hard, blue-hearted with small shell fragments	0.25
LOWER CORNBRASH	
Limestone, rubbly with lumps of bioclastic and micritic limestone in a marly matrix	0.20
Clay, brown and grey with streaks of bioclastic debris and some limestone nodules	0.20
Limestone, rubbly, as above	0.25
Limestone, hard, massive, pale brown	0.60

Loose blocks from the Lower Cornbrash contained well-preserved *Meleagrinella echinata* and a fragment of *Clydoniceras* cf. *discus*.

A 30 m tufa-covered cutting beside the old railway line north of Throop Road exposes the following section low down in the Upper Cornbrash at the southern end [7101 2272]:

	Thickness m
Limestone lenticles in a sandy marl matrix, with *Pholadomya*	0.50
Limestone, buff, sandy	0.15–0.30
Limestone lenticles in a granular marly matrix	0.35
Limestone, buff, massive	0.30
Limestone lenticles and thin beds in a sandy marl matrix, with a dark greyish brown clay bed 0.1 m thick in the middle	0.60
Limestone, buff, sandy, in two beds	c.0.50

North of Templecombe, a quarry [7010 2288] exposes about 0.75 m of hard, coarsely bioclastic and bioturbated Lower Cornbrash. The quarry [7019 2294 to 7024 2297] to the east shows fine-grained, sparsely bioclastic, fossiliferous, sandy, sparry limestone of the Upper Cornbrash.

An exposure [6993 2274] of Lower Cornbrash in the outlier at Northside Wood showed 0.5 m of hard, creamy buff, intensely burrowed and bioturbated, slightly bioclastic, micritic limestone. The burrows, filled with ochre-coloured sand, open on to the bedding surface; they have no particular orientation.

In the stream [7033 2290] north of Abbas Combe, about 1 m of thinly (0.1 to 0.2 m) and irregularly bedded, variably bioclastic, burrowed and bioturbated, micritic and sparry limestones, slightly oolitic or peloidal in the more shelly beds at the top, is exposed. The dip is about 4° NNW.

FOUR

Upper Jurassic

Upper Jurassic strata (see footnote p.16) crop out in an arc through the western, central and northern parts of the district. They comprise an alternating sequence of limestones and clays, with the latter forming the dominant part of the succession. The formations and members of the Upper Jurassic are shown in Table 5. Zones and subzones in the Callovian, Oxfordian and Kimmeridgian are treated as chronozones, and the nominal taxa are recorded in roman type.

KELLAWAYS FORMATION

The Kellaways Formation crops out in three tracts of low-lying ground along the western margin of the Blackmoor Vale. The eastern tract is a 7 km long, north-north-west-striking belt between Horsington and Stalbridge, and is mostly bounded by faults. The central tract, between the Stalbridge Park and Poyntington faults, comprises a series of faulted blocks on either side of the Caundle Brook. A tract west of the Poyntington Fault extends from Ashcombe Farm, south-east of Haydon, through Caundle Marsh toward Burton Hill Wood.

The Kellaways Formation of the district consists of a thin (c.2.5 m) basal sandstone or muddy sandstone overlain by c.6.5 m of silty mudstones and then a c.28 m unit of sandy mudstones and muddy sandstones with some small cementstone nodules and a distinctive bed of calcareous sandstone near the base (Figure 32); however, no formal lithostratigraphical subdivision has been made.

The original geological survey of the district included the Kellaways Formation in the Oxford Clay. However, because of similar weathering-profiles, deep auger holes are required to differentiate the Kellaways Formation from the Oxford Clay. The lower part of the formation weathers to yellow or orange-brown clayey sand with a little granular gypsum. The middle and upper parts weather deeply to a mottled orange, brown and pale yellow, jarositic (hydrated iron sulphate) sandy clay and clayey sand.

The sandy nature of the lowest beds, together with the presence of small limestone clasts, indicates an erosional base, and the mapped junction with the underlying

Table 5 The Upper Jurassic* stratal nomenclature of the district (not to scale).

	Stage	Group	Formation	Member
UPPER JURASSIC	Kimmeridgian		Kimmeridge Clay	
	Oxfordian	Corallian	Ringstead Waxy Clay	See Tables 7 and 8
			Sandsfoot	
			Clavellata Beds (including Coral Rag)	
			Stour	
			Hazelbury Bryan	
			Oxford Clay	Stewartby and Weymouth (undivided)
	Callovian			Peterborough / Mohuns Park
			Kellaways	
			Cornbrash	Upper Cornbrash
MIDDLE JURASSIC	Bathonian			Lower Cornbrash

* See p.16.

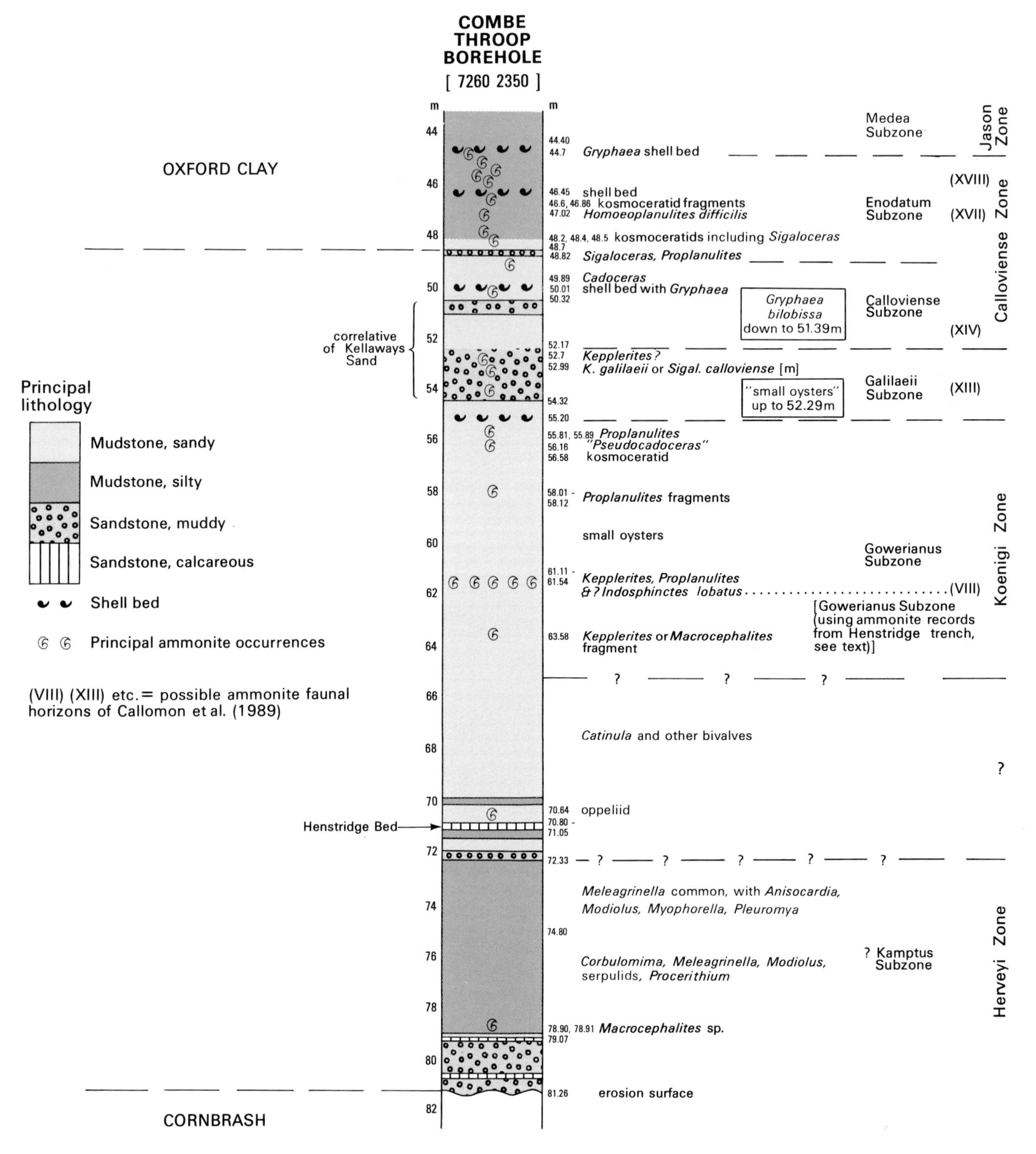

Figure 32 Summary of the Kellaways Formation sequence in the Combe Throop Borehole.

Cornbrash is sharp. No such erosion surface occurs in sections in the Wiltshire type area (Cave and Cox, 1975; Page, 1989), where the corresponding basal calcareous and sandy beds are included in the Cornbrash. The top of the Kellaways Formation is marked by a rapid change from sandy clay to uniform silty clay of the Peterborough or Mohuns Park members.

There are possible correlatives of the Kellaways Clay and Kellaways Sand of the Wiltshire type area in the Shaftesbury district. The 6.5 m thick medium olive-grey mudstone interval near the base of the Kellaways Formation in the Combe Throop Borehole probably corresponds to a shale unit forming the basal part of the Kellaways Clay, first differentiated by Callomon (1955, 1968) and later named the 'Cayton Clay Formation' by Page (1989) after Cayton Bay on the north Yorkshire coast. The unit can also be recognised from geophysical logs in the Mappowder Borehole. A c.4 m thick interval of sandstones and sandy mudstones in the upper part of the Kellaways Formation in the Combe Throop Borehole may be the correlative of the Kellaways Sand in Wiltshire.

Within the Kellaways Formation in the district, a c.300 mm thick laminated calcareous sandstone (the Henstridge Bed) has been mapped around Henstridge (Barton, 1990), and has been identified in the Combe Throop Borehole, 10.2 m above the base of the formation. It is thought to occur 13.7 m above the base of the formation in the Mappowder Borehole and 10 m above it in the Fifehead Magdalen Borehole. An identical sandstone was seen in a temporary section [7198 3052] near Wincanton and, according to Professor J H Callomon (written communication, 1991), a laminated sandstone occurs approximately 7 m above the Cornbrash in the Kellaways Formation near Frome; the Henstridge Bed thus forms a widespread marker bed in the Kellaways Formation. The clay-poor, cemented nature of the sandstone, together with its lateral continuity and sheet-like geometry, are consistent with a storm-deposit origin.

The Combe Throop Borehole proved 32.56 m of Kellaways Formation, the thickest recovered sequence to date (Figure 32). Geophysical log correlations based on Combe Throop give comparable thicknesses in the Fifehead Magdalen, Mappowder, Winterborne Kingston and Holnest boreholes, and show a general eastward-thinning across the district (Figure 33). The apparent thinning northwards to the Gillingham Borehole [7959 2661], and compensating thickening of the Oxford Clay, may in part be an effect of inadequate data (Whitaker and Edwards, 1926) used to recognise the top of the formation.

The Kellaways Formation ranges from the Herveyi Zone to the Calloviense Zone of the Lower Callovian (Callomon, 1964; Page, 1989); a more refined scheme of ammonite faunal 'horizons' is also available for this interval (Callomon et al., 1989).

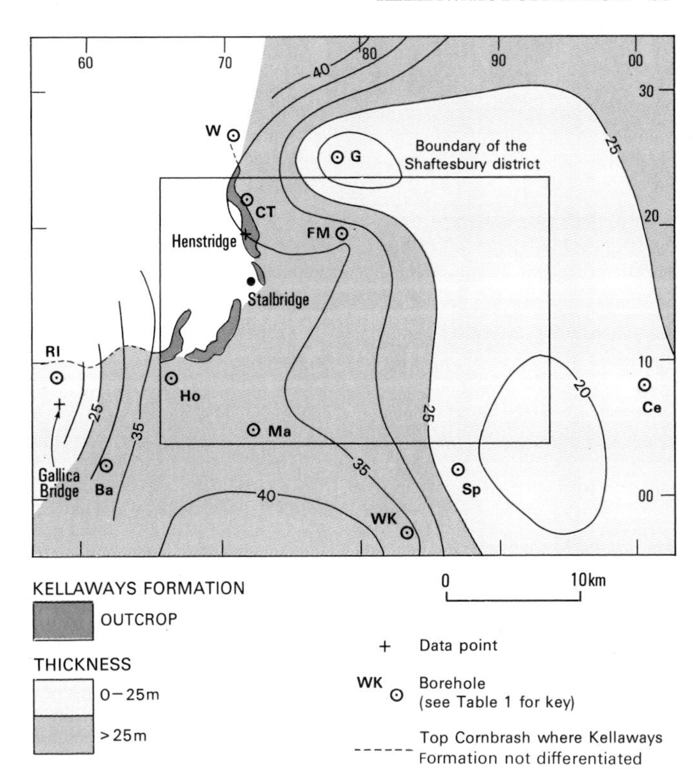

Figure 33 Isopachyte map for the Kellaways Formation.

Details

Combe Throop Borehole

In the Combe Throop Borehole, the Kellaways Formation rests on the eroded top of the Cornbrash at 81.26 m depth (Figure 32). The basal 2.36 m of beds, below 78.90 m, consist of pale olive-grey, muddy, shelly sandstones with thin, more calcareous layers in the lower part. The fauna includes *Corbulomima macneillii, Discomiltha lirata, Goniomya literata, Modiolus bipartitus* and *Myophorella* cf. *irregularis*. These sandstones are overlain by 6.57 m of pale to medium olive-grey, variably shelly, silty mudstones which are finely micaceous towards their base and sparsely sandy above 74.80 m. A mottled appearance and a variable smooth to silty texture resulting from extensive bioturbation is present throughout. Nodules of calcareous sandstone or cementstone are present at two levels. The fauna is sparse and mainly consists of serpulids, *Procerithium* and the bivalves *Anisocardia?, Corbulomima, Meleagrinella* and *Modiolus*; the ammonite *Macrocephalites* occurs at the base (78.90 m).

The succeeding 1.53 m thick, sandy and calcareous interval above 72.33 m contains, in its uppermost part, a 250 mm thick marker horizon of very pale olive-grey, fine-grained, 250 mm thick, laminated, calcareous sandstone, the Henstridge Bed (Figure 32). The basal 50 mm and the upper 30 mm of the sandstone contain 2 to 3 mm diameter subvertical burrows (?*Rhizocorallium*). The uppermost 100 mm of the bed has small-scale cross-lamination, while the lower part has a coarse, parallel lamination and a flaggy appearance. In thin section, the bed consists of 90 per cent subangular quartz (and very rare albite) grains in a ferroan calcite cement; quartz grain-size averages 0.1 mm (range 0.03 to 0.2 mm). The foraminifer *Lenticulina ectypa* makes its first appearance at 72.18 m, a species that has not previously been recorded below the Calloviense Zone.

The predominant lithology of the 22.1 m of strata above the Henstridge Bed at 70.80 m is a medium olive-grey, variably fissile and shelly, sandy mudstone that contains both hair-like pyritous horizontal burrows and 7 to 10 mm diameter subvertical burrows with a coarse sandy infill. There is an interval of medium olive-grey, bioturbated muddy and weakly calcareous sandstones between 50.32 and 54.32 m (approximately 2 to 6 m below the top of the formation). Ammonites are relatively common above 61.54 m, at which depth there are several specimens including *Kepplerites, Proplanulites* and *?Indosphinctes lobatus*. The assemblage belongs to the Koenigi Zone, Gowerianus Subzone and may represent Horizon VIII of Callomon et al. (1989).

The base of the Galilaeii Subzone is taken at a brown-stained silty mudstone between 54.95 and 55.20 m (Figure 32); the bed contains bivalves, including *Catinula, Entolium, Meleagrinella, Mesosaccella* and *Myophorella*. The base of the Calloviense Zone is taken at 52.17 m at the closest bed boundary above the highest *Kepplerites*. *Gryphaea (Bilobissa) dilobotes* first appears at 51.39 m; below this depth (up to 52.29 m) small oysters, including *Catinula*, predominate.

Caundle Marsh

North of Burton Hill Wood, pale brown-weathering ferruginous, extremely sandy clay [around 656 116] is bounded on its south side by the Coker Fault. Farther east, and higher in the Kellaways Formation, an auger hole [6631 1193] showed 1.5 m of fine-grained clayey sand overlying 0.8 m of pale brown, sandy, very ferruginous clay. Near the junction with the Peterborough Member, an auger hole [6675 1198] showed, beneath 0.8 m of head, 1.5 m of brownish grey to pale brown, fine-grained, sandy clay with much ferruginous material and jarosite, on 1.5 m of medium grey, sandy to very sandy clay with shell debris.

An auger hole [6548 1225] in the lower part of the Kellaways Formation near Broke Wood proved thin beds of soft sandstone in a clayey fine-grained sand to extremely sandy clay. Soft sandstone or calcareous sandstone also occurs, together with pale greyish brown sandy clay with jarosite, near the junction with the Cornbrash [6548 1256], and below Alluvium.

Bishop's Caundle to Stourton Caundle

The Kellaways Formation in this area is usually weathered deeply to a brown, sandy clay with common jarosite and gypsum. An auger hole [7098 1227] showed 1.5 m of grey clayey sand, on 0.6 m of greyish brown, ferruginous, jarositic, sandy clay, with grey sandy clay below.

The basal facies of the Kellaways Formation [7021 1298] exhibits a sequence of mottled orange-brown to grey, silt and sand, 2.3 m thick, above blue-grey silt and sand. Fragments of *Gryphaea* and *Catinula* frequently come up in the auger, both in ground north-east of Caundle Brook and in the upper beds adjacent to the faulted contact with the Peterborough Member.

Subdued relief typical of Kellaways Formation elsewhere contrasts with the incised outcrop along the Caundle Brook. A typical augered section in this ground [7224 1498], east of Stourton Caundle, showed 1.6 m of mottled, grey-brown, clayey silt, on 0.5 m of mottled dark brown-yellow, jarosite-bearing silt, on 0.5 m of silty, dark, steely grey clay.

Stalbridge–Henstridge

The Kellaways Formation in a trench at Stalbridge [7425 1772] consisted of pale grey clayey silt and very silty clay, and yielded abundant small *Gryphaea* and *Catinula*, together with *Gryphaea (Bilobissa) dilobotes*, a *Cylindroteuthis* fragment and part of a *Cadoceras* macroconch (Bristow et al., 1989).

A trench along the edge of the disused railway at Stalbridge [7392 1816] exposed mottled, bluish grey silty and sandy clay, in part overlain by alluvial sand and gravel, with abundant small *Gryphaea* and *Catinula*.

Temporary sections in 1986–1989 included the Henstridge pipe trench (Figure 34), dug between the sewage works [7320 1998] and a point [7278 1986] adjacent to the dismantled railway, from where further trenches were taken to Henstridge village at Oak Vale [7260 1996] and near Steel Well House [7274 1970]. In the Henstridge trench, the Kellaways Formation beneath the Henstridge Bed consisted of medium grey to bluish grey silty sand and sandy clay with numerous small *Gryphaea* and a few scattered cementstone nodules. The Henstridge Bed, a buff to pale orange-weathering flaggy sandstone bed, 300 mm thick, was intersected at two points along 300 m of strike length [7275 1974 and 7262 1997] (Figure 34).

A macroconch body-chamber fragment of *Proplanulites*, preserved in cementstone, was recovered [7283 1996] (Figure 34, locality 1) from about 5.5 m above the Henstridge Bed. The principal fossiliferous interval (Figure 34, locality 2), was a 30 m long section of the trench [midpoint 7288 1990], about 7 to 8.5 m above the Henstridge Bed, which exposed bluish grey, weathering, mottled orange and grey, very silty sand with some clay. Ammonites include body-chambers of *Kepplerites (Gowericeras) gowerianus* and *Proplanulites*, both preserved in cementstone, several *Proplanulites* ex gr. *koenigi* also preserved in, or associated with, cementstone concretions, and a fragment of a more coarsely ribbed ammonite, either *Kepplerites* or *Macrocephalites*. The inner or middle whorls of some ammonites are infilled with buff phosphate. The concretions have a maximum diameter of 190 mm; some are weakly septarian, while others have adherent flat oysters; a single *Camptonectes* has been recorded. The fauna indicates the Koenigi Zone, Gowerianus Subzone.

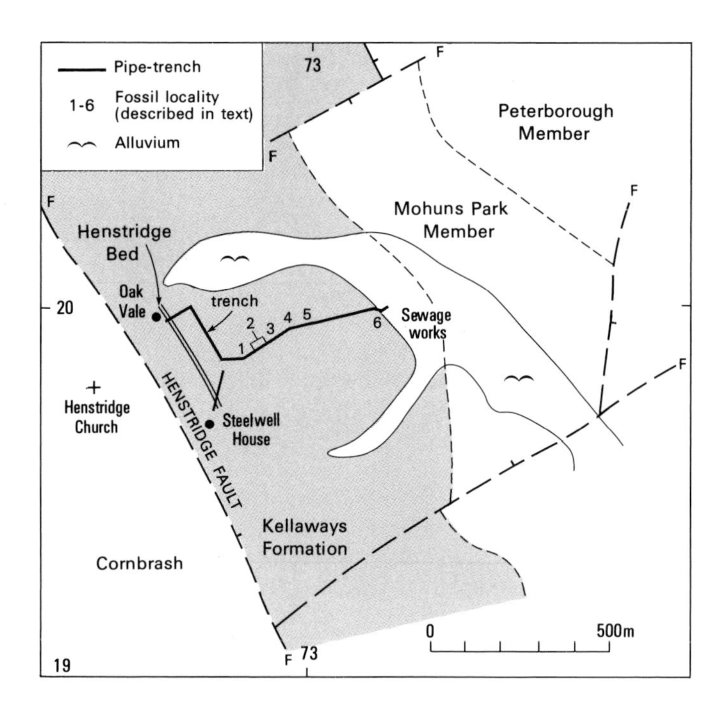

Figure 34 Sketch map of the geology of the Henstridge pipe trench showing localities in the Kellaways Formation referred to in text.

Higher beds (Figure 34, localities 3 and 4) c.9 to 11.5 m above the Henstridge Bed, are pale to medium grey, silty or sandy mudstones with abundant small juvenile *Gryphaea* or *Catinula*, together with *Bositra buchii*, *Grammatodon concinnus?*, *Meleagrinella braamburiensis* and *Thracia*. An inner whorl impression in cementstone of *Proplanulites* was also found.

Spoil from the upper part of the Kellaways Formation (Figure 34, locality 5), about 13 to 15 m above the Henstridge Bed, consists of mottled grey, very silty mudstone, with selenite crystals in part. The fauna includes small *Gryphaea*, uncrushed *Modiolus bipartitus*, *Oxytoma expansa*, an indeterminate kosmoceratid impression and one uncrushed whorl fragment of *Kepplerites* ex gr. *gowerianus* lightly encrusted with thin-shelled oysters. These beds probably fall within the Gowerianus Subzone.

Small (?juvenile) *Gryphaea* and *Catinula*, one *M. bipartitus* and a fragment of a large flat oyster (?*Liostrea*) were recorded in spoil excavated from a deep pit at the Sewage Works [7320 1998] (Figure 34, locality 6). A similar fauna occurs up to c.19 m above the Henstridge Bed in the Combe Throop Borehole.

Henstridge to Horsington

The lithological change from sandy to silty clay, at the Kellaways Formation–Oxford Clay boundary, beneath 1 m of Head, is shown by an auger hole [7353 2002] 250 m south-west of Higher Marsh House:

	Thickness m
HEAD: sand, fine, ochreous-buff, slightly clayey, becoming clayey sand	1.0
MOHUNS PARK MEMBER	
Clay, silty, plastic, light grey, ochreous-mottled, becoming medium grey, brown-mottled below 2.2 m	1.1
KELLAWAYS FORMATION	
Clay, fine-grained, sandy, medium grey, with coarse shell fragments below 2.9 m	0.4
Clay, silty, firm, with a little thin shell debris	0.3
Clay, sandy, shelly, medium grey	0.7
Clay, slightly sandy, shelly	0.2

The mainly sandy strata at the base of the Kellaways Formation were augered [7055 2309] between Horsington and Templecombe:

	Thickness m
Soil, sandy, brown	0.3
KELLAWAYS FORMATION	
Sand, fine-grained, ochreous, becoming slightly clayey below 0.5 m, and more clayey, rusty ochreous-brown below 1.0 m	0.9
Clay, fine-grained sandy, pale grey, ochreous- and buff-mottled, noncalcareous, but with some race at 1.8 m and shell fragments at 1.9 m that include small *Catinula*	1.1
Sand, fine-grained, clayey, calcareous, with some shell debris	0.4
Sand, slightly clayey, calcareous, very firm	0.3
Sand, clayey, slightly shelly and calcareous, ochreous-brown	0.1
Clay, fine-grained sandy, grey, ochreous-mottled, with serpulids	0.1
Sand, clayey, brown	0.1
CORNBRASH: limestone	touched

OXFORD CLAY

The Oxford Clay crops out in an arc from Holnest [655 100], in the south-west of the district, to Kington Magna [760 230] in the north. It forms the low-lying ground of the Blackmoor Vale.

The bulk of the formation comprises some 85 to 120 m of medium grey, calcareous mudstones. Beneath these, there is a lower unit, up to 40 m thick, of bituminous mudstones known as the Peterborough Member (formerly the Lower Oxford Clay). Locally, there is a basal lenticular unit of grey, calcareous, silty, shelly mudstone, the Mohuns Park Member (Tables 5 and 6). Augering to depths of up to 3 m to unweathered clay is required to enable these differentiations to be made. In the type area in Oxfordshire, the Lamberti Limestone divides the upper part of the Oxford Clay into the Stewartby Member (formerly the Middle Oxford Clay) and Weymouth Member (formerly the Upper Oxford Clay). In the present district, this marker bed is absent, except possibly in a limited area between Bagber [753 157] and Manor Farm [768 158]; the Oxford Clay above the Peterborough Member is therefore shown as a single unit (Stewartby and Weymouth members (undivided)) on the map.

The formation spans parts of the Callovian and Oxfordian stages (Table 5; footnote, p.16). Its thickness ranges between 120 m and 145 m in the district, the thinnest sequence being in the south-east, and the thickest successions occurring in the north and south-west (Figure 35). East of the district, the Oxford Clay thins to around 110 m at Fordingbridge; to the south-west and north, it reaches 160 m (but see p.55).

MOHUNS PARK MEMBER

The Mohuns Park Member crops out in a fault-fragmented outcrop from just east of Stalbridge, northwards to the district margin at Combe Throop. There is a smaller outcrop [around 715 115] near Holwell. The member consists of up to 19 m of medium grey, shelly, weakly calcareous, slightly pyritous, silty mudstone. In places, shell debris gives it a white-mottled appearance. In the Combe Throop Borehole, interbedding with brownish grey and greyish brown clay occurs. It is distinguished from the underlying Kellaways Formation by being almost sand free. In the Combe Throop Borehole, two thin sandy clay beds occur within the basal 5 m of the Mohuns Park Member. The top (at 48.7 m) of a sand below these has been taken to define the base of the member; it coincides with a marked change in the gamma-ray log signature (Figure 36). The ammonites *Homoeoplanulites difficilis* and kosmoceratids, including *Sigaloceras*, indicate the Enodatum Subzone within the lowermost 4 m of strata. The base of the Medea Subzone is taken at a *Gryphaea*-shell bed at 44.7 m.

A band of large septarian cementstone concretions near the top of the member generally forms a broad, distinctive feature, on which Mohuns Park Farm [7315 2125] is situated.

In the south-west of the district, the Holnest Borehole penetrated 14 m of the Mohuns Park Member, but it contains a higher proportion of bituminous brown shaly mudstones than the unit does at outcrop in the north. In the Holwell area, the Mohuns Park Member comprises 10 m of grey shelly clay. The Mohuns Park Member is cut out by faulting between Holwell and Stalbridge.

The top of the unit is diachronous; in the Combe Throop Borehole, it lies in the Coronatum Zone, ?Obductum Subzone (Figure 36), whilst a trench section near Stalbridge (Bristow et al., 1989) suggests that the top can be no younger than the Jason Zone in that area.

Table 6 Summary of the stratigraphy of the Oxford Clay of the district.

Chronostratigraphical units				Lithostratigraphical units	Lithology
Stage	Substage	Zone	Subzone		
OXFORDIAN	Lower Oxfordian	Cordatum	Cordatum	Hazelbury Bryan Formation	Orange-brown, bioturbated muddy sand and sandy mudstone
			Costicardia	OXFORD CLAY FORMATION; STEWARTBY AND WEYMOUTH MEMBERS (UNDIVIDED): Red Nodule Beds	Bluish grey, calcareous, shelly mudstone with small (10 cm) cementstone nodules
			Bukowskii		
		Mariae	Praecordatum		
			Scarburgense		
CALLOVIAN	Upper Callovian	Lamberti	Lamberti	(Lamberti Limestone)	
			Henrici		
		Athleta	Spinosum		
			Proniae		
			Phaeinum	PETERBOROUGH MEMBER: 5; Comptoni Bed 4	Brown, silty, bituminous shales with some beds of grey mudstone; five beds of large cementstone nodules
	Middle Callovian	Coronatum	Grossouvrei		
			Obductum	?3, ?2, ?1 cementstone beds	
		Jason	Jason		
			Medea	Mohuns Park Member	Grey, silty mudstone
	Lower Callovian (pars)	Calloviense	Enodatum		
			Calloviense	Kellaway Formation (pars)	Clayey sandstone, silty and sandy mudstone

PETERBOROUGH MEMBER

The Peterborough Member forms low-lying ground in a fault-fragmented outcrop, up to 1.5 km wide, extending in an arc principally from Holwell [705 105] in the south, to just east of Templecombe [720 235] in the north. West of Bishop's Down [670 120], most of the Peterborough Member is cut out by the Coker Fault. A small faulted outcrop occurs near Caundle Marsh [678 132].

Within the district, the Peterborough Member consists of between 18 and 40 m of predominantly bituminous shaly clays and mudstones with, towards the top, several impersistent levels of large septarian nodules (Bristow et al., 1989). Where the Peterborough Member rests on the Mohuns Park Member, there is a transitional boundary with greyish brown, silty mudstones alternating with grey silty mudstones. The mudstones weather to a distinctive dark chocolate-brown; there are some interbeds of less fissile, medium to dark grey and brownish grey mudstone. The chocolate-brown colour may be partly the result of hydration of the bituminous mudstones in the lower part of the weathered zone, because the brown colouration is less intense in fresh material. Nearer the surface, the mudstones weather to pale grey, ochreous-mottled clays, commonly with fine-grained granular gypsum. The zone of weathering is usually 1.5 to 2 m thick, and the transition to hard, unweathered mudstone commonly takes place within 0.5 to 1 m. Deeper weathering-profiles (more than 3 m thick) are present locally, particularly above cementstone doggers, for example south-east of Thornhill House [7431 1456], adjacent to Hargrove Lane [7465 1581] south-east of Poolestown, and around Henstridge Marsh [around 745 195], where mottled grey-brown, silty clay, approximately 3 m thick, overlies chocolate-brown clay.

The septarian cementstone concretions that occur towards the top of the member east of Stalbridge form mappable features (Bristow et al., 1989), one of which (Bed 4) is probably the Comptoni Bed–Acutistriatum Band. This bed was recovered in the Combe Throop Borehole (see Details).

At the top of the Peterborough Member, dark chocolate-brown, shaly mudstone, 0.1 to 0.2 m thick, rests on

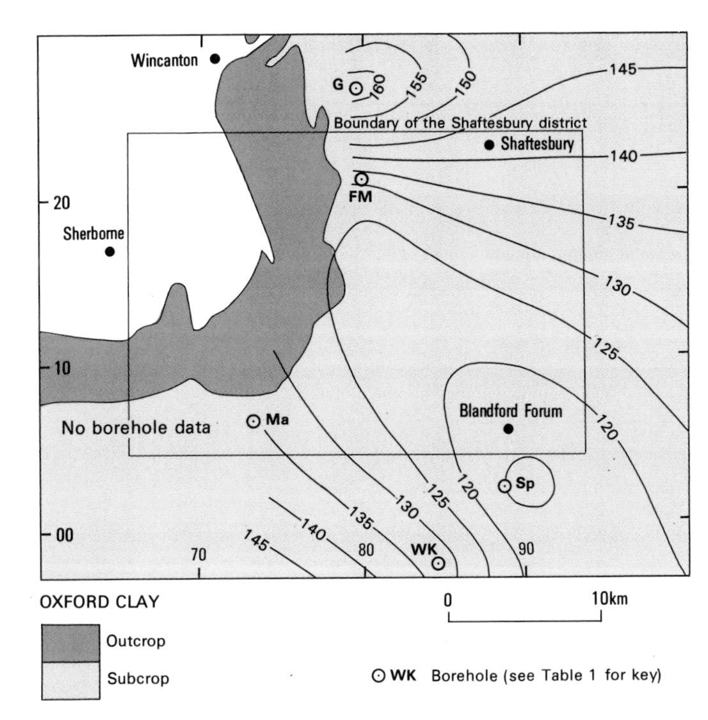

Figure 35 Isopachyte map for the Oxford Clay of the Shaftesbury and surrounding districts (contours at 5 m intervals).

about 0.5 m of greyish brown mudstone, which in turn rests on dark brown, shaly mudstone. This zone of variable lithologies at the top of the member has been recognised at several localities, including the Combe Throop Borehole.

The Peterborough Member ranges from the Calloviense Zone, Enodatum Subzone to the Athleta Zone, Phaeinum Subzone.

STEWARTBY AND WEYMOUTH MEMBERS (UNDIVIDED)

The Oxford Clay above the Peterborough Member consists of medium grey, variably silty mudstones, with some very fine-grained sand. In the weathering zone, the beds are yellow or greyish yellow, and gypsum is commonly abundant, with crystals up to 10 mm long. Small cementstone nodules, up to 0.3 m, occur sparsely at some horizons. Near the top of the sequence, small, red, ferruginous nodules, equivalent to the Red Nodule Beds of the Dorset coast, have been recorded at Marnhull [7771 2025] (Bristow et al., 1989), on Stour Hill [7708 2202] (Bristow, 1990a), near Fifehead Magdalen [7748 2127-7745 2138] (Ross, 1987; Bristow, 1990a) and near Middlemarsh [c.655 069] (Wright, 1981). Very small red nodules have also been noted in trial pits [around 6640 0944] near Holnest.

The Lamberti Limestone or its equivalent has not been recorded in the district, but the only good section seen at this stratigraphical level, in the Stalbridge–Marnhull trench [7425 1770 to 7660 1810] (Bristow et al., 1989), may have been incomplete due to faulting. A prominent feature nearby, near Marsh Farm [758 164], lies close to the Callovian/Oxfordian stage boundary, as determined by microfauna, and may be caused by a harder bed at the level of the limestone.

The Stewartby and Weymouth members range in thickness from around 100 m in the south-western part of the district to 85 m in the Fifehead Magdalen Borehole. In the Dorchester district, they are 98.5 m thick in the Winterborne Kingston Borehole. They range from the Upper Callovian Athleta Zone, Proniae Subzone, to the Lower Oxfordian Cordatum Zone, Costicardia Subzone. The interval is commonly very shelly, with some of the commonest fossils being thick-shelled gryphaeid oysters, with *Gryphaea lituola* characterising the lower part (to the top of the Lamberti Zone) and *G. dilatata* the upper part. Due to lack of exposure, few ammonites have been recovered, but all the zones, and most of the subzones of the Upper Callovian and Lower Oxfordian are recognised (see Details).

Details

MOHUNS PARK MEMBER

Holwell area

The Mohuns Park Member consists of a thin (less than 10 m) bed of grey, very shelly clay. Two auger holes [7103 1160; 7148 1181] proved up to 2 m of sticky grey, very shelly clay. The first yielded foraminifera, bivalves, gastropods, cephalopod hooks and echinoid spines. The presence of *Miliammina jurassica* and *Frondicularia franconica* indicate a Mid Callovian age.

Henstridge to Templecombe area

Auger holes and exposures record the weathering profile and lithology of the Mohuns Park Member in this tract (Taylor, 1991). A representative hole [7351 2028] near Higher Marsh Farm showed 0.3 m of soil, on 0.9 m of pale grey, ochreous-mottled clay, on 2.1 m of pinkish brown, ochreous-mottled clay with jarosite; becoming gypsiferous below 2.55 m, passing down into 0.15 m of silty, grey, ochreous-mottled clay, on 0.15 m of silty, medium grey, hard, crumbling mudstone.

Near Mohuns Park Farm, a slurry pit [7310 2129] showed about 1 m of pale grey-weathering clay. Large septarian nodules, 0.75 to 1.3 m in diameter, have been dug from the pit, which lies at the crest of the feature formed by the Mohuns Park Member.

Beside the road [7179 2263], east of Templecombe, an auger hole proved 2.5 m of weathered ochreous-mottled clay, on 4.25 m of silty, medium bluish grey, slightly shelly clay. Foraminifera from 3.4 m depth included *Epistomina* sp. cf. *stellicostata* and *Planularia eugenii*, and from 5.85 m, abundant *Epistomina parastelligera* and *E. nuda* which together indicate the Jason and Coronatum zones.

Combe Throop Borehole

The Mohuns Park Member occurs between 29.6 m and 48.7 m in the borehole (Figure 36). The base is taken at the change from a dominantly sandy sequence below to a silty and sandy mudstone sequence above. There is a gradual decrease upwards in the sand content over a 4 m interval, above which the unit consists of grey silty clay passing into subordinate beds of brownish grey and greyish brown clay. Fossils are common and include ammonites, bivalves and gastropods. *Gryphaea dilobotes*

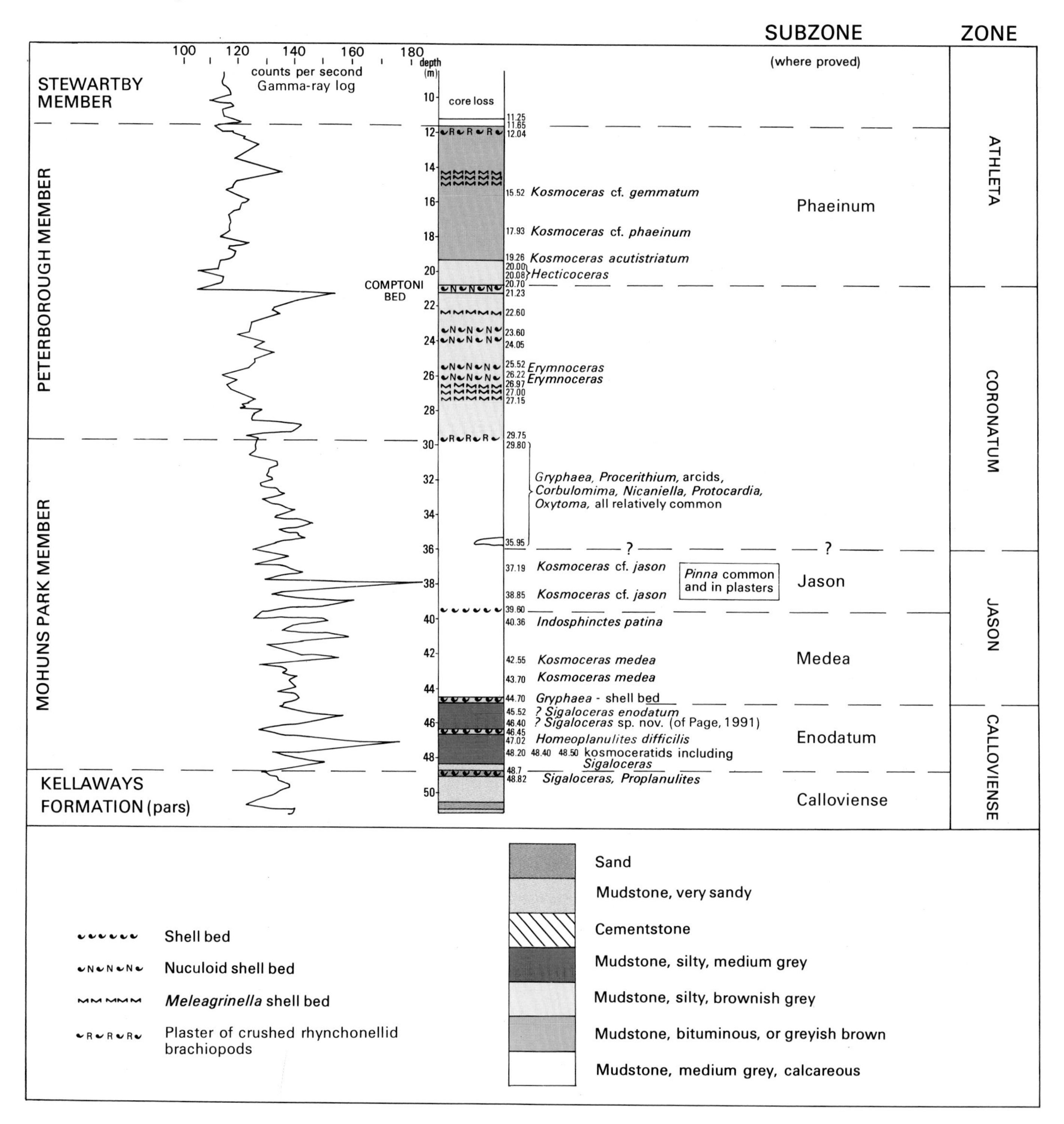

Figure 36 Litho- and biostratigraphy of the Oxford Clay in the Combe Throop Borehole.

is relatively common, together with *Corbulomima macneillii*, *Nicaniella phillis*, *Oxytoma inequivalve*, *Protocardia striatula* and the gastropod *Procerithium damonis*. The Enodatum Subzone is indicated by the occurrence of *Homoeoplanulites difficilis* at 47.02 m, and a possible *Sigaloceras enodatum* at 45.52 m. Diagnostic *Kosmoceras* identify the Medea and Jason subzones, but, above 37 m, specimens are determinable only at generic level (Figure 36).

PETERBOROUGH MEMBER

Bishop's Down to Caundle Marsh

In some auger holes [e.g. 6755 1209] south of Bishops Down, the depth of weathering is shallow, and brown carbonaceous clay was found within 1 m of the surface.

Some 5 m of the Peterborough Member are preserved against a north–south fault at Caundle Marsh. A large nodule with *Gryphaea dilobotes* (oral communication, P Ensom, 1987) was reported at a depth of 2 m in an excavation [6790 1335] in silty, orange-mottled clay.

Holwell to Bishop's Caundle to Stalbridge

Near Holwell, a bed of silt, less than 2 m thick, occurs at the base. Less than 3 m of chocolate-brown silty clay occurs against a NNW–SSE fault to the west of Bishop's Caundle.

Isolated cementstone doggers, approximately 1 m in diameter, occur as brash [7258 1488] near Brunsell Knap Farm, where they form the feature that caps the small knoll adjacent to the farmhouse. Farther south, in Stock Gaylard Park, cementstone doggers give rise to topographical features at two levels. The upper and better developed feature can be followed from the walled garden at Stock Gaylard House, around Church Park above Breach Pit Pond [720 126] to Stock Cottage [7235 1260].

East of Stalbridge, almost continuous exposures in a trench [7454 1778 to 7509 1794] showed fissile, fossiliferous, brown, silty shales, with five layers (Beds 1 to 5) of large cementstone nodules (Bristow et al., 1989). The nodules are up to 1.2 m by 0.3 m and are of two types: the first type has a reticulate-patterned surface of ferruginous-stained calcite veins spaced between 3 and 10 cm apart, which weather to give a knobbly appearance to the surface; the second type, from which the first type may be derived, has a ferruginous skin in concentric layers 1 to 2 cm thick, overlying ferruginous-veined pale grey limestone.

The fauna from below Bed 1 includes [at 7454 1778] fragments of *Kosmoceras* ?ex gr. *jason*; it probably indicates the Jason Zone. The surfaces of Bed 3 doggers yielded *Kosmoceras* ?ex gr. *obductum*, indicative of the ?Coronatum Zone, Obductum Subzone. Mudstones between Beds 4 and 5 yielded *Bositra buchii*, nuculoids, *Parainoceramus*, *Procerithium*, *Binatisphinctes comptoni*, *Kosmoceras* spp., including *K. phaeinum* and *K. (Spinikosmoceras)* sp.; the ammonites indicate the Athleta Zone, Phaeinum Subzone. Bed 4 may be the Comptoni Bed — Acutistriatum Band (Bristow et al., 1989).

Beds 3 and 5 form weak to good features. Bed 5 extends eastwards over a 80 m section of trench [7498 1793 to 7506 1795] and forms a good dip slope inclined eastwards. It is overlain by shelly, brown, silty paper-shales, about 4 m thick, which are exposed along the next 120 m of trench. Fossils from the middle of this length [7505 1793) include *Bositra buchii*, *Meleagrinella* (common), *Procerithium* and poorly preserved *Kosmoceras* spp. (including loop-ribbed form) indicating the Athleta Zone, Phaeinum Subzone.

Deep weathering-profiles (more than 3 m thick) are present locally, particularly above cementstone doggers, for example south of Henstridge Marsh [around 745 195].

Details of auger holes recording the weathering profile of the Peterborough Member and its upper boundary are included in Taylor (1991). A representative hole [7262 2271], east of Templecombe, showed:

	Depth m
Soil, sandy, buff	0.30
ALLUVIUM	
Sand, clayey, ochreous-buff	1.20
STEWARTBY AND WEYMOUTH MEMBERS, UNDIVIDED	
Clay, pale to medium grey, ochreous-mottled	2.00
Clay, silty, calcareous, brownish grey, becoming uniform medium grey; sharp change to	2.75
PETERBOROUGH MEMBER	
Clay, shaly, dark brown	3.00
Clay, silty, medium grey	3.40
Clay, shaly, dark brown	3.45

Combe Throop Borehole

The Peterborough Member was proved between 11.65 and 29.6 m, where it consists of brown, in places bituminous, silty mudstones interbedded, particularly at the top, with grey silty mudstones. The browner mudstones, especially the bituminous ones, are fissile, while the grey clays have a more blocky fracture. The most bituminous section, in which lignitic plant material is common, lies between 21.23 and 29 m, below which the mudstone is greyish brown.

Many plasters of nuculoids and *Meleagrinella* occur up to 21.23 m; the ammonite *Erymnoceras* was obtained from 25.52 and 26.22 m. Other fauna includes *'Dentalium'*, *Procerithium damonis*, *Bositra buchii*, *Corbulomima macneillii*, *Discomiltha lirata*, *Oxytoma inequivalve* and *Thracia depressa*. The Comptoni Bed was recognised between 20.7 and 21.23 m; it is characterised by abundant nuculoid bivalves (mainly *Mesosaccella morrisi)* and the ammonite *Binatisphinctes comptoni*. The mudstones above the Comptoni Bed (11.65 to 20.7 m) contained ammonites such as *Kosmoceras* cf. *gemmatum* (15.52 m), *K.* cf. *phaeinum* (17.93 m), *K. acutistriatum* (19.26 m) and *Hecticoceras* (20.00 and 20.08 m), and lie within the Phaeinum Subzone.

The appearance of the foraminifer *Lenticulina polonica* at 19.02 m and *Planularia eugenii* at 14.09 m indicates the Athleta Zone. *Triplasia acuta* and *T. kimmeridensis*, recorded at 11.5 m, are confined to the early Athleta Zone.

STEWARTBY AND WEYMOUTH MEMBERS, UNDIVIDED

Folke–Holnest area

At Osehill Green, debris from an excavation [6692 0915] included grey silty clay with small ferruginous nodules containing *Dicroloma trifida*, *Modiolus bipartitus*, *Pinna* sp., *Pleuromya alduini*, broken *Cardioceras (Scarburgiceras?)* sp. and *Euaspidoceras*. The same horizon was almost certainly exposed in a nearby trial pit [6640 0944], which showed 5.8 m of medium grey silty clay with two bands of small, red, sideritic nodules, c.0.5 m apart, c.1.5 m from the top. At the base of the pit, a bed of ferruginous siltstone, over 0.2 m thick, contained *Euaspidoceras*. The clays above contained debris of cardioceratids and *Euaspidoceras*. Other nearby pits at the same topographical level yielded further loose fossils, including *Euaspidoceras*, also in ferruginous siltstone, and fragments of large inflated cardioceratids *(Goliathiceras)* and other smaller cardioceratids, indicative of the Mariae Zone, Scarburgense Subzone.

The ammonites *Quenstedtoceras woodhamense*, '*Q.*' *scarburgense* and *Peltoceras* sp., indicative of the Mariae Zone, Scarburgense Subzone, were reported (written communication, Mr H C Prudden, 1989) from a trench [6520 1070] north-west of Holnest.

Middlemarsh–Glanvilles Wootton–Pulham

Wright (1981) recorded *Gryphaea dilatata* and small ferruginous nodules suggestive of the Red Nodule Beds in Lyon's Wood [657 067 to 661 062].

Grey silty clay with ferruginous and limestone nodules containing ammonites of the Cordatum Zone, Bukowskii Subzone, were found in a trench [6830 0881] near Stock Hill (written communication, Mr H C Prudden, 1989). Elsewhere in the trench, ammonites indicate the Cordatum Zone, Costicardia Subzone [6794 0915] and Mariae Zone, Praecordatum Subzone [6827 0900]. The last locality yielded *Goliathiceras (?Pachycardioceras)* aff. *alphacordatum*. North-east of Glanvilles Wootton, debris from a ditch [6937 0925] included fragments of cementstone nodules with *Modiolus bipartitus*, *Nanogyra* sp. and juvenile gryphaeids.

Debris from an excavation [7246 0940] near East Pulham comprised yellow and grey silty clay with *Gryphaea dilatata*. The presence of the foraminifera *Paalzowella feifeli* and *Gaudryina sherlocki* in medium grey, stiff, silty clay at a depth of 2.3 m in an auger hole [7170 0934] indicates the Cordatum Zone. A borehole [7036 0826] at West Pulham showed around 44 m of silty, grey, shelly clay overlying more than 26 m of the Peterborough Member.

Hazelbury Bryan–River Lydden–King's Stag–Poolestown

An exposure [7339 0882] in a river diversion showed dark grey, shelly clay below river terrace gravel. The co-occurrence of the foraminifera *Epistomina tenuicostata* and *Haplophragmoides* cf. *canui* indicates a level high in the Cordatum Zone, as do foraminifera from dark grey, silty, shelly clay in a nearby auger hole [7344 0927]. In Humber Wood, a trench [7275 0805] yielded yellow to medium grey, shelly, silty clay, with *Gryphaea dilatata*.

At King's Stag, pits for brick clay were worked until 1911. The site [724 105] is now flooded or swampy and there is no exposure. Material in the Dorset County Museum at Dorchester includes *Kosmoceras duncani*, *K. spinosum* (Plate 5, 4), *Quenstedtoceras brasili* and relatively common *Q. lamberti* (Plate 5, 5) (Bristow et al., 1989), indicative of the Lamberti Zone and possibly the uppermost Athleta Zone.

Pits [7100 1027 and 7308 1030] and ditches [741 121] south of the Coker Fault expose weathered orange to grey silty clay with *Gryphaea dilatata*.

Bagber area–Sturminster–Stalbridge area

At Bagber, flooded and overgrown brick pits [7555 1330] formerly yielded *Gryphaea dilatata* (Woodward, 1895). North of Bagber, a prominent feature, which extends north-eastwards for about 1 km from the River Lydden [751 140], appears to occur in Oxford Clay of Athleta Zone age. An auger hole [7511 1424] south-west of Mullins Farm, sited just below the feature, proved 2.6 m of calcareous, pale to medium grey, shelly, selenitic clay on limestone. Foraminifera from 2.6 m include *Lenticulina subalata*, indicative of the Coronatum or Athleta zones (Barnard et al., 1981). A second auger hole [7509 1453], about 6 or 7 m lower, yielded *Ophthalmidium dyeri*, *Citharinella nikitini*, *Lenticulina polonica* and *Spirillina infima*, and the ostracods *Nophrecythere cruciata intermedia*, *Eucytherura* cf. *gruendeli* and *Monoceratina scrobiculata*, indicative of an age no older than the Athleta Zone. The ostracod *Lophocythere Karpinskyi* obtained from a medium grey clay in a borehole [7662 1493] south-west of Manor Farm also suggests the Athleta Zone.Farther north-east, there is a strong north-west-trending feature which passes through Manor Farm [769 156]. Micropalaeontological data suggest that it is formed by strata which lie close to the Callovian/Oxfordian boundary. An auger hole [7695 1555], south of Manor Farm and just below the feature, proved:

	Depth m
Clay, mottled, orange/grey, firm	1.30
Clay, as above, with *Gryphaea* at 1.4 m	1.65
Marl, earthy, shelly with *Gryphaea*, red-brown	1.75
Clay, stiff, grey, mottled yellow-brown, with scattered shells; becoming deeper grey with depth	3.00

A sample at 3 m included the foraminifera *Citharinella nikitini* and *Verneuilinoides tryphera* which together indicate the Athleta Zone.

An auger hole [7615 1637] which commenced just above the feature, 1.1 km north-west of Manor Farm, yielded the foraminifera *Brotzenia* cf. *mosquensis*, *Lenticulina quenstedti*, *Citharinella nikitini* and *Nodosaria corallina*, and the ostracod *Lophocythere dorni* indicative of the Mariae Zone.

South-east of Manor Farm, an excavation [7783 1464] exposed medium to dark grey shaly clay with *Gryphaea dilatata* and microfossils, including *Nophrecythere cruciata oxfordiana*, indicative of an early Oxfordian age.

Ryalls Farm [757 160] stands on a prominent hill on the north-eastern extremity of a north-easterly trending ridge. Foraminifera from auger holes about 5m below the feature included *Nophrecythere cruciata oxfordiana* [7564 1560; 7543 1612], *N.c. intermedia*, *Verneuilinoides* cf. *tryphera* and *Procytherura tenuicostata* [7595 1561] indicative of the Lower Oxfordian. An hole [7539 1632], lying below the feature, proved stiff, bluish grey, shelly clay at a depth of 2 m with *Lenticulina subalata*, *Triplasia kimmeridensis* and *Citharinella nikitini*, indicative of the Athleta Zone.

A well-developed feature east of Bungays Farm descends northwards to the Bibbern Brook. An auger hole [7541 1716], about 4 m below the edge of the feature, proved stiff, mottled orange and grey clay with selenite at a depth of 1.4 m. A sample from the base of the clay yielded *?Lenticulina polonica* and

Plate 5 Fossils from the Oxford Clay, Corallian Group and Kimmeridge Clay (all natural size except Figure 12).

1a,b. *Cardioceras (Vertebriceras) quadrarium* S S Buckman: Weymouth Member, 'Red Nodule Beds'; near Marnhull (BGS BRI3044.

2a,b. *Gryphaea (Bilobissa) lituola* Lamarck; Stewartby Member; near Stalbridge (BGS BRI2058).

3 *Cardioceras (Scarburgiceras) praecordatum* Douvillé; Weymouth Member; Marnhull Sewerage works, Marnhull (BGS BRI1881).

4. *Kosmoceras (Kosmoceras) spinosum* (J de C Sowerby); Stewartby Member; King's Stag (Dorset County Museum G2869).

5. *Quenstedoceras (Lamberticeras) lamberti* (J Sowerby); Stewartby Member; King's Stag (Dorset County Museum G2934).

6. *Hibolithes hastatus* Montfort; ?Stewartby Member; near Stalbridge (BGS BRI2053).

7. *Nucleolites scutatus* (Lamarck); Cucklington Oolite; near Stour Provost (BGS BRI2960).

8. *Torquirhynchia inconstans* (J Sowerby); Lower Kimmeridge Clay, Inconstans Bed; near Hazelbury Bryan (BGS CF1680).

9. *Nanogyra nana* (J Sowerby); Lower Kimmeridge Clay, Nana Bed; near Buckland Newton (BGS CF2227a).

10. Assorted valves of *Nanogyra nana* (J Sowerby); Lower Kimmeridge Clay, Nana Bed; Knackers Hole Borehole (48.85 m depth) (BGS BKC6025).

11. Nannocardioceras Cementstone with *Amoeboceras (Nannocardioceras)* and *Isocyprina minuscula* (Blake) in solid preservation; Lower Kimmeridge Clay; Darknoll Brook, Okeford Fitzpaine (BGS BRI3360).

12. Scanning electron micrograph (× 1700) of coccolith-rich limestone with overgrown coccoliths dominated by the genus *Watznaueria*; Upper Kimmeridge Clay, ?White Stone Band; Gear's Mill, near Shaftesbury (BGS BRI3029).

13. Shelly mudstone including a) ventral fragment of *Aulacostephanus*, b) *Nanogyra virgula* (Defrance) and c) *Protocardia morinica* (J de C Sowerby); Lower Kimmeridge Clay (?KC31); Darknoll Brook, Okeford Fitzpaine (BGS BRI3365).

1a
1b
2a
2b
3
4
5
8
9
10
11
12
13
10PM

Saracenaria cornucopiae, indicative of an age no older than the Athleta Zone.

Two auger holes [7535 1703; 7540 1717] east of Bungays Farm proved weathering profiles in excess of 4.2 m.

Stalbridge–Marnhull area

North of the Bibbern Brook, an auger hole [7558 1790] encountered, beneath 1.5 m of river terrace deposits, medium grey clay, with some greyish brown and brown mottling. The rich microfauna from a depth of 3.4 m includes *Lenticulina polonica* and *Verneuilinoides tryphera*, which indicate the Athleta Zone.

Grey shaly mudstone with small (up to 0.3 m across) cementstone nodules was seen in the trench section [7509 1795] west of Gomershay Farm; broken specimens of *Gryphaea (Bilobissa) lituola* were present [7524 1802] (Plate 5, 2a–b). A microsample [7542 1805] from farther east included the ostracods *Polycope sububiquita, Glabellacythere reticulata, Terquemula flexicosta lutzei* and *Lophocythere caesa britannica*, indicative of the ?late Athleta to early Lamberti zones. The belemnite *Hibolithes hastatus* (Plate 5, 6) and a fragment of a coarse pectinid bivalve, suggesting the Upper Callovian, were found in greyish brown silty clay [7573 1812].

Farther east [7598 1817], a bluish grey mudstone [7598 1817] with *Gryphaea dilatata* and an uncrushed, pyritised cardioceratid ammonite *(Goliathiceras* sp. or inflated *Quenstedtoceras* sp.), indicative of the Mariae Zone, were found. From this point eastwards, valves of *Gryphaea dilatata* were common on the surface [as far as 7612 1812].

On the west bank of the Stour [7630 1809], medium grey mudstones yielded *Gryphaea dilatata*, some with encrusting serpulids, *Hibolithes hastatus*, crushed fragmentary and partially pyritised ammonites, including *Cardioceras (Scarburgiceras)* ex gr. *praecordatum* and *Creniceras rengerri*, indicative of the Mariae Zone, Praecordatum Subzone. East of the Stour, 1.8 m of head and river terrace deposits overlie 5 m of bluish grey mudstone [7658 1810] with *Gryphaea dilatata, 'Astarte', Entolium, Cardioceras (Scarburgiceras) praecordatum* (Plate 5, 3), *Peltoceras (Parawedekindia) arduennense* and *Hibolithes hastatus* which also indicate the Praecordatum Subzone.

Cementstone nodules form a small feature [747 194] south-east of Rhode's House Farm. Auger holes [7472 1940; 7484 1952] above and close to this feature proved gypsiferous, mottled, greyish brown clay to depths of 4 m or more.

An auger hole [7530 1879] in the bottom of a 0.4 m deep ditch near Triangle Farm, close to the base of the Stewartby Member, proved 3.55 m of clay, mottled orange-grey to about 2.5 m, becoming brown below; selenite appears at a depth of about 1.5 m. A sample from the base yielded *Saracenaria oxfordiana, Lenticulina subalata, L. muensteri, L. ectypa, Brotzenia mosquensis, Glabellacythere reticulata, Nophrecythere cruciata intermedia, Lophocythere caesa britanica* and a pyritised ammonite nucleus (loop-ribbed *Kosmoceras)* which probably indicates the Proniae Subzone.

An auger hole [7516 1983] north of Gibbs Marsh Farm proved 1.2 m of soft, mottled grey/orange-brown clay, on 1.28 m of bluish grey clay with some orange mottling (layer of oolitic race at 1.5 m; selenite below 1.5 m), on a 2 cm thick, brick-red siltstone, on 0.6 m of mottled olive-grey/brown clay. A rich microfauna from the basal clay included *Citharinella nikitini*, indicative of an age no older than the Athleta Zone.

Marnhull–Fifehead Magdalen

Red Nodule Beds were seen in a temporary excavation [7771 2025] near Marnhull. There, two layers of red sideritic nodules, 0.7 m apart, occurred in a 2 m section of pale to medium grey clay. The nodules were ellipsoid or round, up to 45 mm across, or flattened, c.20 × 70 mm. Of several hundred nodules examined, only two ammonite fragments were found; they were *Cardioceras (Vertebriceras) quadrarium* (Plate 5, 1a–b) and *Goliathiceras* sp. (Bristow et al., 1989), indicative of the Costicardia Subzone. Specimens of *Gryphaea dilatata* with an encrusting fauna of serpulids, oysters and foraminifera occurred in the clays. Because of faulting, the position of the nodule beds relative to the top of the Oxford Clay could not be determined. The site lies just south-east of an old clay pit [7758 2035] in which Woodward (1895) recorded *Gryphaea dilatata*.

The concentration of clay-ironstone nodules, some up to 0.5 m diameter, noted by Ross (1987) on the west side of Fifehead Magdalen [7748 2127 to 7745 2138] may also be from the Red Nodule Beds. Their position, some 28 m below the top of the Oxford Clay, may have been affected by solifluction.

Limonitic nodules up to 30 cm across in a ditch [7708 2202] on the south-west side of Stour Hill, about 20 m below the top of the Oxford Clay, may also be from the Red Nodule Beds.

A specimen of *Peltoceras (Parawedekindia)* sp. from near Coking Farm, Fifehead Magdalen [7666 2150], indicative of the Praecordatum or Bukowskii subzones, came from a level about 35 m below the top of the Oxford Clay.

Beside the A30 [7412 2115], west of Bow Brook, an auger hole [7412 2115] proved 1 m of river terrace deposits, on 1.3 m of gypsiferous, grey, brown and ochreous-mottled clay, on 1m of silty, noncalcareous, uniform, medium to dark grey clay, with some shell debris and a lignitic fragment at 2.8 m. A sample at a depth of 3.3 m yielded *Trochammina globigeriniformis, Epistomina stellicostata, E. regularis* and *Saracenaria oxfordiana*, and rare *Citharinella nikitini* and *Planularia eugenii*, indicative of the Coronatum or Athleta zones; the lithology suggests the latter.

A temporary exposure [7518 2352] north-west of Kington Magna, about 45 m below the top of the Oxford Clay, proved medium grey clay with common *Gryphaea dilatata* and partially pyritised pentacrinoid columnals.

An auger hole [7260 2349] alongside the Combe Throop Borehole, duplicates the upper part of the borehole sequence, for which there was little recovery:

	Depth m
Soil, clayey, silty, medium brown	0.2
TERRACE	
Clay, finely sandy, silty, buff-ochre	1.0
STEWARTBY MEMBER	
Clay, calcareous, grey, ochre-mottled, with ferruginous clayey-weathered cementstone nodule at 1.1 m; granular gypsum below 1.3 m; passing into	1.5
Clay, calcareous, medium grey, with some pale brownish ochre mottling; granular gypsum below 2.0 m	3.1
Clay, calcareous, brownish buff, with some gypsum, some grey mottling and 5 to 6mm-long gypsum crystals below 3.3 m	3.9
Clay, calcareous, buff to greyish buff, with scattered irregular gypsum crystals up to 12 mm across; some patches of unaltered medium grey mudstone below 4.2 m	4.8
Mudstone, calcareous, medium grey, with some white shell fragments; resting on cementstone at	8.3

CORALLIAN GROUP

The Corallian Group represents an interlude of shallow-marine, mixed carbonate and siliciclastic sedimentation within a long period of argillaceous deeper-water shelf sedimentation that started with the Kellaways Formation and ended with the Kimmeridge Clay. The group comprises a siliciclastic lower part with very sandy, shelly limestones (Hazelbury Bryan Formation), an interval of commonly oolitic, calcareous clays and oolitic, pisolitic and bioclastic limestones (Stour Formation, Clavellata Beds and Coral Rag), and a sandy and argillaceous upper part (Sandsfoot Formation and Ringstead Waxy Clay). This tripartite division is maintained throughout Dorset, although from Marnhull northwards, the middle part becomes increasingly sandy. The evolution of the terminology and the stratigraphical sequence currently recognised in the district is shown in Table 7. In part, the terminology follows Wright (1981), but mapping has resulted in the recognition of an additional member (Hinton St Mary Clay) and the correct stratigraphical position of the Newton Oolite (= Cucklington Oolite (Bristow, 1989b)). The Ringstead Waxy Clay has been mapped with the Kimmeridge Clay because of the difficulty of mapping the junction (see p.74).

The Corallian Group crops out in a tract from West Stour in the north of the district, southwards towards Sturminster Newton and then south-westwards towards Hazelbury Bryan, south of which its outcrop is disrupted by a major east–west fault, the Merriott–Hardington Fault (Figure 37). The outcrop continues westwards, much disturbed by both east–west and north-north-west trending faults, towards Glanvilles Wootton and Lyon's Gate, with the upper units being progressively cut out westwards by overstep of Cretaceous strata.

The group is thickest in the northern and southern parts of the district (90 and 80 m respectively), and thinnest between Marnhull and Fifehead Neville (40 m), over the Cranborne–Fordingbridge High. Regionally, the thickness varies considerably (Figure 37); it is at least 100 m thick in the Winterborne Kingston Trough and in south Dorset, and c.90 m thick in the Mere Basin.

In the mid and late Oxfordian, ammonite provincialism became acute and different standard zonations have been developed for each province. The zonation traditionally applied to the Corallian Group of Dorset is based predominantly on perisphinctids, but a more boreal zonation based exclusively on cardioceratids, used elsewhere in England, can also be applied in part (Table 8).

The Corallian Group ranges from the Lower Oxfordian (Cordatum Zone, Cordatum Subzone) to the Upper Oxfordian (Pseudocordata Zone).

Table 7 Evolution of the terminology of the Corallian Group in north Dorset.

Old Series one-inch geological map (1875)	Blake and Hudleston (1877)	Woodward (1895)			Arkell (1933)		Gutmann (1970)		Wright(1981)		This account	
											Formation	Member
KIMMERIDGE CLAY												
	Sandsfoot Grit	Sandsfoot Beds			Upper Calcareous Grit		*T. clavellata* bed or *Trigonia* Beds			Passage Beds (Ringstead Clay)	Ringstead Waxy Clay	
									Sandsfoot Formation	Sandsfoot Grit	Sandsfoot	Sandsfoot Grit
	Marl and clay				Sandsfoot Clay					Sandsfoot Clay		Sandsfoot Clay
	Calcareous sandstone and limestone with *Trigonia* (Calcareous Group)				*Trigonia clavellata* Beds	Glos Oolite Series				*Trigonia clavellata* Beds (or Formation)	Clavellata Beds (including Coral Rag)	Eccliffe
Coral	Rag-stone (Florigemma Beds)		*Trigonia* Beds and Osmington Oolite							Coral Rag		
Rag	False-bedded limestone	Coralline Oolite		Upper Corallian	Marnhull and Todber Freestone	Osmington Oolite Series		Todber Freestone (False-Bedded Limestone)	Osmington Formation	Todber Freestone		Todber Freestone
	Marl and Pisolite (Series)				Littlemore Clay Beds (pisolite at base)		Osmington Oolite (Series)	Pisolite Facies				Newton Clay
										Sturminster Pisolite		Sturminster Pisolite
										Newton Oolite	Stour	Hinton St Mary Clay
										Nothe Clay		
					?	Berkshire Oolite Series			Stour Formation	Cucklington Oolite		Cucklington Oolite
										Woodrow Clay		Woodrow Clay
Sand	Sand	Lower Calcareous Grit		Lower Corallian	Lower Calcareous Grit					Lower Calcareous Grit Formation	Hazelbury Bryan	
OXFORD CLAY												

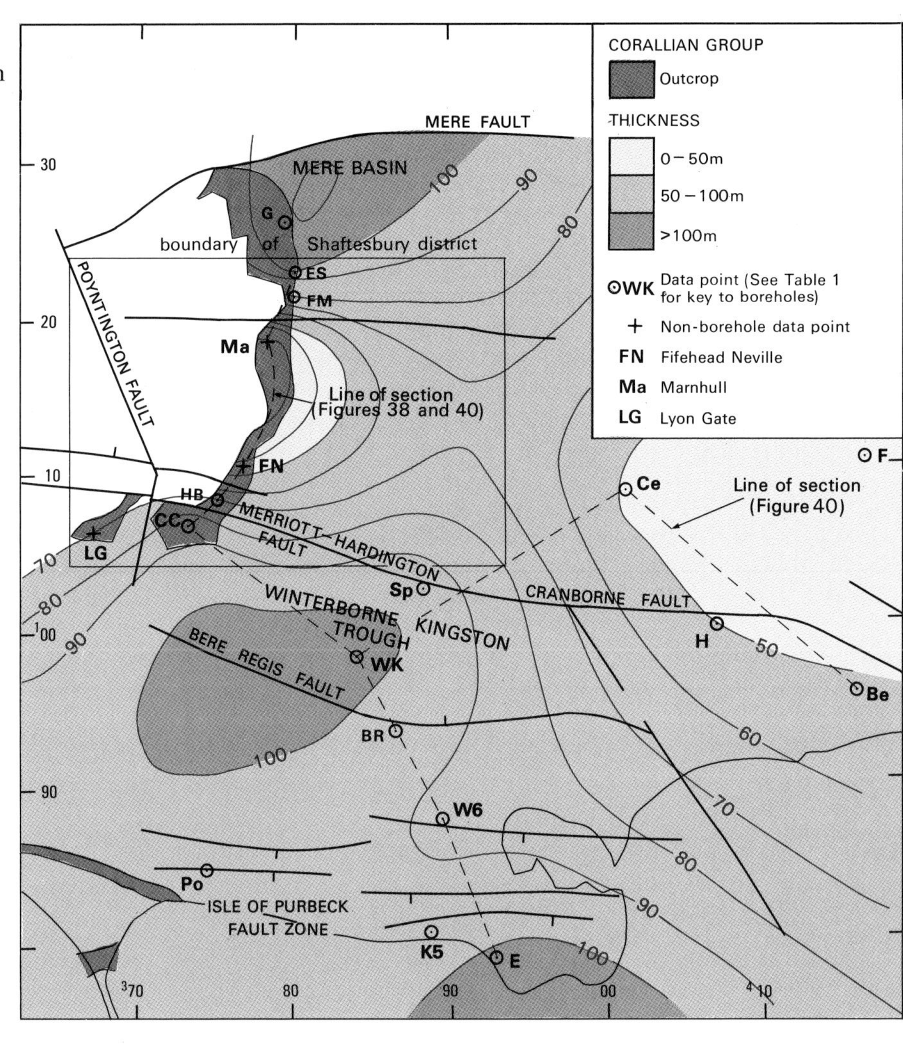

Figure 37 Isopachytes of the Corallian Group and major fault pattern for the Shaftesbury and adjacent districts.

Hazelbury Bryan Formation

The formation consists of clays, clayey sands and sands, mostly in coarsening-upward sequences. Thin sandy limestones occur locally in the upper part of the coarsening-upward sequences. The sands generally have pronounced spring lines at their bases. Where the Hazelbury Bryan Formation is thickest, there are four such sand units named, in ascending order in this account, A, B, C and D (Figures 38, 39 and 40). Sands A and B occur only in the northern and southern parts of the district; they are missing in the tract between Marnhull and Fifehead Neville. South of the district, these lower sands are well developed in the Winterborne Kingston Borehole and as far south as the Wareham C6 Borehole (Figure 40), but farther south, in the Isle of Purbeck, and east and south-east of the district, in the Cranborne, Fordingbridge, Hurn and Bransgore boreholes, the Nothe Grit (which equates with Sand C) rests on Oxford Clay. Sand D does not seem to be present in south Dorset.

The base of the formation is defined by the base of the lowest sand in the sequence, which is Sand A in the southern and northern parts of the district and Sand C in the central area.

The sands in the Hazelbury Bryan Formation range in mean grain size from very fine to medium grained (1.95 to 3.90ø), the sorting from 0.5 to 2.45ø and the inclusive graphic skewness from −0.03 to 0.89. Some are calcareously cemented and oolitic. Glauconite occurs at one level in the Cannings Court Borehole, but was not noted elsewhere. Samples from the East Stour, Hazelbury Bryan and Canning's Court boreholes show smectite to be the dominant clay mineral, particularly in the last two boreholes. Kaolinite is also common, mainly in the sandier parts of the sequence.

The formation is thickest in the north of the district (52 m) and is relatively thin (10 to 16 m) from Marnhull to south of Sturminster. Farther south, it thickens to c.45 m between Hazelbury Bryan and Middlemarsh (Figure 37).

The macrofauna of the Hazelbury Bryan Formation is dominated by bivalves including *Chlamys, Gervillella, Isognomon, Lopha, Modiolus, Myophorella, Pinna* and *Thracia*; gastropods, such as *Dicroloma* and *Procerithium*, and cardioceratid ammonites, are also present.

The occurrence of *Cardioceras cordatum?* at the base of Sand B in the Hazelbury Bryan Borehole, indicates the Cordatum Subzone. Temporary sections [7432 2977 to 7435 2479] on the A303 in the Wincanton district yielded *Cardioceras (C.)* ex gr. *ashtonense - persecans* (12.5 m from the top of the formation), and the same species group, together with *Cardioceras (C.* trans. *Vertebriceras)* and *C. (Plasmatoceras)* ex gr. *plasticum*, in a sand (probably Sand B) c.15.5 m from the top. This fauna also indi-

Table 8 Chronostratigraphical subdivision of the Oxfordian Stage.

Predominantly perisphinctid zonation: Subzone	Predominantly perisphinctid zonation: Zone	Substage	Cardioceratid zonation: Zone	Cardioceratid zonation: Subzone	Lithostratigraphy of the Corallian Group of the Shaftesbury district
Evoluta	Pseudocordata	Upper Oxfordian	Rosenkrantzi		Ringstead Waxy Clay
Pseudocordata					Sandsfoot Grit
Pseudoyo			Regulare		
Variocostatus	Cautisnigrae				
Cautisnigrae			Serratum	Serratum Koldeweyense	Sandsfoot Clay Eccliffe Mb
Nunningtonense	Pumilus		Glosense	Glosense Ilovaiskii	Clavellata Beds
Parandieri		Middle Oxfordian	Tenuiserratum	Blakei Tenuiserratum	Coral Rag
Antecedens	Plicatilis		Densiplicatum	Maltonense	Todber Freestone Newton Clay Sturminster Pisolite
Vertebrale				Vertebrale	Hinton St Mary Clay Cucklington Oolite Woodrow Clay
Cordatum	Cordatum	Lower Oxfordian	Cordatum	Cordatum	Hazelbury Bryan Formation
Costicardia				Costicardia	Oxford Clay (pars)
Bukowskii				Bukowskii	
Praecordatum	Mariae		Mariae	Praecordatum	
Scarburgense				Scarburgense	

The zonations are based on ammonite faunas — one exclusively cardioceratid (for the boreal and subboreal areas) and one predominantly perisphinctid (for parts of the subboreal area) (Sykes and Callomon, 1979, emend. Birkelund and Callomon, 1985).

cates the Cordatum Subzone. At Revels Farm [6773 0560], Buckland Newton (p.78), the Lyon's Gate Bed, which sits on top of Sand C, and which there occurs c.15 to 20 m below the top of the formation, yielded *Cardioceras (Scoticardioceras)* sp., indicative of the Vertebrale Subzone. The Cordatum–Plicatilis/Densiplicatum zonal boundary therefore occurs within, or at the top of, Sand C. A specimen of *Cardioceras (Plasmatoceras)* ex gr. *tenuicostatum* also indicative of the Vertebrale Subzone, was collected (Holmes and Melville, MS 1952) from Sand D in the Kingsmead railway cutting [7760 2473], in the southern part of the Wincanton district. Vertebrale Subzone faunas also occur in the overlying Cucklington Oolite which indicates that the Hazelbury Bryan Formation above Sand D must belong with this Subzone.

Microfauna from the middle of the formation includes the ostracod *Schuleridea triebeli* from 50.2 m in East Stour Borehole, indicative of an age no older than the Cordatum Zone, the foraminifer *Ophthalmidium strumosum*, which becomes extinct in the Tenuiserratum Zone, and 'flood' proportions of *Ammobaculites coprolithiformis* (Wilkinson *in* Bristow et al., 1989).

Lyon's Gate Bed

The Lyon's Gate Bed, approximately 18 m below the top of the Hazelbury Bryan Formation (Figures 38, 40), forms a fairly continuous outcrop in the south-west of the district around Lyon's Gate and Cosmore, where it consists of very calcareous, shelly, ferruginous sandstone or very sandy limestone, which commonly causes a pronounced spring line. It is usually overlain by a highly ferruginous clay. It occurs in the Hazelbury Bryan Borehole between 42.02 and 43.56 m, where it consists of a mixture of calcareous, shelly, muddy, extremely poorly sorted sandstone and very sandy limestone (Plate 6E,F). The bed is highly ferruginous and the sand is very coarse grained, some quartz grains being up to granule size. Other coarser clasts, mainly of rounded shell debris and sandy limestone, also occur; they indicate reworking of older sediments. The bed is not obvious in the Cannings Court Borehole, but it may be represented by some rather gritty glauconitic clayey sand at c.41 m.

The bed yields a macrofauna dominated by bivalves, such as *Gervillella*. Comparison of the north and south Dorset sequences (Figure 40), suggests that it correlates with the Preston Grit.

Stour Formation

The term Stour Formation (Bristow, 1989a) is a modification of Wright's (1981) formation of the same name. The change was necessitated by the recognition of an additional clay member (Hinton St Mary Clay), and the elimination of one of Wright's (1981) members (Newton Oolite = Cucklington Oolite (Bristow, 1989b)). The formation as now recognised comprises six members (Table 7), not all of which occur in any one area. The members form a unified sequence of alternating calcareous, commonly oolitic and locally sandy clays, and oolitic and pisolitic limestones.

Woodrow Clay

The name Woodrow Clay was introduced by Wright (1980) for 2 to 4 m of grey clay at the base of the Stour

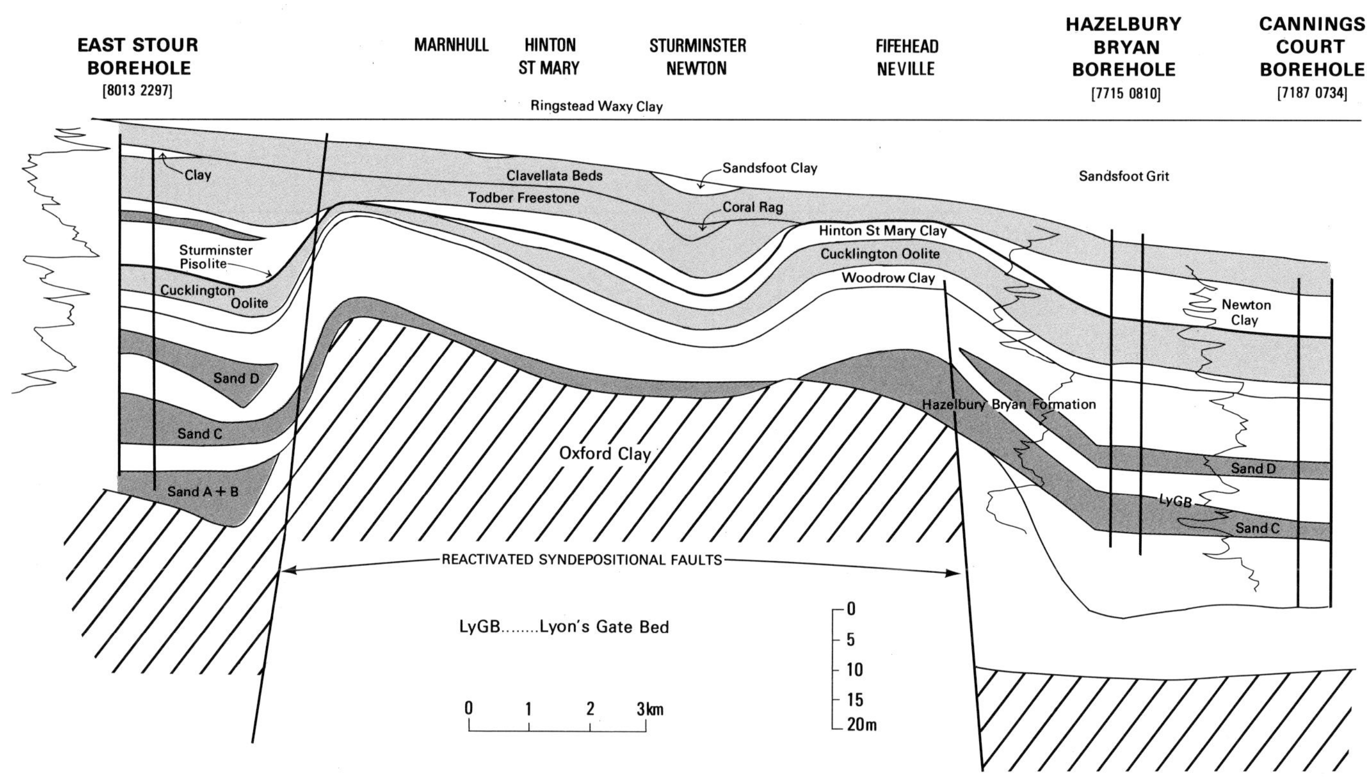

Figure 38 North–south section showing Corallian Group stratigraphical relationships and thickness variations (for position of section see Figure 37).

Formation. The type locality is the hamlet of Woodrow [759 108] in the south of the district.

The member consists of up to 5 m of grey, generally sand-free, calcareous, locally oolitic clay. In the Hazelbury Bryan and Cannings Court boreholes, the Woodrow Clay consists of olive-grey, slightly sandy to sandy clay, with wisps and disrupted laminae of very fine-grained sand or silt. The clay is locally very shelly; scattered pisoliths may occur (p.78). In the East Stour Borehole, 0.4 m of very shelly, sparsely oolitic mudstone with clasts of pale and medium grey mudstone occur at the base of the member. The mudstone is succeeded by a shelly, sparsely oolitic micrite with abundant clasts of medium grey mudstone; this micrite forms a very distinctive peak on the gamma-ray log because of its high mudstone clast content (Figure 38). The overlying bed is a 0.44 m thick, fine-grained, sandy mudstone with scattered ?chamosite ooliths and clasts of mudstone, and the highest bed is 1.4 m of extremely shelly, slightly sandy, oolitic (?chamositic) mudstone. In the northern part of the district, the member becomes less oolitic and more sandy.

The clay mineralogy of the member varies considerably; smectite dominates over illite and kaolinite in the East Stour Borehole, and kaolinite and illite occur without smectite in the south in the Hazelbury Bryan and Cannings Court boreholes.

The macrofauna from the boreholes consists mainly of the bivalves *Chlamys*, *Corbulomima*, *Pinna* and *Pseudolimea*, with one perisphinctid ammonite (ex gr. *Alligaticeras* - *Properisphinctes*).

Cucklington Oolite

The name Cucklington Oolite was proposed by Wright (1980) for oolitic limestones at Cucklington [756 275] in the Wincanton district.

The Cucklington Oolite is dominantly a flaggy, coarse-grained, shelly oolite or oobiosparite to oobiomicrite (Plate 6D). The ooliths are commonly micritised and ferruginous, and set in a micritic matrix; a sparry ferroan calcite cement is also present. The nuclei of the ooliths commonly consist of quartz or feldspar grains. Scattered pisoliths occur locally at the base.

The Cucklington Oolite is best developed from Hazelbury Bryan in the southern part of the district, where it is 3.2 m thick, northwards through Sturminster, where it is over 5 m thick, towards Hinton St Mary. There it is 3 to 4 m thick but, just to the north, it pinches out, reappearing north of Marnhull with a thickness of up to 5 m.

South-west of Fifehead Neville, the limestone passes [around 764 106], over a distance of a few tens of metres laterally, into a fine-grained oolitic sandstone; a similar oolitic sandstone occurs farther south-west at Kingston [750 098] (p.78). In the Hazelbury Bryan Borehole, the upper half of the member is oolite, and the lower half is a slightly oolitic sandstone. Farther west, in the Cannings Court Borehole, it is a much

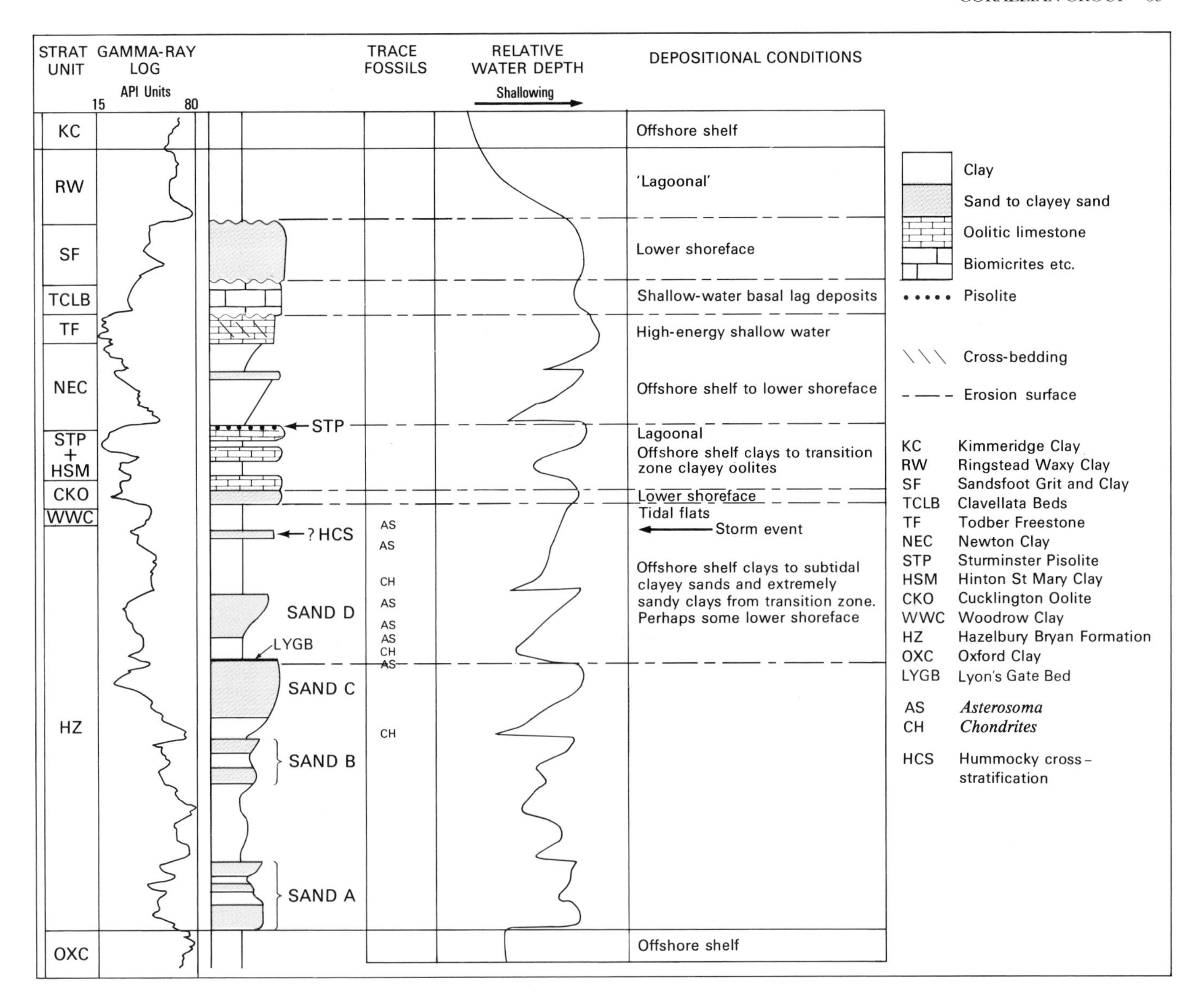

Figure 39 Generalised stratigraphy and conditions of deposition for the Corallian Group of north Dorset.

more variable sequence with many beds of oolitic clay.

The Cucklington Oolite contains bivalves, including *Camptonectes, Grammatodon, Nanogyra nana, Pleuromya uniformis, Pseudolimea* and *Thracia*. Gastropods and echinoids are common, but are often too fragmentary for identification; the echinoid *Nucleolites scutatus* (Plate 5, 7) is the only common well-preserved fossil. Ammonites occur locally; Wright (1982) recorded *Perisphinctes (Arisphinctes) kingstonensis*, indicative of the Vertebrale Subzone, from the Cucklington Oolite near Stoke Trister [740 286] in the Wincanton district. Poorly preserved ostracods from the East Stour Borehole include *Nophrecythere oertlii*, known only from the upper part of the Densiplicatum Zone in England, and *Procytheropoteron cryptica*, which first appears in the Densiplicatum Zone.

Hinton St Mary Clay

Lithologically, the member is highly variable. Between Hazelbury Bryan and Hinton St Mary, and at Cuckrow Copse, it is an oolitic clay and marl with a little oolite. In the Hazelbury Bryan Borehole, it is mainly an oolitic limestone, and in the Cannings Court Borehole, the proportion of oolite is also high.

The member occurs only in the southern half of the district, and even there, it has been mapped only between Hazelbury Bryan and Hinton St Mary. South and west of this area, it has not been mapped, although it can be defined geophysically in the Cannings Court Borehole.

The member is 3.3 m thick at Cuckrow Copse, 5.5 m in the Hazelbury Bryan Borehole and 3.1 m in the Cannings Court Borehole. Between Sturminster Newton and Hinton St Mary, it is about 5 m thick.

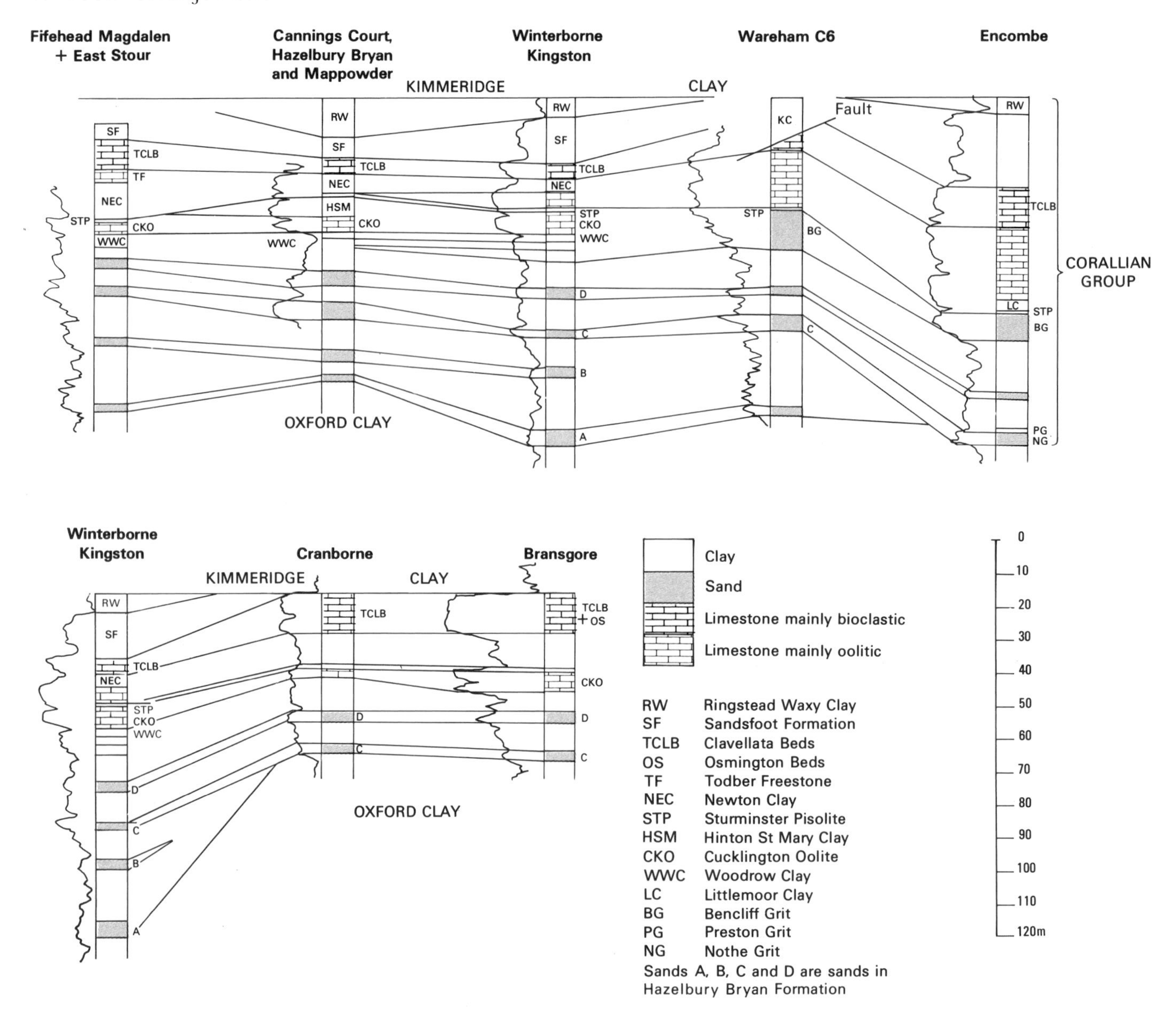

Figure 40 Correlation of the Corallian Group across Dorset and west Hampshire (see Figure 37 for line of section).

The only fossil found was *Nucleolites scutatus*. An Antecedens Subzone fauna from the overlying Sturminster Pisolite suggests that the junction of the Vertebrale and Antecedens subzones lies within the Hinton St Mary Clay or close above it.

Sturminster Pisolite

In the northern and central parts of the district, this member is a pisolith concentrate with pisoliths, up to 8 mm across, and in contact. South of Sturminster Newton, the pisoliths, up to 20 mm across, are matrix supported, commonly in a clayey, oolitic micrite. Although some of the pisoliths are spherical, most are flattened, with their height in extreme cases being only one quarter of their diameter (Plate 6C; see also Bartlett and Scanes, 1917, pl. 23b). In hand specimen, broken pisoliths show a clear concentric structure (see also Edmunds, 1938, fig.11b). Thin sections (Plate 6C) show the pisoliths set in a matrix of poorly sorted, patchily micritised ooliths in a fine-grained, ferroan, calcite spar cement. The pisoliths are irregularly laminated, commonly with a bioclast as their nucleus. Under high magnification, the pisoliths show a vermiform internal structure. Wright (1981, p.24) noted that the larger (7.5 to 10 mm), ovoid pisoliths are oncoliths of algal origin, whereas the smaller (3 to 5 mm) spherical bodies, developed around ooliths, are probably true pisoliths. Their nucleus may be shell debris, grains of sand or aggregates of ooliths.

At Cucklington [756 275], in the Wincanton district, a thin unit of shelly, coarse-grained, flaggy oolite occurs above the pisolite. A similar oolite above pisolite occurs at Stour Provost (see pp.80–81). Because of its uncertain lateral extent, this oolite is regarded here as part of the Sturminster Pisolite.

The Sturminster Pisolite is less than 0.8 m thick in the western part of the district near Lyon's Gate [6556 0638], 0.56 m thick in the Cannings Court Borehole and 0.35 m thick in the Hazelbury Bryan Borehole. Nearby, at Cuckrow Copse [7620 0935], Wright (1981) noted 2.1 m of pisolitic clays and pisolitic oolite. At the type section [7830 1348] at Sturminster Newton, the member is 1.2 m thick; farther north, south-east of Hinton St Mary, it forms a 2 m thick limestone mainly consisting of oolite, but with pisolite both at the top and bottom. In the north of the district, in the Todber and East Stour area, it is 0.6 m thick when a pisolite, but up to 3 m thick when a pisolitic, oolitic clay.

Fossils are not abundant in the Sturminster Pisolite, but commonly include *Nucleolites scutatus*. Gutmann (1970, p.129) recorded over 20 species of gastropods, bivalves and echinoids from the pisolite near Todber [c.7965 1989]. Ammonites, mostly poorly preserved perisphinctids, are more common in the Sturminster Pisolite than in the other Corallian members. The presence of *Perisphinctes (Arisphinctes)* aff. *helenae* and *P. (Dichotomosphinctes)* cf. *antecedens* (Arkell *in* Mottram, 1957), is indicative of the Antecedens Subzone.

Newton Clay

The name Newton Clay was introduced by Bristow (1989b) for a widely developed clay unit above the Sturminster Pisolite; it approximates to Arkell's (1933) Littlemore Clay in north Dorset. The type locality is the road cutting [7825 1347] at Newton. This member is only mappable north of the Merriott–Hardington Fault and east of Hazelbury Bryan, although farther west, it can be recognised in the Cannings Court Borehole, where it is 6 m thick. It is 7.2 m thick in Hazelbury Bryan Borehole, but is absent between Fifehead Neville and just south of Sturminster Newton. North of this, it maintains a thickness of c.5 m to beyond Hinton St Mary. The Newton Clay is missing in a limited area north of Marnhull, but where it reappears north of there, it is up to 15 m thick.

The dominant lithology is an oolitic, grey, calcareous, commonly sandy clay, although in the northern part of the district, around East Stour, sand and sandstone are also present. Marly limestones with scattered pisoliths occur in the basal bed, and at one point [7923 1719], north of Hinton St Mary, a coarse-grained oolite with scattered pisoliths occurs low in the clay. Wright (1981) recorded sandy oomicrite at the type locality.

In the northern part of the district, the sequence consists of a clay/sand cycle which increases in sand content upwards, although there are minor oscillations within the cycle (Figure 38). The sequence is dominantly a silty and sandy clay, with bedding planes coated with fine-grained sand, together with thin interbedded units of fine-grained sand. Locally, in the Wincanton district, beds of fine-grained sand which are thick enough to map [e.g. around 774 245], probably correspond to the mid-cycle sand in the Newton Clay of the East Stour Borehole. There, the sands consist of subrounded quartz grains, micritised ooliths and common foraminifera set in a mixed ferroan and non-ferroan calcite cement. Marly limestones are present in the uppermost beds in the north, where they form a transition to the overlying Todber Freestone. Burrowing and bioturbation is common throughout; disseminated lignite and scattered ooliths occur locally.

In the Cannings Court and East Stour boreholes, smectite is the dominant clay mineral, apart from the basal part at East Stour, where kaolinite and illite are more important. In the Hazelbury Bryan Borehole, illite predominates.

Fossils are common at certain levels in the Newton Clay. Blake and Hudleston (1877, p.276) recorded gastropods, bivalves and echinoids from the Sturminster railway cutting. The fauna from the Cannings Court Borehole consists exclusively of bivalves including *Camptonectes, Isognomon* and *Pseudolimea;* that from the East Stour Borehole includes the bivalves *Chlamys, Nanogyra nana, Pholadomya* and *Thracia depressa,* and the echinoid *Nucleolites scutatus.* None of this fauna is diagnostic, but ammonites from both the underlying Sturminster Pisolite and overlying Todber Freestone are indicative of the Antecedens Subzone (Mottram, 1957, p.165; Wright, 1981, p.27), thereby indicating that the Newton Clay also belongs to that Subzone. Collectively, the ostracods from several sites are indicative of the upper part of the Densiplicatum Zone. Ostracods from the East Stour Borehole include *Macrodentina tenuistriata* at 19.1 m, a species which extends from the upper part of the Densiplicatum Zone into the Upper Oxfordian, and the 'Dorset Form' (sensu Fuller, 1983, MS) of *Pseudohutsonia tuberosa* (at 18.2 m) and *Galliaecytheridea sp. (= G. posterospinosa* of Fuller, 1983, MS) (at 16.2 m), which are restricted to the Tenuiserratum and Glosense zones. Ostracods and foraminifera from the type section at Newton indicate an age no older than the Densiplicatum Zone.

Todber Freestone

The Todber Freestone, named by Mottram (1957), is the uppermost member of the Stour Formation. It extends northwards from Sturminster Newton [785 140] to East Stour [800 230], beyond which it has not been recognised. Wright (1981; 1985) considered that the Todber Freestone north of East Stour passed into a spicular shelly micrite, but, because the micrite is so different from the Todber Freestone and more closely resembles the Clavellata Beds, it is here included within the latter.

The Todber Freestone consists of up to 5 m of cross-bedded, fine-grained oosparite (Plate 6B) with some pelmicrite. Recognisable bioclastic fragments are generally scarce, but *Nucleolites scutatus* is commonly found intact. At Todber [796 198], the ammonite *Perisphinctes* cf. *antecedens,* found loose, was believed to have come from the Todber Freestone (Mottram, 1957, p.165).

A lenticular bed of buff marly clay, up to 2 m thick, can be mapped within the Todber Freestone round the south-west side of East Stour [800 230]. In the East

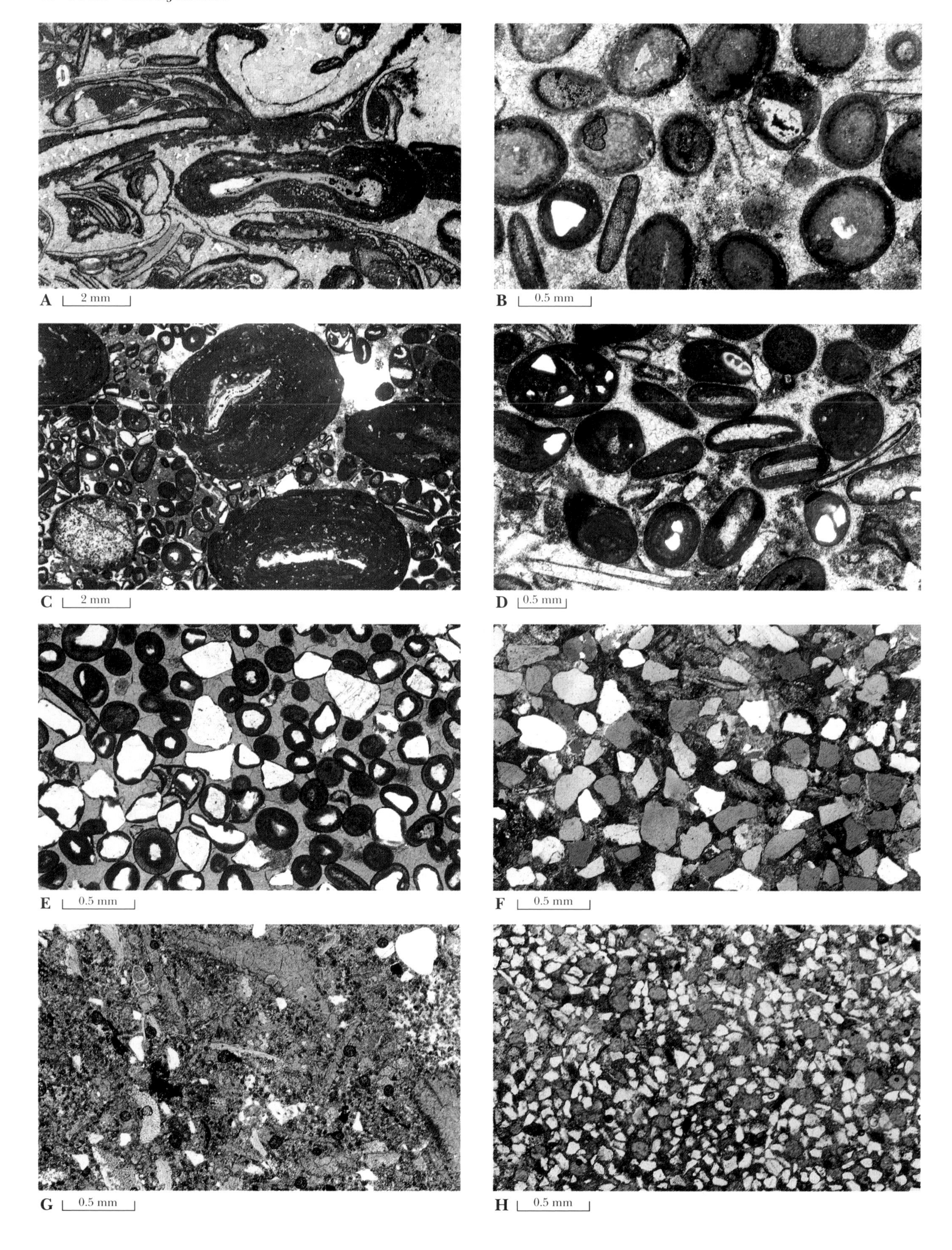
A 2 mm
B 0.5 mm
C 2 mm
D 0.5 mm
E 0.5 mm
F 0.5 mm
G 0.5 mm
H 0.5 mm

Plate 6 Photomicrographs of limestones and sandstones of the Corallian Group (all plane polars, dual carbonate stain, unless otherwise stated).

A. Clavellata Beds, Todber
Framework grains are dominated by coarse bioclastic material set in a slightly ferroan, sparry calcite cement (pale blue). Some of the larger bioclastic fragments (pink centre field) have an irregularly laminated outer coating of probable algal origin. Other bioclasts have been completely replaced by ferroan spar leaving only a micritic envelope. Oolitic grains are moderately common in the section.

B. Todber Freestone, Todber
Well-sorted, coarse sand grade, ferruginous ooliths dominate the framework grains, with subordinate bioclastic debris set in a slightly ferroan, sparry calcite cement. Quartz/feldspar grains and bioclastic fragments commonly form oolith nuclei.

C. Sturminster Pisolite (plane polars), near Marnhull. Ferruginous irregularly laminated pisoliths in a matrix containing smaller, ferruginous, micritised ooliths and abraded bioclasts. Under high magnification the pisoliths show a vermiform structure of possible algal origin.

D. Cucklington Oolite, near Hinton St. Mary. Moderately well-sorted ferruginous ooliths dominate the framework grains in a ferroan sparry calcite cement. The oolith nuclei are either calcitic bioclast fragments or quartz grains.

E. Hazelbury Bryan Formation (?Lyon's Gate Bed)
Fine- to medium-grained, sand-sized quartz, oolitic and coated grains (superficial ooliths) in a strongly ferroan sparry calcite cement (blue stain).

F. Hazelbury Bryan Formation (Lyon's Gate Bed)
Medium-grained quartz/feldspar detrital sand grains set in a ferroan sparry calcite cement (blue). Abraded non-ferroan calcite shell debris commonly occurs.

G. Hazelbury Bryan Formation
The framework grains are a mixture of coarse and finely a braded carbonate bioclasts and sparse quartz grains in a ferroan sparry calcite cement. The bioclasts either show a non-ferroan calcitic, prismatic structure (pink-stained) or are more commonly replaced by strongly ferroan (blue) sparry calcite. Throughout the section small brown rhombs of siderite are abundant.

H. Hazelbury Bryan Formation
Well-sorted, very fine quartz sand (white) and ferroan calcite (blue) calcite-filled spheres (?replacing siliceous *Rhaxella* spicules) in a sparse, ferruginous, micritic matrix. The spheres often coalesce to give the appearance of a patchy sparry cement.

Stour Borehole, a peak on the gamma-ray log between 3.3 and 4.4 m probably represents this unit.

Clavellata Beds

The Clavellata Beds, especially the basal bed, typically consist of tough, coarsely bioclastic, oolitic limestone; beds of fine- to medium-grained oolite, in units up to 0.3 m thick, and shelly, sporadically oolitic, spicular micrite also occur. Beds of sandy marl are common and, locally, they can make up almost half the succession. The limestones range from biosparites (Plate 6A) and oobiosparites to pelmicrites and biopelmicrites. The bioclasts are either ferroan calcite fragments with micritised envelopes or unaltered, non-ferroan calcite fragments. Calcite-filled spheres, possibly after *?Rhaxella*, are often present and foraminifera are common. Very fine-grained siliciclastic sand is also common and often forms the nuclei of ooliths.

Wright (1981, p.25) reinterpreted the basal 2.7 m of shelly rubbly limestone rich in gastropods at Sturminster Newton (Bed 4 of Blake and Hudleston, 1877) as the **Coral Rag**. This bed rests on a waterworn and irregular surface of the Todber Freestone. Wright (1981) also regarded 1 m of flaggy, shelly micrite with scattered ooliths, common comminuted shell fragments and *Nerinea* sp. at two localities [7847 1354; 7860 1346] south of the River Stour, as part of the Coral Rag. These beds were not sufficiently distinctive to be differentiated from the Clavellata Beds in the field and are included in that formation on the published map.

From East Stour northwards, a unit of flaggy, fine-grained oolite, the Eccliffe Member (Bristow, 1990a), up to 5 m thick, occurs at the top of the Clavellata Beds. It closely resembles the Todber Freestone, with which it has been confused (Wright, 1981, p.25).

The Clavellata Beds are c.7 m thick in the Glanvilles Wootton–Buckland Newton area. The Hazelbury Bryan Borehole proved 6.1 m, and it is between 3 and 4.5 m thick in the Sturminster Newton to Marnhull area. In the north of the district, it thickens to between 7 and 12 m, the greater thickness being in areas where the Eccliffe Member is present.

The fauna of the Clavellata Beds is dominated by bivalves, gastropods and echinoids (Blake and Hudleston, 1877; Woodward 1895; Gutmann, 1970); material collected by Woodward and Rhodes in the Sturminster Railway cutting has been re-examined (p.82). Gutmann (1970) recorded the ammonite *Pseudarisphinctes* sp. from the Clavellata Beds at Todber [796 198], and Mottram (1957) recorded *Perisphinctes (Arisphinctes)* aff. *osmingtonensis*, also believed to have come from that locality. Arkell (*in* Mottram, 1957) identified '*Decipia* aff. *lintonensis, D.* cf. *decipiens, Perisphinctes (Pseudarisphinctes)* cf. *shortlakensis, P.* sp. indet. cf. aff. *decurrens* and *Perisphinctes* sp.', and Wright (1985) recorded many ammonites, including *Perisphinctes* spp. and *Amoeboceras glosense*, from Whistley Farm [7815 2850] in the Wincanton district. These faunas allow a good correlation between the Clavellata Beds of north Dorset and those of the coast, which belong to the Cautisnigrae Subzone (Wright, 1986).

Sandsfoot Formation

The Sandsfoot Formation is locally divisible into two parts, the Sandsfoot Clay, which is grey, oolitic, locally sandy and ferruginous clay, and the Sandsfoot Grit, which varies from a fine-grained sand, commonly ferruginous, to a fine-grained sandy clay.

Sandsfoot Clay

The Sandsfoot Clay is only locally developed in the district. In the Mappowder area [7277 0722] it consists of 0.1 m of oolitic clay (p.84). Blake and Hudleston

(1877) recorded 2.59 m of clay and marl in the Sturminster railway cutting [7860 1427]. In the Marnhull–Sturminster Newton area it is a medium grey, locally oolitic, sand-free clay, up to 3 m thick (Bristow, 1989b). In the north of the district, up to 3 m of fine-grained sandy, shelly clay at the base of the Sandsfoot Formation is regarded as Sandsfoot Clay.

SANDSFOOT GRIT

The Sandsfoot Grit typically consists of fine-grained sandy clay and clayey fine-grained sand; fine-grained sandstone, ferruginous fossiliferous sand and sandstone, and ferruginous oolite occur locally. The Sandsfoot Grit is overstepped by the Gault in the south-west of the district, near Glanvilles Wootton [678 082], where the outcrop is much fragmented by faulting. North of the Merriott–Hardington Fault, the member forms a fairly continuous outcrop between Hazelbury Bryan and Marnhull; farther north, it is again faulted.

In the western part of the district, the thickness cannot be determined because of faulting. South of Mappowder, it appears to be about 12 m thick, but to the north, the Fir Tree Farm Borehole [7359 0714] shows only 6.8 m of Sandsfoot Grit. The member is at least 9 m thick around Droop [753 082], but its top is cut out by a fault. At Knackers Hole [7791 1188], it is 7.7 m thick. Between Hinton St Mary and Fifehead Neville, the Sandsfoot Grit varies between 5 to 8 m. North of Hinton St Mary and towards Marnhull, the maximum thickness probably does not exceed 3 m and may be as little as 2 m, but in the north of the district, the Sandsfoot Grit is up to 10 m thick.

In many areas in the district, small oysters are common in the Sandsfoot Grit (see p.85); serpulids, bivalves, ammonites and belemnites also occur (Blake and Hudleston, 1877; Woodward, 1895; White, 1923) (p.85). The Sandsfoot Grit on the Dorset coast has yielded perisphinctids and *Amoeboceras*, indicative of the Pseudoyo and Pseudocordata subzones (Wright, 1986).

Ringstead Waxy Clay

On the Dorset coast, the Ringstead Coral Bed and the Ringstead Waxy Clay separate the Sandsfoot Formation and the Kimmeridge Clay. In the Shaftesbury district, where a distinction is not easily made by augering, the Ringstead Waxy Clay has been mapped as part of the Kimmeridge Clay. A few exposures of pale grey silty clay with small, red, sideritic nodules were noted (p.86). Subsequently, the Knackers Hole Borehole [7791 1188] proved 3.28 m of medium to dark grey mudstones with bivalves, gastropods and a few ammonites overlying 4.46 m of pale grey mudstone and siltstone with incipient clay ironstone nodules and a restricted fauna of bivalves and perisphinctid nuclei, overlying 1.61 m of interburrowed sandy clays and clayey sands with phosphatic nodules, which forms the basal unit of the Ringstead Waxy Clay. An important erosion surface (at 59 m) occurs at the base of this unit which is marked by a distinctive signature on the geophysical logs. Although the topmost unit of medium to dark grey mudstone strongly resembles the Kimmeridge Clay, its position below the Inconstans Bed, which at present defines the base of that formation, assigns it to the Ringstead Waxy Clay.

Ammonites collected from the Knackers Hole Borehole and temporary sections for the A303 [7815 3061] in the Wincanton district will enhance dating and correlation within the Pseudocordata Zone. On the Dorset coast, the Ringstead Waxy Clay is assigned to the Evoluta Subzone (Wright, 1986).

History of sedimentation

The sedimentation in the Corallian Group has long been described in terms of cyclic sequences. Arkell (1933) noted three clay-sandstone-limestone cycles in the Corallian of south Dorset; other authors (Wilson, 1968 a, b; Talbot, 1973; Fursich, 1976; Wright, 1986) have described a variety of cycles, mostly coarsening upward. Sun (1989) reinterpreted the sequence in terms of four rather symmetrical transgressive–regressive cycles with an erosional base to each transgressive unit.

After the deposition of the Oxford Clay, the Corallian Group sequence in north Dorset shows an abrupt regression, from offshore shelf muds to subtidal clayey sands, followed by a series of minor, fairly symmetrical, transgressive–regressive cycles culminating in a major regressive sand (Sand C; Figure 39). This sand is poorly sorted, fine grained and silty, and shows intense bioturbation, with common *Asterosoma*; it was probably deposited below the lower shoreface. Above an erosion surface, the sand is succeeded by the Lyon's Gate Bed, the probable equivalent of the Preston Grit of south Dorset. According to Sun (1989), the Preston Grit represents a transgression. The presence of ooliths, abraded bivalve debris and coarse sand grains indicates reworking of pre-existing shelly oolitic sands into subtidal shoals. The silty clay above the Lyon's Gate Bed probably indicates a reversion to deeper water shelf conditions before coarsening up to Sand D. The top of Sand D is a very silty, very fine-grained, poorly sorted, bioturbated sand with common *Asterosoma* traces, again probably deposited below the lower shoreface. The succeeding clay unit up to the base of the Woodrow Clay shows little variation other than a variable input of sand; it probably indicates a fluctuating position between the lower shoreface and the offshore shelf. Discrete sand or sandstone beds occur in the Woodrow Clay; these are mostly highly bioturbated, but a 0.2 m-thick bed in Cannings Court Borehole exhibited dome-shaped laminae, possibly hummocky cross-stratification, suggesting a storm event.

The low sand content of the Woodrow Clay, compared to that of the Hazelbury Bryan Formation, can be explained by a cut-off in sand supply due to reduction in the relief and greater distance of the hinterland. The lamination indicates gentle reworking of the fine-grained sediments, possibly in a tidal-flat setting; the bivalves suggest well-oxygenated waters.

Above the Woodrow Clay, the Stour Formation is predominantly a clay–limestone sequence, with some sand in the Newton Clay. It exhibits a series of upward-coarsening units with a transgressive oolitic clay at the base, passing up into an oolitic limestone at the top. The Cucklington Oolite over part of the district shows a decrease in clay content upwards, as does the Hinton St Mary Clay, which terminates in oolitic limestones. The overlying Sturminster Pisolite was probably deposited during low-energy, possibly 'lagoonal' sedimentation, although Wright (1981) refers to frequent agitation of the sediment. These Stour Formation cycles are similar to those in the siliciclastic Hazelbury Bryan Formation, except that the coarse top consists of poorly sorted, commonly strongly bioturbated oolite or clayey oolite, rather than sand. These oolites probably represent sediment derived from shoals upslope, redeposited subtidally.

Sharply defined transgressions occur at the bases of the Newton Clay, Clavellata Beds, Sandsfoot Grit and Ringstead Waxy Clay. Sand deposition recommenced in the Newton Clay at the top of the lower of two upward-coarsening sequences. A sample from this sand in the East Stour Borehole was moderately well sorted. Farther north, outside the district, exposures in a new road cutting [7558 2997] showed a cross-bedded sand at the same level, suggesting that shoreface conditions developed north of the district.

The Clavellata Beds have a marked erosion surface at their base; on the structural high between Newton and Fifehead Neville, they rest on the Sturminster Pisolite. The Clavellata Beds are distinguished by the relative abundance of coarse shelly debris, consisting mainly of broken and abraded bivalves and gastropods. Sun (1989) suggested that the Clavellata Beds in south Dorset are basal-lag deposits, related to a transgressive phase.

The Sandsfoot Grit, which contains phosphatic nodules above an erosive base, is a very poorly sorted, bioturbated sand and was probably deposited in the transition zone between offshore shelf and shoreface, or in the lower shoreface zone. The commonly ferruginous clay matrix suggests that it was deposited in an area of lower than normal salinity, perhaps close to a delta or estuary. The bivalve faunas are consistent with this type of environment.

The top of the Sandsfoot Grit is marked by a further erosional break, with associated phosphatised nodules and fossils in the sandy transgressive base of the overlying Ringstead Waxy Clay. These sandy beds pass up into relatively sand-free, pale grey clay with an impoverished fauna. These clays include small sideritic nodules, possibly indicative of lowered salinities, which would explain the poor fauna. The upper part of the Ringstead Waxy Clay is darker and contains a varied fauna more like that of the Kimmeridge Clay. Phosphatic nodules occur just above the junction of these two units, and presumably marks the transgressive base of more offshore shelf deposits over sediments deposited in a lower salinity environment.

The study of the north Dorset Corallian Group succession suggests that there are probably at least six, variable, regressive–transgressive cycles. Some cycles show an abrupt transgressional base while others are more gradational. On the east–west-trending Fordingbridge structural high, in the middle of the district, the lower cycles and the Newton Clay cycles are missing; to the south and north, all six cycles are present. This lateral variability in the development of the cycles suggests local tectonic control on Corallian Group sedimentation, in competition with the overall eustatic controls operating throughout the Wessex Basin. The local tectonic activity is also probably responsible for the periodic influxes of sand.

Thickness variation within the group shows evidence of synsedimentary fault control (Figure 37). The thickest deposition took place in the Mere Basin, in the Winterborne Trough and south of the Isle of Purbeck Fault Zone in the Channel Basin. The group thins markedly eastwards, losing both the lowest part of the Hazelbury Bryan Formation and the Sandsfoot Formation, probably due to movements on a buried NW–SE-trending fault.

DETAILS

Hazelbury Bryan Formation

Middlemarsh area

Auger holes in a stream bed [6530 0647] showed 2 m of grey, very silty clay to clayey silt and much shell debris, on 0.3 m of grey sandy shelly clay, on sandstone. Farther upstream [6531 0629], several hard, shelly, calcareous sandstone ribs occur.

A temporary section [6730 0619] east of Grange Wood showed:

	Thickness m
DRIFT	
Clay, orange, sandy, with some flints	0.2
HAZELBURY BRYAN FORMATION	
Clay, orange-grey to greyish brown, sandy, with some race	1.1
Sandstone, very shelly	0.1
Clay, sandy to very sandy, grey, with shells	1.3

Holnest area

In Holnest Park, a dip slope, possibly formed by the Lyon's Gate Bed, is inclined 2°N and suggests that the stratigraphical level is still rising northwards towards the Merriott–Hardington Fault. Near the top of the dip slope, an auger hole [6563 0904] proved a calcareous clayey sand with shells below.

East of Holnest, the Hazelbury Bryan Formation forms low hills. An auger hole [6699 0836] drilled near the base of the formation penetrated grey, very shelly, very sandy clay below a stream bed to a depth of 1.3 m. A fragment of *Cardioceras (Plasmatoceras)* sp. from the clay suggests a Cordatum Zone, Cordatum Subzone, age.

Sandhills

The formation forms a salient northwards towards Sandhills [686 105]. Most of the outcrop consists of very sandy clay. On the tops of the three summits of the hill, there are caps of orange-grey or yellow, clayey, fine-grained sand.

Pulham area

West of the Poyntington Fault, a trench exposed calcareous sandstone containing *Cardioceras* sp. [between 7067 0766 and

0788 0757] and gritty oolitic limestone [around 7115 0740]. Debris from the trench also includes an oolitic, shelly, calcareous, medium-grained sandstone. Similar material caps hills to the south-west [7060 0718; 7085 0710; 7042 0673].

Cannings Court area

Narrow outcrops of the Hazelbury Bryan Formation occur between faults east of Cannings Court Farm (Freshney, 1990, p.11). Most of the formation was penetrated in the Cannings Court Borehole between 19.9 m and the base of the hole at 52 m. Two sand units, with tops at 29.3 m (Sand D) and 40.9 m (Sand C), represent the uppermost parts of upward-coarsening sequences. The intervening beds consist mainly of extremely sandy, commonly shelly, bioturbated clays and subordinate shelly calcareous sandstones and sandy limestones. Glauconite and ooliths are common in some of the clays between 39.5 and 40.45 m. The fauna from the borehole is dominated by bivalves: *Myophorella* sp. occurs throughout, together with *Chlamys, Modiolus* and *Pinna. Gervillella aviculoides* was common between 26.90 m and 31.85 m. The only ammonite was *Cardioceras cordatum?* at 36.35 m, 42.30 m and 47.40 m.

Mappowder–Westfields area

West of Mappowder and north of Westfields [724 062], the Hazelbury Bryan Formation occupies a low-lying area. In a trench, sandy clay with sandstone doggers [approximately? 7234 0670 to 7259 0652] and a pinkish white limestone containing a perisphinctid ammonite [7260 0652] were reported. In the west [718 068], a flattish hill top is capped by a thin sand. Another thin sand, partly cemented to sandstone, was seen in a stream [7210 0663]; it yielded oysters and *Myophorella.* Southwards, the land rises gently southward in what appears to be a dip slope. An excavation [7210 0623] produced debris of grey and ferruginous clay, calcareous sandstone and fragments of large oysters. An auger hole [7220 0650] down dip to the north penetrated wet, ferruginous, sandy, grey clay, which overlays a calcareous sandstone or sandy limestone, possibly a correlative of the lower sand of the Hazelbury Bryan area.

Hazelbury Bryan area

East of Kingston [750 097], sand (Sand D) forms an extensive dip slope which dips about 1°SE. Spoil and stream sections show yellowish brown, fine-grained sand with fragments of soft, bioturbated, calcareous sandstone [7543 0970], soft, rubbly fine-grained, calcareous sandstone with bivalves [7580 0967] and 1.5 m of soft, rubbly, fine-grained sandstone [7564 0954]. Higher strata, ranging from silty to extremely sandy grey clay, commonly with race, occur south-east of the stream. An old pit [747 099] worked grey clay with much oyster debris from beneath the sand. A lower sand, the base of which is marked by springs [e.g. 7560 0962], occurs down the scarp slope to the west. An auger hole [7445 0962] on the scarp intersected most of the lowest 10 m of the formation below the lowest sand, and yielded a calcareous microfauna indicative of the Cordatum Zone:

	Depth m
LANDSLIP	
Sand, clayey, with flints	1.0
HAZELBURY BRYAN FORMATION	
Sand, yellow, fine-grained, clayey	1.2
Clay, extremely sandy, brownish grey with orange stain	3.1
Clay, sandy to very sandy, grey, orange-stained with ferruginous material; some shells	3.6
Clay, grey, sandy to slightly sandy; becomes dark grey from 4.4 m, with some shell debris	5.3

Nearer Hazelbury Bryan, another deep auger hole [7420 0865] showed the following section with microfaunas indicative of the Cordatum Zone:

	Depth m
HAZELBURY BRYAN FORMATION	
Sand orange-brown/buff, slightly clayey	0.50
Sand, mottled orange-grey, extremely sandy.	1.20
Clay, brown, becoming pale to medium grey at 1.4 m, extremely sandy; shelly at 2.6 m	2.90
Sandstone, grey, fine-grained, soft	2.95
Clay, pale to medium grey, sandy to extremely sandy, shelly	4.10
Sandstone, fine-grained, clayey, soft	4.20
Clay, sandy, passing gradually downwards into a silty clay with very little sand	5.60

South-west of Kingston, a few exposures of sandstones in Sand C occur at the bottom of ditches. One [7486 0933] showed 0.2 m of yellow fine-grained sandstone, resting on over 2 m of rubbly fine-grained sandstone with pectinid bivalves. North-west of Kingston, an auger hole [7510 0999] proved:

	Depth m
Topsoil, sandy	0.40
Clay, green, very sandy	0.60
Clay, very sandy, mottled orange/grey, becoming more clayey at 1.5 m and interlaminated with fine-grained sand; becomes more sandy at 2.1 m and with scattered shells	2.10
Clay, fine-grained, sandy, not laminated, medium grey	2.95
Sandstone, very fine-grained, ?calcareous, dark grey	3.00
Sand, fine-grained, very clayey, grey	3.40
Clay, extremely sandy (very fine-grained), grey, passing down into	3.70
Clay, very silty	5.10

The main part of Hazelbury Bryan village is built on the lowermost sand and sandstone unit (Sand C) of the Hazelbury Bryan Formation. At [7445 0807], a lenticular sandstone with bivalves, up to 0.3 m thick, was seen in a trench, underlying 1.5 m of yellowish brown sand. The base of the sand is well marked by a spring line on the western side of the village. The beds above the sand consist of extremely sandy clay, commonly with shell debris and race.

Hazelbury Bryan Borehole

Most of the formation was penetrated between 22.06 m and the bottom of the hole at 52.1 m; there was some core loss in the sands. About 50 per cent of the section was clayey to clay-free, fine-grained sand; the other part was mainly extremely sandy clay, with a relatively small amount of clay with little sand. Some sand is cemented by calcite; a few beds of sandy limestone were present. The sand mainly occurred at the top of two upward-coarsening cycles, the top of the lower cycle (Sand C) being at 42.85 m, and that of the upper cycle (Sand D) at 34.85 m. These two sands probably correlate with the two sands mapped around Kingston. The formation is highly bioturbated and commonly shelly, with a fauna mainly of bivalves, including *Myophorella*.

Fifehead Neville–Sturminster area

In this area, the combined Sand C and D has a wide outcrop, with springs commonly marking its base. It consists of up to 10 m of yellow, brown and orange, fine-grained, locally clayey sand in the south-west. North of Haydon [757 120], the outcrop narrows and the clayey sand thins to about 7 or 8 m. Between Puxey [770 125] and Rolls Mill [776 134], there is no basal sand. The beds above the basal sand comprise up to 10 m of mottled orange and grey, silty and finely sandy clay. Locally, medium grey or grey-brown clay occurs.

The outlier at Road Lane Farm [765 133] consists of up to 15 m of dominantly mottled orange and grey sandy clay or very clayey sand. The basal beds are more sandy than the higher strata and commonly give rise to springs.

Hinton St Mary–Marnhull

North of the south-west-trending fault at Hinton St Mary, the formation is about 20 m thick. North of the fault, the formation can be divided into a basal clayey sand (Sand C), 3 to 4 m thick, overlain by about 16 m of silty and finely sandy clay. From Walton Elm [780 178] to Marnhull, the sequence is tripartite, with a relatively sand-free clay in the middle. At the sewage works, Marnhull [785 1747], and 800 m east-north-east of Walton End [7884 1803], slabs of fine-grained calcareous sandstone, with small bivalves and gastropods, are common at the surface. They occur close to the top of the formation.

Four metre-deep sections at Walton End were recorded by Bristow (1989b); one of these [7748 1918], in the middle, dominantly clay unit, showed:

	Thickness m
Topsoil, sandy, dark brown, becoming orange-brown downwards	0.5
Clay, mottled orange and grey; spring at base	2.0
Shell bed of large (10 cm) oysters	0.1
Clay, dark grey, sandy, oolitic, with clay clasts up to 2 cm by 5 cm, dark grey; water comes in at base	0.9

A sample from the upper clay yielded: *Schuleridea triebeli, Ophthalmidium strumosum* (abundant), *Epistomina porcellanea?, Ammobaculites agglutinans,* cf. *Vaginulina striatoides, Nodosaria* cf. *opalini, N. radicula, Lenticulina muensteri, Citharina serratocostata* and *Tristix triangularis*. A sample from the basal clay was poorer in species and included *Schuleridea triebeli, ?Procytherura tenuicostata, Lenticulina muensteri* (abundant), *Ophthalmidium strumosum* (abundant) and *Epistomina porcellanea?*. Collectively, the two faunas indicate the Cordatum to Tenuiserratum zones.

At Pleck, material from a ditch [7710 1814] in Sand C included brown clayey sand with fragments of oysters. To the east, Sand C, which forms a gently dipping dip slope of about 1 or 2°E, passes up into a silty clay, which in turn is overlain by 5 to 10 m of very clayey sand or sandy clay. This unit is then succeeded by 3 to 4 m of mottled orange and grey sandy clay. On the north side of Marnhull [around 771 198], Sand C forms a similar eastward-dipping slope. The sand is overlain by about 5 m of smooth grey or mottled orange and grey clay, which in turn is succeeded by about 10m of mottled orange and grey, silty and finely sandy clay.

A dutch auger hole [7844 2019] in Sand C south of the River Stour proved:

	Depth m
Soil	0.25
Sand, fine- to medium-grained, clayey, with pebbles of sandstone and some manganese dioxide grains; orange mottles	0.85
Sand, laminated, with some clay laminae and 1 to 2 cm clay pellets	1.00
Sand, clayey, mottled, buff; clay pellets locally abundant; lignite fragments at 1.55 m; shell fragments at 2.35 m; abundant clay pellets at 2.75 to 2.95 m	3.10
Gap	c.4.50
Sand, clean, orange-brown	5.55
Sand, beige, compact, alternating with running soft sand; rare clay pellets; clay laminae at 7.3 m	7.55

Sand C is succeeded by a silty, sandy, shelly and, locally, oolitic clay and clayey fine-grained sand that varies in thickness from 5 to 15 m.

Fifehead Magdalen area

The basal sand (Sand A + B) is well developed near Fifehead Magdalen, where it is over 10 m thick, but it thins northwards and disappears near Kington Magna, where silty and fine-grained sandy clay rests on Oxford Clay.

Sand C is well developed around Fifehead Magdalen, where it is up to 8 m thick, but it thins rapidly northwards and disappears north of Kington Magna on the main escarpment, and north of West Stour along the Stour valley. An exposure in a ditch [7716 2244] on Stour Hill revealed a shelly sandstone conglomerate of angular to subrounded pebbles and cobbles of sandstone at the top of the sand, underlying a thin sandy clay. The clasts are set in a sand matrix identical with the clasts. A hard cemented sandstone is exposed on Kington Hill [7683 2275].

On the west side of the Stour valley near West Stour, Sand C, which has a maximum thickness of about 2 m, can be traced by the line of springs issuing from its base.

A lenticular bed of fine-grained sand developed within the clay above Sand B has a maximum thickness of about 4 m, and crops out over some 400 m [7657 2351 to 7660 2385] north of Kington Magna.

In the East Stour Borehole, the Hazelbury Bryan Formation is over 29.45 m thick and consists of bioturbated mudstones, mainly very sandy, and clayey silty sand and sandstone. Two upward-coarsening sequences were seen, each terminating in a sand unit, sands C and D (Figure 38). The borehole finished in a sandy bed, which may be the top of Sand B. Thin limy sandstones and sandy limestones are also present. A thin section (Plate 6H) of one of these limestones shows well-sorted ferroan calcite spar spheres, possibly after *?Rhaxella* spicules, and very fine-grained detrital quartz and feldspar grains set in a sparse micritic matrix (see also Plate 6G).

At its thickest, north of Fifehead Magdalen, Sand D is about 5 m thick. This sand, and the higher part of the formation, were formerly exposed on either side of the Kingsmead railway tunnel, just north of the district (Bristow, 1990a). The ammonite *Cardioceras (Plasmatoceras)* ex gr. *tenuicostatum*, almost certainly indicative of the Vertebrale Subzone, was found in Sand D.

Lyon's Gate Bed

The most westerly occurrence of this bed is to the north-west of Lyon's Gate [around 6520 0666], where an auger hole showed the following section:

	Thickness m
HAZELBURY BRYAN FORMATION	
Clay, yellow, ferruginous, extremely sandy, ranging to clayey sand	0.8
Clay, sandy, gritty, grey with race	0.9
Clay, sticky, sandy, grey with race	2.0
LYON'S GATE BED	
Sand, clayey, very shelly, passing rapidly down into sandstone	0.1

Springs commonly issue from the base of the bed, e.g. above Lyon's Gate Farm [6591 0602]. One auger hole [6675 0594] showed 1.5 m of ferruginous, very sandy clay, overlying 0.2 m of oolitic sandstone of the Lyon's Gate Bed. Near Cosmore, an excavation [6773 0560] exposed 0.4 m of slightly oolitic, sandy biosparite from which *Cardioceras (Scoticardioceras)* sp. indet. was recovered. A section [6785 0575] in a stream exposed 0.2 m of grey oolitic sandstone, on 0.3 m of calcareous sand, which rested in turn on over 2 m of rubbly sandy shelly biosparite with *Gervillella aviculoides*.

Farther north, around Glanvilles Wootton, auger holes [6811 0767] showed 0.9 m of ferruginous sandy clay, resting on oolitic, shelly, ferruginous, extremely sandy clay to a depth of 1.3 m. North-east of Glanvilles Wootton, oolitic sandstone was seen in a ditch [6912 0868].

Stour Formation, undivided

Lyon's Gate area–Glanvilles Wootton–West Pulham

West-north-west of Lyon's Gate, small outcrops of oolitic limestone, up to 0.3m thick, are visible in an old quarry [6555 0634]; Wright (1981) regarded this limestone as the Cucklington Oolite. Auger holes show mainly oolitic clay, which is locally sandy [e.g. 6888 0858] or silty [e.g. 6817 0828]; oolite also occurs [6553 0605]. An auger hole [7011 0800] on a small outcrop adjacent to the Poyntington Fault, starting about 3 m below the Sturminster Pisolite, was as follows:

	Depth m
STOUR FORMATION	
Clay, oolitic	1.0
Oolite, argillaceous, with a few pisoliths	1.8
Clay, with layers of ooliths near the top; layers of fine-grained sand occur below 2.0 m	2.3
?HAZELBURY BRYAN FORMATION	
Clay, medium to dark grey,sandy	2.8
Clay, extremely sandy with layers of brown very fine-grained sand and some shell debris	2.8
Sand, orange-brown, silty, very fine-grained	3.3

Buckland Newton–Cannings Court–Mappowder

Debris from excavations near Buckland Newton included oolitic clay, fragments of oolitic limestone and loose pisoliths [6764 0535], along with *Nucleolites scutatus* and clay, shelly micritic limestone and pisoliths [6768 0531].

A small pit [7230 0754] east of Cannings Court showed 0.3 m of fine-grained, slightly oolitic limestone with scattered shells. South-east of Cannings Court, an auger hole [7252 0715] proved 1 m of yellowish brown and grey, extremely sandy clay, to clayey sand with ferruginous pellets, on 1 m of grey, oolitic clay with thin bands of soft white limestone and race.

South-west of Mappowder, a stream section [7290 0539] revealed the following:

	Thickness m
STOUR FORMATION	
Limestone, muddy, shelly, with some large ooliths	0.40
Marl, shelly	0.15
Limestone, nodular, micritic, with calcispheres	0.1–0.30
?HAZELBURY BRYAN FORMATION	
Clay, dark grey, sandy	over 0.60

WOODROW CLAY

West of Droop [748 080], the Woodrow Clay consists of grey, relatively sand-free, locally oolitic clay [7538 0888]. In the Hazelbury Bryan Borehole [7515 0810] (between 22.06 and 25.5 m), it consists of olive-grey, slightly sandy to sandy clay, with wisps and disrupted laminae of very finely grained sand. A thin sideritic claystone layer occurs between 24.55 and 24.57 m. In the Cannings Court borehole [7187 0734] (between 18.10 and 19.9 m), the member comprised grey to olive-grey, silty to very finely sandy clay, with very fine-grained sand to silt laminae, and layers of ooliths in the upper part. An abundant bivalve fauna included *Chlamys*, *Corbulomima* and *Pinna*.

At the type locality [762 108], Wright (1981) recorded 3 to 4 m of grey clay overlying the Hazelbury Bryan Formation. Similar clay, in places oolitic, was encountered at many points along the outcrop as far as Newton. North-west of Newton Mill, the Woodrow Clay forms a wide south-easterly dipping slope.

Oolitic grey and brown clay, about 5 m thick, occurs on the outlier east of Walton Elm. The new cemetery at Marnhull [7807 1940] is dug into pale to medium grey, very oolitic clay. On the faulted outlier west of Hains, up to 5 m of grey oolitic, locally shelly clay with scattered pisoliths occur. South and west of Marnhull, the Woodrow Clay is only about 2 to 3 m thick.

The Woodrow Clay in the East Stour Borehole is described on p.68.

CUCKLINGTON OOLITE

Hazelbury Bryan area

An old pit [7667 0991] has debris of oolitic limestone around it, and at Cuckrow Copse [7610 0939], much oolitic sandstone occurs (Wright, 1981). The lower sandy part of the Cucklington Oolite forms a small outlier in Kingston [750 098], where about 0.4 m of shelly, rubbly slightly oolitic sandstone occurs in the foundations of a cottage [7504 0973].

Fifehead Neville area

West and north of Fifehead St Quintin, the Cucklington Oolite is about 4 m thick. Near Fifehead Neville, there are several old pits [7677 1064; 7687 1061; 7690 1059; 7687 1078], but no section remains. Wright (1981) saw the following section in the last pit:

	Thickness m
Oomicrite, slightly sandy, shelly, well bedded with *Nucleolites scutatus*	1.80
Clay, oolitic, shelly, with alternations of thin limestone beds and black clay	c.1.50
Limestone, oolitic, flaggy, slightly cross-bedded	0.75

South-west of Fifehead Neville, the oolitic limestone passes into a fine-grained sand over a 50 m tract [around 764 106]. Wright (1981; personal communication) noted blocks of calcareous sandstone, sandy oolite and ferruginous oolite in an old quarry [7659 1088] west of Fifehead Neville, and flaggy, calcareous, oolitic sandstone with '*Nanogyra nana* and *Pseudolimea* sp.' in a roadside exposure [7651 1076] to the south-west.

Near the ford, the junction of flaggy, very sandy oolite and calcareous sandstone, with oolitic grey clay, can be found in the roadbank [7716 1109]. Oolitic shelly limestone occurs in the stream bank [7726 1103] and in ditch sections [7729 1117] east of the stream. Fine-grained flaggy oolite floors the tributary stream [7731 1197] just south of the ford, and farther south [7715 1062] nearer Lower Fifehead Farm.

Sturminster Newton area

North of Salkeld Bridge, an exposure in the bank [7712 1223] shows:

	Thickness m
Limestone, shelly, soft, ferruginous	c.1.00
Clay, gritty, grey	0.50
Unexposed	0.50
Limestone, shelly, pisolitic	0.15
FAULT OF UNKNOWN THROW TO SOUTH	
Limestone, shelly, coarse-grained, with *Perisphinctes* sp. and common *Nanogyra*	0.70
Clay, grey	0.30

At Newton, the exposure in the roadbank [7813 1347] is the type locality of Wright's Newton Oolite, where he saw 3 m of flaggy oolite resting on clay. The lowest bed consists of massive, oomicrite, overlain by 2.5 m of flaggy, medium-grained oomicrite with lenses and thin beds of oosparite. Shell fragments abound and include bivalves, gastropods and echinoids.

East and West Stour area

In the East Stour Borehole [8013 2297], the Cucklington Oolite consists of 3.17 m of fine- to medium-grained, shelly oolite, commonly with a micritic matrix. Mudstone partings and clasts are common throughout; scattered pisoliths are present in the top 2.2 m.

A quarry [7799 2290], just west of West Stour, exposes about 0.4 m of coarse-grained, shelly oolite with *Nucleolites*, overlain by 0.1 m of pisolite. The coral *Stylina* found on the surface of a field [786 239] on the west side of the Stour (Ensom, 1985) probably came from the Cucklington Oolite.

HINTON ST MARY CLAY

Cannings Court Borehole

The Hinton St Mary Clay, between 10.25 and 13.35 m, contains a high proportion of oolite and is very sandy, commonly with fine lignite debris; *Nucleolites scutatus* is common.

Hazelbury Bryan–Fifehead Neville area

Oolitic clays and argillaceous oolite form an outcrop around Zoar Lane, north-west of Hazelbury Bryan. Wright (1981) measured a section [7610 0939 to 7620 0935] in Cuckrow Copse, which is reproduced here with some emendations:

	Thickness m
STURMINSTER PISOLITE	
Clay, pisolitic, and marls with bands of pisolitic oolite	2.10
HINTON ST MARY CLAY	
Oomicrite, fine-grained	0.30
Marl, oolitic, soft	0.45
Clay, grey, with scattered ooliths	0.30
Marl, white, with scattered ooliths	0.60
Clay, light grey	0.60
Marl, white	0.45
Marl, oolitic, with grey clay bands	0.60
CUCKLINGTON OOLITE	
Sandstone, oolitic, flaggy, calcareous, fine- to medium-grained	0.80

The beds below the Sturminster Oolite were referred to as Newton Oolite by Wright. In the Hazelbury Bryan Borehole, the Hinton St Mary Clay occurs between 13.6 and 18.7 m, and is dominantly oolitic limestone, with a limestone:clay ratio of 3:2.

Sturminster Newton–Marnhull

In the railway cutting at Sturminster Newton, Blake and Hudleston (1877) recorded 0.9 m of coarse-grained oolitic marl beneath pisolite (Figure 41). On the north-east side of Sturminster Newton, the clay is at least 5 m thick, and north-north-west of the town, it is 3 to 4 m thick; it is absent near Marnhull. The Hinton St Mary Clay reappears as a thin, c.2 m development north of Pilwell [7828 1928], and is probably represented by the 0.46 m of blue marl and clay recorded by Gutmann (1970) beneath pisolite at Todber.

STURMINSTER PISOLITE

Lyon's Gate–West Pulham–Cannings Court–Mappowder

Pisolitic debris was seen at two localities [6553 0605; 6556 0638]; at the latter, there was 0.8 m of pisolitic clay, with pisoliths up to 10 mm in size. An auger hole [7008 0800] at West Pulham showed the following:

	Depth m
STOUR FORMATION	
Limestone, cream coloured, crumbly	1.1
STURMINSTER PISOLITE	
Clay, grey , sandy, oolitic, with scattered pisoliths which become abundant and up 1 cm in size in the bottom 0.2 m	1.5
Oolite to oolitic clay with large pisoliths	1.9

The Cannings Court Borehole [7187 0734] has several levels of pisoliths in a oolitic clay matrix between the depths of 4 and 10 m; near the borehole, there are abundant loose pisoliths in the field bank [7226 0709]. South of Mappowder, cream pisolitic oolite, with individual pisoliths up to 1.5 cm in size, was seen in a trench [7329 0571].

Hazelbury Bryan–Fifehead Neville area

North of Zoar Lane [7677 1000 to 7620 0927], Hazelbury Bryan, the pisolite forms a distinct feature. Wright (1981) recorded 2.1 m of pisolitic clays and marls with bands of pisolitic oolite in a ditch [7620 0935] at Cuckrow Copse.

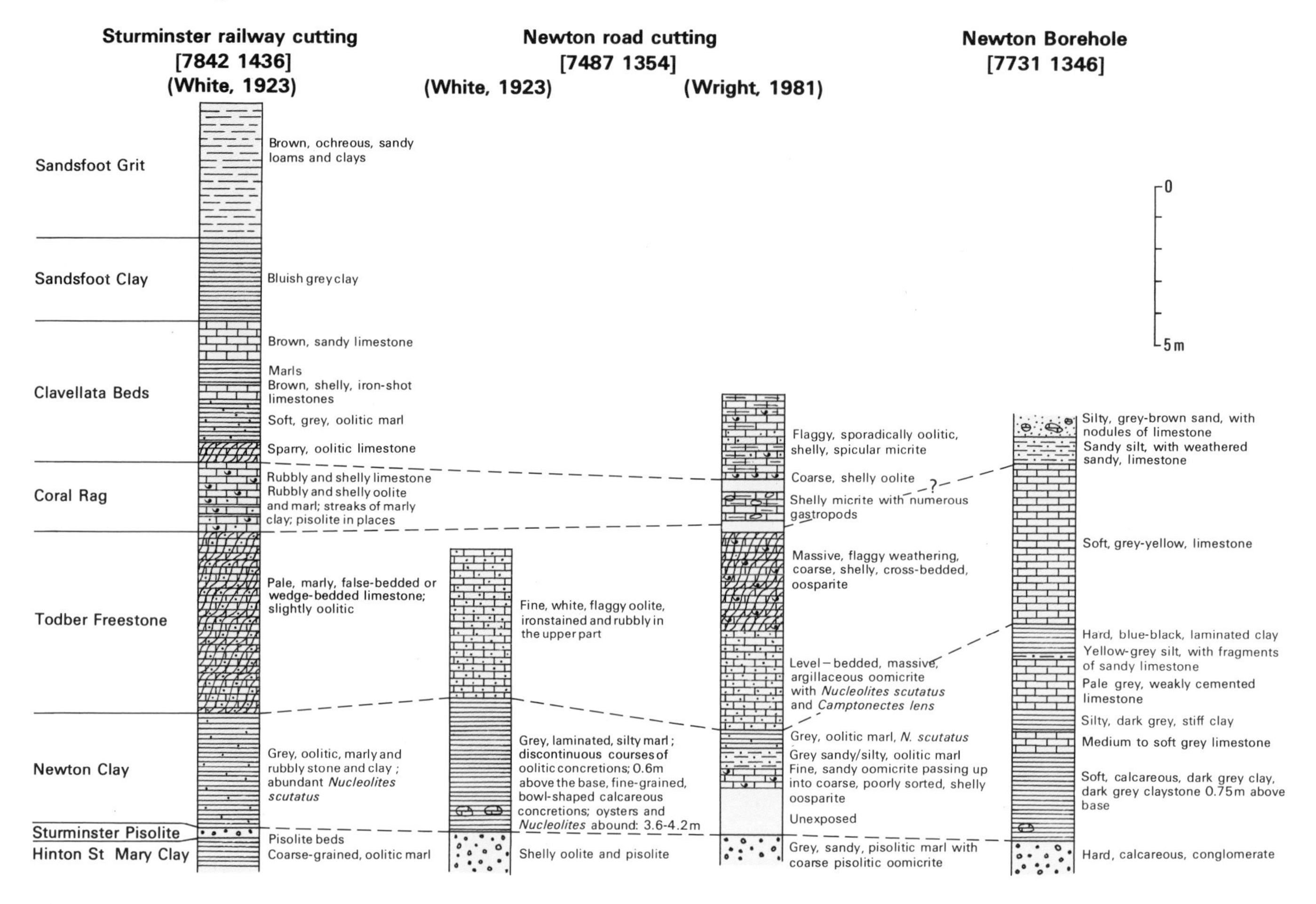

Figure 41 Comparative sections in Corallian strata recorded in the Sturminster railway and Newton road cuttings.

The pisolite in the Hazelbury Bryan Borehole consists of a band of oolitic pisolite between 14.3 and 14.65 m. The overlying clay contains some scattered pisoliths at the base.

Pisolitic marl is common in the fields [around 7704 1007; 7726 1086] south-west of Fifehead St Quintin, below the feature of the Clavellata Beds.

Sturminster Newton–Hinton St Mary area

A section [7812 1327] behind a house at Newton showed 0.7 m of soft, crumbly, pisolitic marl above Cucklington Oolite.

About 1.2 m of pisolitic limestone is exposed as a low wall beside the pavement on Newton Hill [7830 1348]. White (1923) described this section as 'like a retaining wall of pebbly concrete', together with the fossils '*Ostrea*, *Serpulae* and *Nucleolites scutatus*'.

In the railway cutting [7842 1436], Blake and Hudleston (1877) noted a pisolitic bed, 0.3 m thick, resting on coarse-grained oolitic marl. Farther north, springs [7848 1467 to 7852 1482] mark the base of the pisolite low down on the scarp face.

A section [7882 1495] on the northern edge of Sturminster Newton showed 0.5 m of brown clay, on 0.3 m of pisolite, which in turn rested on 0.1 m of grey marly oolite. Pisoliths are common in brash at [7902 1524] and also 130 m farther north-east [792 155]; blocks of pisolite, associated with common *Nucleolites scutatus*, occur at [7923 1548] and [7924 1559]. These occurrences are close to the base of an overlying oolitic limestone. About 200 m north-west of the last locality, pisoliths and pisolite are again common on the surface [7908 1571; 7907 1575], but are at the top of a limestone succession.

On the north side of Spa Coppice, north-east of Hinton St Mary, about 1 m of soft pisolitic limestone is succeeded by about 2 m of pisolitic clay with common small oysters.

On the north side of Chivrick's Brook, blocks of pisolite and shelly oolite are common along a narrow tract [around 7923 1768]. The pisolite is about 1 m thick. Wright (1981) saw a temporary section [7925 1780] of shelly, pisolitic oomicrite with *'Exogyra' nana* and *Lopha* sp., overlying oolitic limestone, at the northern end of this field. West of Chivrick's Brook, towards Marnhull, the pisolite has a wide outcrop and, with the Cucklington Oolite, forms a prominent feature [7885 1815 to 7855 1822].

Todber–East Stour area

Gutmann (1970) saw a section [c.7965 1989] at Todber, where, beneath 1.2 m of hard yellow limestone (Todber Freestone), 0.76 m of shelly, pisolitic, grey marlstone rested on 0.46 m of blue marl and clay (Hinton St Mary Clay). A rich fauna, including '*Pseudomelania heddingtonensis*, *Exogyra nana*, *Chlamys superfibrosa*, *Opis curvirostra* and *Nucleolites scutatus*' was recorded.

At Stour Provost, the village square overlies the pisolite (Blake and Hudleston, 1877); blocks of pisolite are dug out of graves [7939 2152] in the churchyard. South of the church, roadside exposures [7945 2149] reveal coarse-grained shelly oolite, apparently stratigraphically higher than the pisolite. Possibly this is a development of non-pisolitic oolite at the top of the sequence as seen at Cucklington (Bristow, 1990a) and north of Sturminster Newton. At the crossroads east of the village, coarse-grained oolitic pisolite is exposed in the road bank. Shallow excavations [around 793 214] on the south-west side of the village revealed a very shelly, coarse-grained, pisolitic limestone. Perisphinctid ammonites are common and include *Perisphinctes (Arisphinctes)* aff. *helenae.*

In the East Stour Borehole [8013 2297], 0.15 m of oolitic, pisolitic, shelly, micritic limestone rests on an irregular piped surface of the Cucklington Oolite.

North-east and south-east of Kington Magna, long broad dip slopes are developed on the Sturminster Pisolite, on which fragments of pisolite, together with common to abundant pisoliths, occur [773 235; 778 237; 777 228].

Newton Clay

Hazelbury Bryan–Sturminster Newton area

The Newton Clay is represented by 7.2 m of grey, sandy, oolitic clay with argillaceous oolitic limestones and some biomicritic limestones in the Hazelbury Bryan Borehole [7515 0810].

The clay was formerly well exposed in the road bank at Newton, where it is thick enough to cause stability problems (White, 1923, p.33). Sections have been recorded in the road cutting by Blake and Hudleston (1877), White (1923) and Wright (1981), but the quoted thicknesses of the clay vary (Figure 41). Arkell (1933, p.388) recorded abundant *'Nucleolites scutatus'*, together with rarer *'Acrosalenia angulata* and *Hemicidaris intermedia*', from the Newton Clay (his Littlemore Clay Beds) at this locality.

A sample of oolitic, grey, marly clay with thin earthy limestones from a section [7822 1345] at Newton yielded the following fauna of ostracods and foraminifera: *Schuleridea triebeli, Galliaecytheridea postrotunda praecursor, Eripleura parvaesulcata, Lenticulina muensteri, Citharina heteropleura, Frondicularia franconia, Spirillina infima, Planularia suturalis, Cornuspira eichbergensis* and *Trocholina nodulosa.* Collectively, they indicate the Cordatum or Densiplicatum zones.

Wright (1981) described an exposure [7800 1325] of tough, fine-grained, sandy oomicrite on pisolite at Glue Hill, Newton.

In the Sturminster railway cutting, Blake and Hudleston (1877) described the 2.44 m of strata between the Sturminster Pisolite and Todber Freestone as 'black and white rubbly marl, the whiter portions in hard bands'. Woodward (1895) referred to these beds as grey, oolitic, marly, rubbly stone and clay with abundant *Nucleolites scutatus.* Fossils collected by Woodward and Rhodes in the BGS collections include *Ceritella* sp., *Littorina muricata, Procerithium inornatum, P. muricatum,* '*Turbo*' *funiculata, Isocyprina glabra, Limatula elliptica, Nanogyra nana,* pectinids including *Chlamys, Quenstedtia laevigata, Acrosalenia* including *A. decorata, Hemicidaris intermedia* and *Nucleolites scutatus.*

On the east side of the B3092, temporary sections [7876 1472; 7881 1479] show up to at least 1.9 m of marly, locally oolitic and calcareously cemented, shelly, grey clay; to the north, the clay is 5 to 7 m thick. At two places near Hinton St Mary [7847 1517; 7871 1519], pisoliths were found within the clay. Brash north-east of Hinton St Mary [7923 1719] includes a 1 cm bed of very coarse-grained oolite with scattered pisoliths, overlying oolitic clay.

Marnhull–East Stour

The Newton Clay can be followed around Marnhull where it consists of 3 to 4 m of grey, commonly oolitic clay. In the north-east, at Todber, the member is absent (Gutmann, 1970). South of Trill Bridge, the Newton Clay is represented by about 2 m of shelly, oolitic, sandy marl; it thickens north-westwards to about 3 m [around 7975 2125]. Between Stour Provost and East Stour, the thickness is about 5 m.

In the East Stour Borehole, the Newton Clay is 12.24 m thick. Towards the top of the clay, there is a 0.6 m thick bed of sand and sandstone. Northwards from the A30 road, the Newton Clay, which is more sandy and with fewer ooliths, is up to 10 m thick. On the steep east bank of the River Stour, oolitic grey clay and marl, with minor amounts of sandy clay, about 5 m thick, forms a narrow outcrop.

Todber Freestone

Blake and Hudleston (1877) noted that the Todber Freestone in the road cutting at Newton [7723 1345 to 7747 1353] thins rapidly westwards from 3 to 0 m, with the ?Coral Rag resting on Newton Clay in the west. The lower part of the freestone is a level-bedded, massive, argillaceous oomicrite, 3 m thick, with '*Nucleolites scutatus* and *Camptonectes lens*' (Wright, 1981). The upper 3 m of Wright's section consist of massive, flaggy weathering, coarse, shelly oosparite showing strong cross-bedding.

Woodward (1895) noted that the basal bed of the Todber Freestone in the Sturminster railway cutting was pisolitic in places. Fossils collected by Woodward and Rhodes include *Bourguetia* sp., *Procerithium muricatum, Limatula elliptica* and *Pseudodiadema radiatum.*

Sections in the Todber Freestone at Marnhull, Todber and Stour Provost are included in the account of the Clavellata Beds.

In the East Stour Borehole [8013 2297], the Todber Freestone is 4.25 m thick and consists of fine- to medium-grained oolite with scattered shell fragments and with some clay wisps and clasts. Towards the base, there is a 0.17 m bed of silty and fine-grained sandy clay and clayey fine-grained sand. The Todber Freestone has not been identified north of the A30 road.

A temporary section [7971 2287] at Church Farm, East Stour, showed 1 m of flaggy limestone overlying 1 m of yellowish brownish grey clay. In the road cutting [7974 2299] to the north, yellowish buff marly clay was augered. Marly clay was also augered south of the farm [7978 2276].

Clavellata Beds, including Coral Rag

Glanvilles Wootton

A deep ditch [6906 0852] exposed less than a metre of soft, shelly micritic limestone with some high-spired gastropods. Debris at the edge of a field [6833 0819] consisted of shelly micritic limestone containing bivalves, such as *Myophorella clavellata,* and gastropods.

Brockhampton Green area–Cannings Court–Mappowder

In the faulted outcrops near Brockhampton Green, the shelly micrite seen in a small pit [7408 0544] and in debris from a trench appears to be less than 2 m thick.

West of Cannings Court Farm, micritic shelly limestone yielded scattered specimens of *Myophorella clavellata* and other bivalves, and some high-spired gastropods, caps a hill [between 7240 0762 and 7280 0775]. Between Cannings Court and Mappowder, there are many old pits in the the Clavellata Beds [e.g. around 7192 0735].

Small outcrops of shelly limestone occur in old pits [7214 0670; 7298 0656] west of Mappowder, the first yielding *Myophorella clavellata.* Most of Mappowder is built on the

Clavellata Beds. In the centre of the village, Wright (1981) noted an exposure [7340 0600] of micritic oolite. A section [7409 0575 to 7423 0586] recorded by Wright showed:

	Thickness m
CLAVELLATA BEDS	
Limestone, micritic, light brown, impure, bioturbated, with abundant *Rhaxella* spicules	1.2
Limestone, micritic, light grey, shelly, spicular, containing *Liostrea* sp and other bivalves, and *Phasianella* sp.	c.2.4
Limestone, micritic, light blue-grey, spicular, with scattered ooliths	0.6
Marl, soft, blue-grey, sporadically oolitic	0.3

An outcrop nearby [7420 0579] exposed 0.8 m of limestone classifiable in thin section as a bipelmicrite containing some quartz sand and abundant calcite-filled spheres, possibly *?Rhaxella*.

Exposures of hard, sandy, shelly, oolitic limestone occur south of Saunders Dairy House Farm [7420 0545] and nearby in a stream [7460 0577; 7447 0559; 7453 0533; 7455 0528]. A thin section from one locality [7453 0533] showed peloidal micrite with abundant calcitic sparry spheres *(?Rhaxella)*, sparse ooliths and bioclasts, and some fine quartz/feldspar grains.

At Mappowder Quarry [7267 0557], now filled, Wright (1981) saw 0.3 m of tough shelly limestone with *Myophorella clavellata, Mytilus varians* and *Gervillella aviculoides*.

Hazelbury Bryan area

Manuscript material in BGS files records sandy marl and shelly blue-hearted limestone in a pit [?7624 0928] west of Zoar (probably Zoar Farm). Fossils included '*Trigonia, Ostrea* sp, *Pecten fibrosus* and *Astarte*'.

A shallow pit [7618 0881] near Locketts Farm shows up to 1 m of yellowish orange, rubbly, shelly oomicrite, with a poorly preserved trigoniid; rubbly limestone was worked from other overgrown pits nearby [around 7560 0860].

The Hazelbury Bryan Borehole penetrated the Clavellata Beds between the depths of 1.95 and 4.75 m. They consist of 0.75 m thick shelly micrites at the top, with a 0.70 m shelly, slightly oolitic, limestone at the base; the intervening strata are sandy clays with thin shelly micrites.

Fifehead St Queintin–Sturminster Newton

West of Fifehead St Quintin, the basal Clavellata Beds consist of oolitic shelly micrite, but about 2 m above the base, pisoliths are common [e.g. around 7700 1015 and to the south-west].

In the Coombs Valley south of Newton, a section [7835 1290] at stream level showed:

	Thickness m
Flaggy oolite	0.4
Limestone, crumbly, poorly bedded, oolitic	0.5
Flaggy oolite	0.2
Limestone, crumbly, poorly bedded,oolitic; many of the ooliths in the upper part are flattened into discs. In the basal 0.8 m, lenses (up to 10 cm by 1 cm), pods, thin beds (1 to 2 mm thick) and wisps of medium grey clay occur. Gastropods, *'Ostrea' moreana,* some encrusted with *Nanogyra nana,* and an ammonite occur	1.5

At the top of the road section [7847 1354] at Newton, Wright (1981) recorded about 1 m of shelly micrite with scattered ooliths and comminuted shell fragments, including numerous gastropods, set in a fine-grained micritic matrix, which he thought was the Coral Rag. In a quarry [7860 1346] east of the bridge over the Stour, 1.5 m of well-bedded limestone of similar lithology is also exposed below the Clavellata Beds; well-preserved *Nerinea* are common at both localities. A section seen in the quarry in 1986 showed 1 m of rubbly shelly limestone of the Clavellata Beds, resting on about 1.5 m of marly silty clay with a patchy calcareous cement (?Coral Rag), on 0.4 m of fine-grained oolitic limestone (Todber Freestone).

In the Sturminster railway cutting, Blake and Hudleston (1877) described a 2.7 m thick bed (their Bed 4) of rubbly limestone with common abraded fossils. Material collected by Woodward and Rhodes in the BGS collections includes *Bathrotomaria reticulata, Bourguetia buvignieri, Dicroloma deshayesea, Natica dejanira, Nerinea* sp., *Littorina muricatum, Procerithium muricatum, Pseudomelania heddingtonensis,* '*Astarte*' sp., *Isognomon promytiloides, Lima (Acesta) subantiquata, Limatula elliptica, Lopha gregarea, Mactromya* sp., *Modiolus* sp., *Nanogyra nana, Opis (Trigonopis) corallina, Sowerbya triangularis* and *Plegiocidaris florigemma.* Wright (1981 p.25) attributed this unit to the Coral Rag.

A section in Clavellata Beds, above the Coral Rag, in the same railway cutting was described by Blake and Hudleston (1877), Woodward (1895) and White (1923) (Figure 41). Fossils from the brown sandy limestone at the top of the formation collected by Woodward and Rhodes include *Isognomon promytiloides, Pleuromya uniformis, Trautscholdia* [*Astarte*] cf. *curvirostris* and *Perisphinctes* sp. From the basal limestone, they collected a serpulid, a rhynchonellid brachiopod, *Pleurotomaria* sp., *Discomiltha* [*Lucina*] *rotundata, Gryphaea* sp., *Myophorella clavellata, Nanogyra nana, Nucleolites scutatus* and *Pseudodiadema* sp.

Hinton St Mary–Todber

Several small exposures in fine-grained flaggy *Rhaxella* biomicrite [7845 1615; 7895 1650], flaggy shelly fine-grained oolite [7853 1610], some with large gastropods [793 166], and oolitic marl [7810 1632] were noted in and around Hinton St Mary (Wright, 1981; Bristow, 1989b).

North of Chivrick's Brook, there are several old limestone quarries. In the now-filled southernmost [794 178] quarry, Wright (1981) recorded 2.5 m of flaggy shelly oosparite with *Myophorella clavellata.* Gutmann (1970) noted a section in the basal Clavellata Beds similar to that seen north of the road (see below); it yielded *Pseudonerinea clytia* and *Astarte* sp. North of the road, a quarry [7930 1819] formerly exposed the junction of the Clavellata Beds and the Todber Freestone (Gutmann, 1970). A recently opened quarry [794 179] showed the following section in 1992:

	Thickness m
CLAVELLATA BEDS	
Oolite, bioclastic, rubbly weathering, passing down into massive, very shelly, bioclastic oolite	0.50
Oolite, fine-grained, flaggy	0.20
Oolite, moderately bioclastic	0.30
Oolite, friable	0.05 to 0.07
Oolite, hard	0.45
Oolite, friable	0.02
Oolite, hard, with scattered shell fragments	0.45
Oolite, bioclastic, pebbly; with large shells, including bivalves, gastropods and some belemnites; pebbles of rounded oolite up to 10 cm	0.25

Todber Freestone	
Oolite, hard, massive	0.95
Oolite, friable	0.05 to 0.10
Oolite, hard, massive	0.85
Oolite, friable	0.05 to 0.10
Oolite, massive, bioclastic, coarsely bioclastic at base	0.60
Oolite, massive, with scattered shells	1.30
Oolite, friable	0.05 to 0.10
Oolite, massive	0.70

From Marnhull church eastwards, limestone has been worked in three quarries [7855 1885; 7865 1890; 7880 1875]. In the first, a section recorded by Blake and Hudleston (1877) showed 4.2 m of fine-grained, unfossiliferous oolite in beds 0.3 to 0.6 m thick (Todber Freestone), overlain by shelly limestone (Clavellata Beds) with *'Trigonia clavellata, Chemnitzia heddingtonensis, Nerinea fasciata, Pleuromya tellina* and *Nucleolites scutatus'*. In the second quarry, cross-bedded shelly oolite, 0.6 m thick, rests on 1.6 m of fine-grained oolite of the Todber Freestone (Bristow, 1989b, p.43).

Near Todber, there are several limestone quarries. One [7949 1980] is still worked on a small scale; others [796 199; 7975 1990; 7970 1973; 8015 2003; 8030 2005; 8005 1996], although partially filled, still expose sections in Todber Freestone capped by the Clavellata Beds (Bristow, 1989a and b, 1992a; Gutmann, 1970; Blake and Hudleston, 1877; Wright, 1981). All sections are essentially similar and show a hard, bluish grey, coarsely bioclastic limestone up to 1.2 m thick, resting with an undulating base on up to 3.6 m of cross-bedded, flaggy oolite with scattered *Nucleolites scutatus*, in units that vary in thickness between 0.1 and 0.6 m, separated by sandy limestones. The foresets dip at about 10°E (Plate 7).

A representative section is that seen in the easternmost pit [7975 1990]. Blake and Hudleston (1877) recorded the following section (slightly modified), of which the uppermost part is still exposed:

	Thickness m
Clavellata Beds	
Marl, soft, yellow, oolitic, and thin-bedded rubbly oolitic limestone with *'Natica* sp., *Chemnitzia heddingtonensis, Ostrea solitaria, Nucleolites scutatus'*	0.91
Limestone, oolitic, shelly, with soft marly partings, flaggy. *"Ammonites plicatilis"* [= *Perisphinctes* sp.], *Nucleolites scutatus*	2.13

Plate 7 Junction of Clavellata Beds and Todber Freestone near Todber.

	Thickness m
Limestone, blue, solid, composed of comminuted shells, some relatively intact bivalves, belemnites and echinoids, with scattered ooliths; undulating base	1.22
TODBER FREESTONE	
Clay, rubbly	0 to 0.3
Limestone, oolitic, fine-grained, creamy, cross-bedded, blue, hard	3.66
STURMINSTER PISOLITE	
Marl, blue, hard, pisolitic, with oysters including *Nanogyra nana* (unexposed)	5.18

Warren's, the most southerly pit [7970 1975], was described by Gutmann (1970) and Bristow (1989b). From the Clavellata Beds, Gutmann (1970) recorded a fragment of the ammonite *Pseudarisphinctes* sp. and common bivalves including *Trigonia clavellata.* This is probably the pit in which Mottram (1957) recorded the ammonite *Perisphinctes (Arisphinctes)* aff. *osmingtonensis,* found loose and believed to come from the Clavellata Beds; another ammonite, *Perisphinctes (Dichotomosphinctes) antecedens,* was thought to have come from the Todber Freestone.

An old quarry [7931 2026] south of Trill Bridge, exposed:

	Thickness m
CLAVELLATA BEDS	
Oolite, flaggy, fine-grained	0.95
Conglomerate, with tabular clasts of limestone up to 8 cm across, mostly horizontal, but some vertical, in a matrix of sandy oolitic marl	0.10
TODBER FREESTONE	
Marl, sandy, oolitic	0.15
Oolite, fine-grained, with thin (2 to 3 cm) beds of oolitic sandy marl; shelly 0.85 cm below top	0.95
Oolite, fine-grained, soft	0.15
Oolite, fine-grained, well-bedded	0.30

The rocks dip 2 to 3° to the east.

Stour Provost–East Stour

At Stour Provost, several old quarries [7955 2175; 7958 2185; 797 217] formerly exposed the junction of the Todber Freestone and Clavellata Beds (see Bristow, 1990a). In the quarry west of the road, the following section [7960 2182] was seen in 1988:

	Thickness m
?SANDSFOOT FORMATION	
Clay, oolitic, greyish brown, shelly	0.80
Clay, oolitic, very shelly, brown	0.10
CLAVELLATA BEDS	
Limestone, shell-fragmental, oolitic, hard	0.20
Limestone, oolitic, passing into crumbly oolite, orange-brown	0.20
Limestone, shelly, hard	0.25
Oolite, crumbly, orange	0.02 to 0.06
Oolite, roughly flaggy	0.20
Limestone, shell-fragmental, oolitic, hard	0.40
TODBER FREESTONE	
Limestone, oolitic, crumbly	0.30

If the Sandsfoot Beds are correctly identified at the top of the section, then the Clavellata Beds are only 1.3 m thick, a similar thickness to that quoted by Arkell (1933).

From the Clavellata Beds at Stour Provost and East Stour, Arkell (1933) recorded 'abundant *Trigonia clavellata*', many other bivalves and '*Nerinea* sp.'

The old quarry [795 231] on the west side of East Stour, now filled, may be the one referred to by Arkell (1933, p.388), in which he described the Clavellata Beds as dark, soft, shelly beds, only 0.76 m thick, sharply demarcated from the underlying beds of white oolite of the 'Osmington Oolite Series' (= Todber Freestone). However, the description is more likely to apply to the Stour Provost Quarry (see above), where he also saw the junction with the Sandsfoot Formation (which does not crop out near the quarry at East Stour). The Todber Freestone has not been recognised north of the A30 road.

North of East Stour, on the valley sides of a tributary of the River Stour, shelly to very shelly (including *Natica* and *Pleuromya*), sparsely oolitic sparite crops out on the lower valley sides [7975 2365; 794 237; 794 231] and is capped by fine-grained flaggy oolites of the Eccliffe Member. At the last locality, the undifferentiated Clavellata Beds are about 5 m thick.

ECCLIFFE MEMBER

Near East Stour, brash consists of sparsely oolitic, sparsely shelly sparite [around 796 233]. Up to 1.2 m of flaggy fine-grained oolite is exposed in the road bank of the old A30 [7962 2308].

Sandsfoot Formation, undivided

Glanvilles Wootton–Sharnhill Green

A typical auger hole [6842 0820] proved 0.6 m of orange-grey, extremely sandy clay with oysters, on 0.7 m of orange-grey to medium grey, clayey, fine-grained sand with oysters.

Near Sharnhill Green, auger holes showed lithologies ranging from grey sandy clay, particularly in the lower part, to clayey fine-grained sand. Auger holes near Beaulieu Wood [7047 0625] revealed grey sandy clay on limestone, presumably the Clavellata Beds. A nearby auger hole [7027 0630], higher in the sequence, showed orange-grey, ferruginous clayey sand to extremely sandy clay with oysters. Some of the ferruginous material was in thin lateritic layers.

Mappowder area

North and north-west of Mappowder, orange to yellowish grey, sandy to extremely sandy, locally shelly, ferruginous clay and clayey fine-grained sand were proved. One auger hole [7277 0722] showed 1.9 m of yellowish and brownish grey, extremely sandy clay with ferruginous pellets, resting on 0.1 m oolitic grey clay (probably Sandsfoot Clay), which rested in turn on the Clavellata Beds. North-east of Mappowder, springs [7375 0661 to 7422 0630] are prominent at the contact of the Sandsfoot Formation and the Clavellata Beds. Auger holes in the stream bed [7481 0587] showed dark greyish green, extremely sandy clay; in the stream, there was abundant debris of large oysters and occasional belemnites. To the west, a nearby auger hole [7453 0553] showed:

	Depth m
SANDSFOOT FORMATION	
Clay, sandy, brownish grey and ferruginous	0.8

	Depth m
Clay, sticky, greyish yellow-stained, becoming ferruginous at base, with oysters	2.0
Clay, very ferruginous	2.2
CLAVELLATA BEDS	
Limestone, oolitic	2.3

An auger hole [7486 0540] in a ditch showed a sequence a few metres higher in the succession of 1.5 m of orange-grey, clayey sand, with oysters at the base, on 0.2 m of soft, cream-coloured limestone, on 0.8 m of brownish grey and orange-stained, extremely sandy clay.

Hazelbury Bryan area

In one locality [7657 0892], an auger hole revealed yellow clay resting on hard clayey silt. A nearby auger hole [7658 0691] showed highly ferruginous, patchily brick red, sandy clay to clayey sand. The Hazelbury Bryan Borehole [7515 0810] penetrated 2 m of Sandsfoot Formation, but there was no recovery. Auger holes beside the rig and on top of a hill nearby proved the following succession:

	Thickness m
SANDSFOOT FORMATION	
Clay, extremely sandy, ranging to clayey fine-grained sand, orange, brown and grey; race 3 m below top; brown and olive-grey below	5.30
Gap (between base of highest auger hole and one next to rig)	1–2
Clay, very sandy, with shells, probably oysters, below 1.8 m	2.00
CLAVELLATA BEDS	
Limestone, oolitic, rubbly, pale yellow	0.05

SANDSFOOT CLAY

North-north-east of Broad Oak, oolitic grey clay, 1 to 2 m thick [7894 1340 to 7902 1335], is overlain by orange-brown sandy clay of the Sandsfoot Grit. In the Sturminster railway cutting, Blake and Hudleston (1877) recorded 2.59 m of clay and marl with small oysters between the Sandsfoot Grit and Clavellata Beds.

South of Hinton St Mary, and underlying the village, there are outliers [787 157; 786 162] of orange or greyish brown, only locally oolitic clay resting on Clavellata Beds. North of Chivrick's Brook, 0.5 m of oolitic grey-brown clay rests on Clavellata Beds in an old quarry [7931 1810].

South of Todber, sand-free oolitic grey clay rests on Clavellata Beds and is overlain by Sandsfoot Grit [around 797 196].

SANDSFOOT GRIT

Fifehead Neville–Sturminster Newton

Near Fifehead Neville, the Sandsfoot Grit consists of about 5 m of orange or mottled orange and grey, sandy clay. North-eastwards from Rivers' Corner, the outcrop increases in width to about 350 m and the deposits become more sandy, so that fine-grained clayey sand and, locally, clay-free fine-grained sand occur. The low hill [782 132] on the south side of Newton is about 8 m high and is underlain by orange-brown sandy clay.

In the Knackers Hole Borehole [7791 1188], the Sandsfoot Formation is 7.7 m thick and consists mainly of fine-grained clayey sands and weakly cemented sandstone, intensely burrowed, resting on sandy shelly, burrowed, oolitic mudstone. The Sandsfoot Clay is probably represented by a basal unit, 0.18 m thick, of shelly oolitic mudstone. Above this, there is a bed of sandy oolitic mudstone, 0.35 m thick, which contains common matrix-supported limonitised ooliths. This may be the lateral equivalent of the oolitic iron ore recorded in the Broad Oak section (see below). The fauna is dominated by bivalves that include *Gervillella, Isognomon,* oysters *(Liostrea, Lopha, Nanogyra nana),* pectinids, *Pholadomya, Pinna* (including *P. sandsfootensis)* and *Pseudolimea.*

At Newton, on the east side of Coombs Valley, at least 2 m, and possibly up to 3 m, of fine-grained clayey, patchily oolitic sand overlie the Clavellata Beds [7835 1291; 7843 1286]. From Coombs eastwards, the Sandsfoot Grit becomes more ferruginous. Blake and Hudleston (1877) noted clays and sands with, at the top, ferruginous, fossiliferous sand and concretions, beneath 'Kimmeridge Clay' in the road [c.7865 1325] to Broad Oak. Woodward (1895) gives a slighty more detailed section which was also noted by White (1923) (a revised faunal list is included):

	Thickness m
RINGSTEAD WAXY CLAY (see p.74)	
Clay	no thickness
Sandy layer	no thickness
Clay with selenite, *Deltoideum delta*	0.91
Nodular bed	no thickness
SANDSFOOT GRIT	
Ferruginous beds like the oolitic iron-ore of Abbotsbury and Westbury, with band of blue-hearted sandy limestone; serpulids, *Chlamys (Aequipecten) midas, Deltoideum delta, Oxytoma* sp., *Pinna sandsfootensis, Pleuromya uniformis, Protocardia* sp., perisphinctids and belemnites	3 to 4.5
?SANDSFOOT CLAY	
Clays, ochreous; *Nanogyra nana, ?Pachyteuthis abbreviata*	no thickness

White (1923) noted that the ferruginous calcareous oolite at the top of the section is about 0.45 m thick. During the Second World War, an analysis of the oolite was carried out to see if it could be exploited for iron ore. The results of a sample [7869 1320] collected by Mr G A Kellaway are: Fe (27.1 per cent), SiO_2 (22.4 per cent), Al_2O_3 (4.8 per cent), CaO (10.1 per cent), P_2O_5 (1.5 per cent), S (0.2 per cent), ignition loss (18.75 per cent), H_2O (4.3 per cent). Its low iron content (cf. 50.3 per cent at Westbury, 44 per cent at Abbotsbury (Lamplugh et al., 1920), thinness and limited areal extent preclude it from economic exploitation.

Wright (1981) recorded the following section [7869 1329] in the road bank:

	Thickness m
Sandstone, iron-rich, with scattered limonitic ooliths and phosphatic nodules; *Liostrea* sp.	0.75
Sandstone, calcareous, with white altered ooliths	1.20
Sand, argillaceous, red	3.00

Wright noted that ooliths within the phosphatic nodules of the uppermost bed were still recognisable as chamosite.

In the railway cutting at Sturminster Newton, Blake and Hudleston (1877) recorded 4.27 m of argillaceous, yellow sand, on 2.59 m of Sandsfoot Clay. At Rixon, a dutch auger hole [7949 1475] proved 3.8 m of orange-brown, clayey, fine-grained sand. A sample of the sand proved, despite the high clay content (25 per cent by weight), to be moderately sorted and very positively skewed.

Hinton St Mary–Marnhull

North-east of Hinton St Mary, 1 to 2 m of orange-brown silty clay are overlain by 2 to 3 m of ferruginous fine-grained sand and sandstone [around 796 170]. Some 2 m of orange-brown, very sandy clay overlies the Sandsfoot Clay on the north-east side of Chivrick's Brook [around 7965 1800]. An auger hole [7984 1780] proved:

	Depth m
TOPSOIL	0.50
?RINGSTEAD WAXY CLAY	
Clay, pale to medium grey with race	0.90
SANDSFOOT GRIT	
Sand, fine-grained, very clayey, grey with scattered shells	1.05
Clay, very sandy, fine-grained, mottled orange and grey, shelly, with small (up to 5 mm) ?phosphatised black nodules	1.50
Clay, very sandy, grey	1.60
Sandstone, clayey, grey	1.70
Clay, very sandy, medium grey	2.30
Stone (?Clavellata Beds)	touched

To the south-east, yellow-brown, sandy, locally oolitic clay [around 7990 1745] rests on Clavellata Beds and is overlain by ?Ringstead Waxy Clay. On the east side of Marnhull [around 790 190], yellow-brown or grey, locally shelly clay with very little sand forms an outlier on the Clavellata Beds.

Todber–East Stour

From Todber [around 798 203] to East Stour, the Sandsfoot Grit consists of about 2 m of fine-grained, very sandy clay.

In the faulted wedge-shaped inlier [around 7950 2345] north-west of East Stour, grey and brown sandy clay occur. An auger hole [7953 2343] proved 0.8 m of mottled orange and brownish grey sandy clay, above 1.4 m of laminated, fine-grained, sandy clay and clayey fine-grained sand.

Ringstead Waxy Clay

In the south-west of the district, small exposures in a stream [7167 0477; 7167 0474] showed up to 1.3 m of pale grey silty clay, with a few small red sideritic nodules at the second locality.

South of Mappowder, a small exposure and auger holes [7350 0531] proved grey sandy clay with oysters and red sideritic nodules.

Wright (1981, pp.26–27) saw a poor section [787 132] above the Sandsfoot Grit in the road bank to Broad Oak. There, some 3 m of clay with red claystone nodules showing grey, clay-filled *Chondrites* burrows were seen. The top of the Ringstead Waxy Clay was taken at the top of a line of red nodules. Specimens collected by Woodward (1895, p.102) from this clay (his 'Kimmeridge Clay') include *Deltoideum delta*.

The best section in Ringstead Waxy Clay was in Knacker's Hole Borehole [7791 1188] (See p.74).

KIMMERIDGE CLAY

The Kimmeridge Clay is a succession of marine mudstones which, at various levels, are calcareous, kerogen-rich (bituminous mudstones and oil shales), silty or sandy; there are associated thin siltstone and cementstone beds. Throughout southern Britain, these lithologies occur in a complex sequence of small-scale rhythms, many of which can be correlated over distances of tens of kilometres. At its type locality on the Dorset coast, c.30 km south of the district, the formation is over 500 m thick and has traditionally been divided into Lower and Upper Kimmeridge Clay. Although, these units have no formal lithostratigraphical status under modern rules of nomenclature, it is convenient to retain them for descriptive purposes; they can be readily distinguished on the basis of ammonite genera (see below). In the lower part of the Lower Kimmeridge Clay, an idealised rhythm consists of grey siltstone or silty mudstone overlain by dark grey mudstone and then by pale grey calcareous mudstone. In the upper part of the Lower Kimmeridge Clay and the greater part of the Upper Kimmeridge Clay, brownish grey 'bituminous' mudstone or oil shale takes the place of the basal silty lithology. On a broader scale, the whole of the Kimmeridge Clay can be regarded as a single rhythm when compared with other Upper Jurassic clays, changing upwards from rhythms with silt to rhythms with oil shale, and then becoming almost uniformly very calcareous (Cox and Gallois, 1981).

On the Dorset coast, the base of the Kimmeridge Clay is taken at a minor erosion surface which marks a lithological change from the mixed lithologies of the Corallian Group to the more uniform mudstones of the Kimmeridge Clay. The Inconstans Bed at the base is a dark grey, intensely bioturbated clay with wisps and burrow-fills of silt, fine-grained sand and scattered limonite ooliths in the lower part; phosphatic pebble beds occur, at and close above, the base. As well as the eponymous brachiopod *Torquirhynchia inconstans*, there is a rich molluscan fauna together with serpulids (Cox and Gallois, 1981). In the Shaftesbury district, the boundary sequence was proved in the Knackers Hole Borehole [7791 1188]; there is field evidence of the Inconstans Bed near Mappowder [7464 0642] (see p.90).

However, generally it has not been possible to map the basal boundary and, as a result, the Kimmeridge Clay and the underlying Ringstead Waxy Clay of the Corallian Group have been mapped together.

On the Dorset coast, the mudstones in the topmost part of the Kimmeridge Clay ('Rhynchonella Beds' and above) are silty and sandy, and form part of a gradational transition into the clayey sands of the basal Portland Beds (Arkell, 1947; Townson, 1975). In the Shaftesbury and adjacent districts, there are more substantial developments of arenaceous strata lower down in the Upper Kimmeridge Clay, but their precise lower limit has not everywhere been established. In the Church Farm No. 2 Borehole [8555 2223], Shaftesbury (Figure 42, locality 29; see p.91), sand laminae and burrowfills have been recorded in silty mudstones as low as the Pectinatus

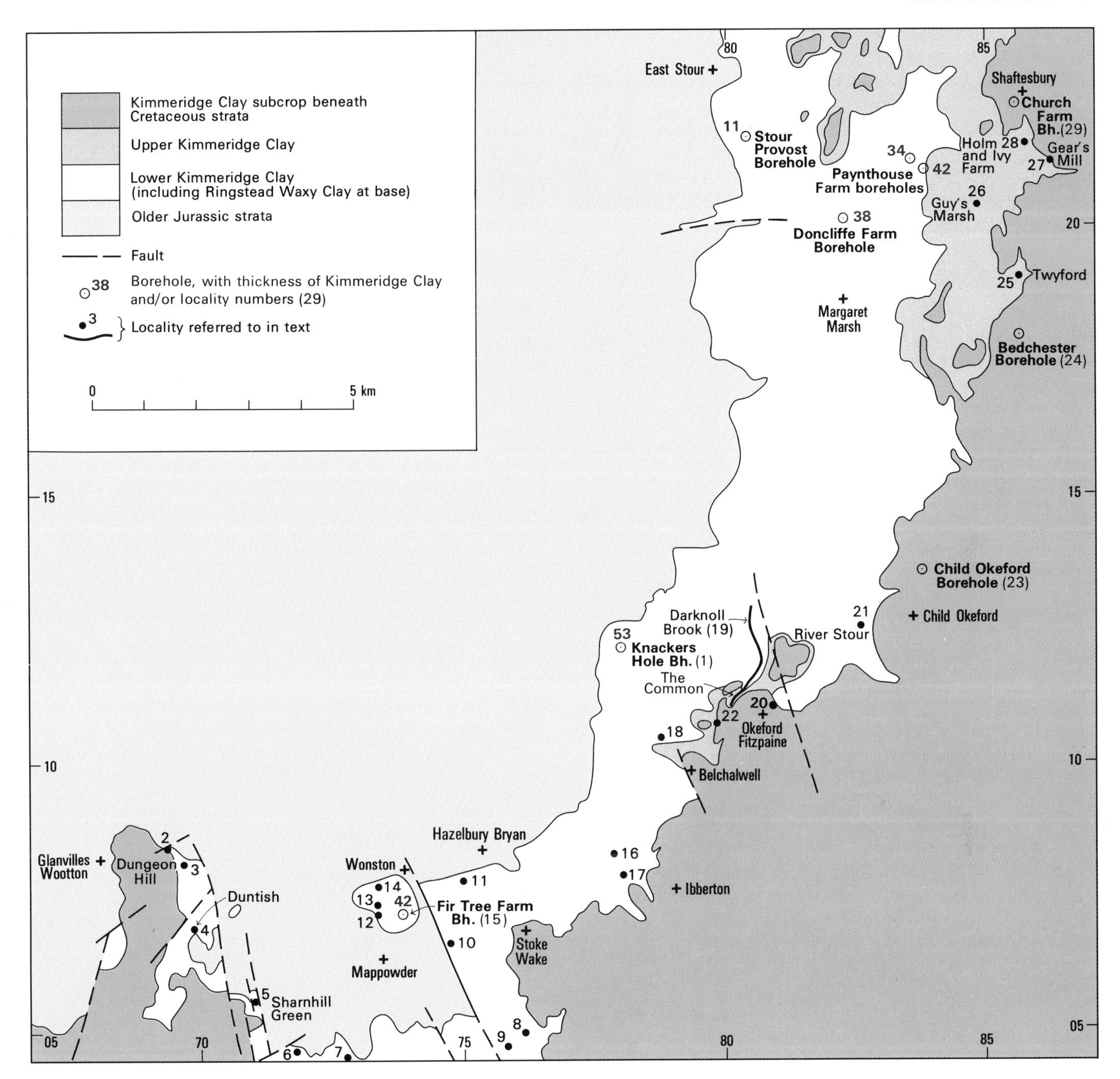

Figure 42 Sketch map of the outcrop of the Kimmeridge Clay of the district showing localities referred to in text.

Zone; there, they are unconformably overlain by Lower Greensand. In the north-eastern part of the district, Portland Beds overlie Kimmeridge Clay at depth (Whittaker, 1985, map 22) and, in the Dengrove Farm Borehole [9253 2417], Donhead St Andrew, some 750 m north of the district, sand and sandstone have been recognised in the Kimmeridge Clay, just beneath the Portland Beds (Bristow, 1990b). In the Tisbury Borehole [9359 2907], at the eastern end of the Vale of Wardour, sandy and silty strata extend down to the top of the Pectinatus Zone (Gallois, 1979).

In the district, the Kimmeridge Clay crops out in an arc from Buckland Newton in the south-west to just west of Shaftesbury in the north. West of Buckland Newton, it is cut out beneath the unconformable Gault. On a broad scale, the Lower Kimmeridge Clay forms gently undulating ground, with the Upper Kimmeridge Clay rising more steeply from the clay vale. This change in topography may be due to the presence of a greater proportion of calcareous mudstones and oil shales in the Upper Kimmeridge Clay. By combining palaeontological evidence with a weak feature break, a tentative line is drawn on Figure 42 to show the approximate boundary between the two divisions.

The preserved thickness of the formation thins southwards from a maximum of about 300 m in the Mere Basin, to the north-east of the district, to about 160 m where the outcrop emerges from beneath Cretaceous strata near Shaftesbury, to less than 100 m adjacent to the Cranborne Fault on the southern edge of the district. South of the Cranborne Fault, thicknesses of up to 130 m are locally present in the Winterborne Kingston trough. South and south-east of the district, much reduced thicknesses of Kimmeridge Clay are recorded in the boreholes at Winterborne Kingston (47.7 m), Spetisbury (42.4 m) and Shapwick (58.5 m). Much of this thinning is a result of pre-Aptian erosion, with different thicknesses of Kimmeridge Clay preserved in fault-bounded blocks beneath the Lower Greensand or Gault. Provisional work shows that there is also some original depositional thinning. This is evident from seismic data in the area near Shaftesbury, where there is a thin, probably Lower Kimmeridge Clay succession to the west at outcrop (on a 'shelf'), and a thicker succession of both Lower and Upper Kimmeridge Clay, preserved beneath Cretaceous strata, to the east (in a 'basin').

In the northern part of the district, the Lower Kimmeridge Clay appears to have a maximum total thickness of c.70 m, compared to c.120 m in the Okeford Fitzpaine area farther south (Figure 43). Southwards from there, the thickness of the lower part of the Lower Kimmeridge Clay remains remarkably constant (Figure 43). The preserved thickness of the Upper Kimmeridge Clay is about 95 m in the Shaftesbury area.

Equivalents of the Yellow Ledge Stone Band, Grey Ledge Stone Band and White Stone Band, three of the marker stone bands recognised on the Dorset coast, have almost certainly been identified. The last-named is the thickest (nearly a metre), best developed and most widespread of the coccolith-rich bands which occur at a number of levels in the English Kimmeridge Clay (Gallois and Medd, 1979); many of these are less than 10 mm thick, but all are characteristically very pale brown in colour, finely laminated and light in weight. Other marker horizons from the Dorset coast which have been recognised in the district include the Inconstans Bed, the Nana Bed, the Supracorallina Bed and the Nannocardioceras Cementstone (Figure 43; see p.90).

The Kimmeridge Clay is highly fossiliferous (Plate 5) and at most levels contains abundant ammonites (used as the basis of the standard zonation) and bivalves. Other molluscan groups, echinoderms, brachiopods, serpulids and crustaceans are also present amongst the invertebrate fossils, and there is a rich microfauna and microflora, including foraminifera, ostracods, coccoliths and dinoflagellate cysts (e.g. Riding and Thomas, 1988). Marine vertebrates are also present; vertebrae and other bones have been recovered from the Lower Kimmeridge Clay at Stoke Wake [76 06] (Delair, 1966). By combining faunal markers and the rhythmic variation in lithology within the framework of the ammonite zonation, the formation can be subdivided into small-scale stratigraphical units. Although originally established in cored borehole sequences in eastern England (Gallois and Cox, 1976; Cox and Gallois, 1979), these units (hereinafter referred to as KC 1, KC 2, etc.) can be recognised widely in sections and boreholes, including downhole geophysical logs, throughout southern Britain (Cox and Gallois, 1981; Penn et al., 1986). They are shown, together with the ammonite-based chronostratigraphy, in Figure 43 and are used in the descriptive details that follow.

Details

LOWER KIMMERIDGE CLAY

Much of the Lower Kimmeridge Clay sequence was proved in the Knackers Hole Borehole [7791 1188] (Figure 42, locality 1), where the highest recovered core material, at a depth of 8.30 m, belongs to a low level of the Eudoxus Zone. A number of marker beds, including the Supracorallina Bed and Nana Bed (Plate 5, 10), are well developed; correlatives of other marker beds, including the Black Head Siltstone, Wyke Siltstone and Inconstans Bed, were also identified (Figure 43).

A partly faulted outcrop of the Kimmeridge Clay appears from beneath the Gault to the north-east of Dungeon Hill. A trench section [6890 0845] (Figure 42, locality 2), noted by Cope (1972), showed bituminous shales of the Lower Kimmeridge Clay (Eudoxus Zone) faulted against oolites of the Corallian Group. An auger hole [6896 0842] c.3 m above and to the south-east of the latter locality showed black to very dark grey silty clay which yielded abundant ostracods, including *Dicroygma reticulata, Galliaecytheridea dorsetensis, G. mandelstami, G.* cf. *punctata, Macrodentina (P.) proclivis* and *Mandelstamia rectilinea,* indicative of the Mutabilis Zone. This suggests that either there is a fault between this locality and the last, or the dip is fairly steep in that direction. Farther south-east in the trench [6950 0810] (Figure 42, locality 3), very dark grey clays and bituminous shales yielded an ammonite fauna of the Eudoxus or Autissiodorensis zones. Older Kimmeridge Clay, assigned to the Cymodoce Zone, was also seen south-east of the trench as far as [c.6962 0805] (written communications, R Brand, R D Clark, M R House, H C Prudden, H S Torrens, 1989–1991).

Some of the clay in the Duntish area is more sandy than the Lower Kimmeridge Clay of nearby areas. One auger hole [6970 0688] (Figure 42, locality 4) showed 1.2 m of grey, silty to sandy, very shelly clay with secondary gypsum, overlain by 1.1 m of flinty head deposits. Long-ranging foraminifera, including *Ammobaculites agglutinans* (abundant), *Lenticulina subalata, L. muensteri, L. varians, Vaginulina secunda, Citharina serratocostata, Tristix suprajurassica* and *Planularia* sp., were recovered, together with the ostracods *Dicroygma reticulata, Mandelstamia rectilinea, Macrodentina steghausi, Galliaecytheridea elongata* and *Eocytheropteron bispinosum* s.s. These faunas indicate the Eudoxus Zone, probably the higher part. West of Duntish, an auger hole [6910 0636] revealed 1.4 m of sticky yellow clay on bituminous brown clay to 1.8 m. South of Duntish, an auger hole [6921 0517] showed 1.5 m of extremely sandy clay with streaks of green sand on sticky, dark grey, silty clay to 2 m.

The outcrop at Sharnhill Green, characterised by sticky, commonly shelly clay, is fragmented by faulting. A sample from an auger hole [7086 0566] (Figure 42, locality 5) yielded a sparse fauna of foraminifera, including *Vaginulina prima, Citharina serratocostata* and *Lenticulina* sp., and the ostracod *Galliaecytheridea* cf. *elongata,* which tentatively suggest the Eudoxus Zone. The larger outcrop on the western side of the Poyntington Fault was worked for brick clay [701 077], but no section remains.

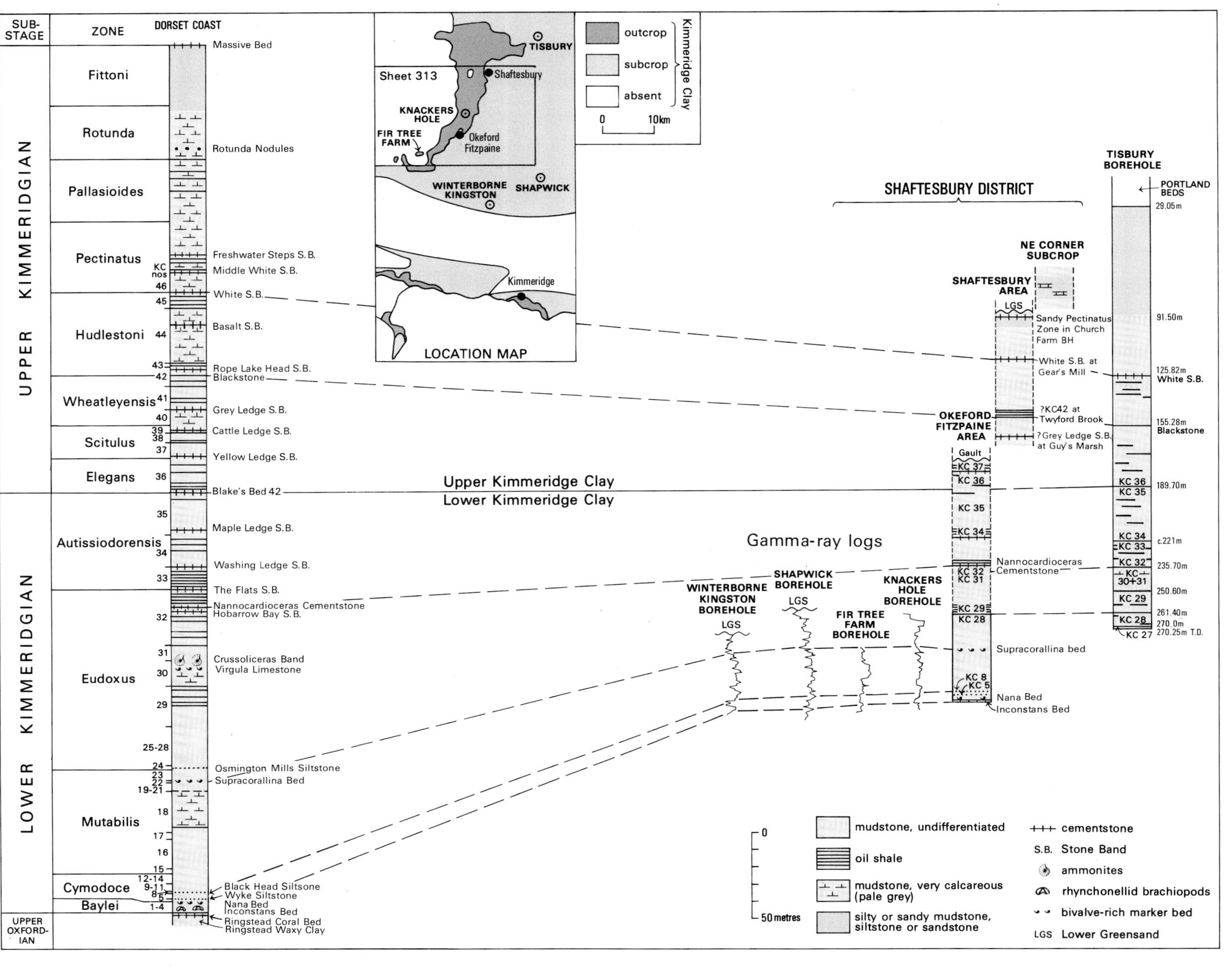

Figure 43 Correlation of the Kimmeridge Clay of the Shaftesbury district with that of the Dorset coastal sequence (Cox and Gallois, 1981) and the Shapwick, Winterborne Kingston (Rhys et al., 1982) and Tisbury (Gallois, 1979) boreholes.

South-east of Sharnhill Green, a stream section [7167 0467] (Figure 42, locality 6) near Noake Farm showed 1.6 m of silty, pale to medium grey, very shelly mudstone with abundant *Nanogyra nana* (Plate 5, 9), overlying 0.2 m of cementstone, also with small oysters. The mudstone yielded a trigoniid bivalve and a small 'raseniid' ammonite. This exposure is almost certainly in the Nana Bed (KC 2) (Figure 43).

Farther east, debris from a trench [7278 0448] (Figure 42, locality 7) yielded pieces of cemented lumachelle rich in small shells and shell fragments including *Aulacostephanus* ex gr. *eulepidus*, '*Astarte supracorallina*' [= *Nicaniella (N.) extensa*] and small oysters; these indicate the Supracorallina Bed (KC 22) (Figure 43).

A small stream section [7611 0497] (Figure 42, locality 8) to the south of Stoke Wake showed 0.8 m of brownish grey, highly fissile, shelly mudstone with shell plasters, mainly of *Aulacostephanus eulepidus*, together with '*Astarte*', *Isocyprina minuscula, Protocardia*, fish fragments and faecal pellets. This lithological and faunal assemblage indicates KC 16 or KC 20 in the Mutabilis Zone. A sample from an auger hole in a stream bed [7558 0453] (Figure 42, locality 9) yielded foraminifera (including common *Ammobaculites cobbani*) and ostracoda (including *Macrodentina (M.) cicatricosa*); these are suggestive of the upper part of the Mutabilis Zone or the basal part of the Eudoxus Zone.

One of the few exposures in the Mappowder area occurs in a stream bank [7464 0642] (Figure 42, locality 10), where about 0.5 m of grey clay with a cementstone layer crops out below 1.2 m of flinty clayey head. The cementstone yielded *Torquirhynchia inconstans* (Plate 5, 8), which is indicative of the Inconstans Bed at the base of the Kimmeridge Clay (Figure 43). About 100 m north-north-west, another exposure [7461 0653] in the stream shows 0.5 m of grey clay with some cementstone nodules and abundant oysters.

An auger hole [7495 0785] (Figure 42, locality 11) just south of the Merriott–Hardington Fault, south-west of Hazelbury Bryan Church, yielded the following foraminifera: *Epistomina ornata* (abundant), *Citharinella* sp. cf. *disjuncta* of Lloyd (1962), *Marginulina radiata, Lenticulina muensteri, Citharina serratocostata, Vaginulina prima, Dentalina* sp., *Ramulina* sp., and ostracoda: *Macrodentina (P.) proclivis* s.s., *Schuleridea triebeli, Mandelstamia rectilinea, Dicrorygma reticulata* and *Exophthalmocythere fuhrbergensis*. Taken together, these microfaunas indicate the Eudoxus or, more probably, the Mutabilis Zone.

West of a north-north-west-trending fault, an outlier of Kimmeridge Clay south-west of Wonston occurs at a higher topographical level than the main outcrop to the east. Debris from a ditch [7330 0710] (Figure 42, locality 12) yielded the ammonite *Pictonia* sp., preserved in cementstone and indicating the Baylei Zone. Evidence of the succeeding Cymodoce Zone is provided by an auger hole [7335 0731] (Figure 42, locality 13), samples from which yielded rare foraminifera, together with the ostracods *Galliaecytheridea dissimilis* (common), *G. dorsetensis* (rare) and *Mandelstamia rectilinea* (common). The ostracods are indicative of the *G. dissimilis* ostracod Zone, the upper boundary of which coincides with the top of the Cymodoce Zone (KC 14). The presence of early Mutabilis Zone strata on the northern part of this outlier is tentatively suggested by the foraminifera *Trochammina squamata, Ammodiscus* sp. and *Cornuspira* sp. (of Lloyd, 1962) in a sample from another auger hole [7351 0776] (Figure 42, locality 14). The Fir Tree Farm Borehole [736 071] (Figure 42, locality 15), on the south-eastern part of this outlier, proved 31 m of Kimmeridge Clay above 11 m of Ringstead Waxy Clay. Comparison of the gamma-ray log with that of the Knackers Hole Borehole (Figure 43) establishes that the former borehole commenced in the Supracorallina Bed.

The Supracorallina Bed (KC 22) was also recorded in a small stream exposure [7778 0832] south-west of Belchalwell (Figure 42, locality 16), where pale grey, 'tough' clay, with a fauna including '*Astarte supracorallina*' [= *Nicaniella extensa*] and *Aulacostephanus eulepidus* formed a sill in the stream bed. Higher up the stream [7790 0775] (Figure 42, locality 17), grey clay yielded a sparse, early Kimmeridgian microfauna indicative of a stratigraphical level probably no younger than early Mutabilis Zone.

A temporary exposure [7862 1039], 500 m south-south-west of The Common (Figure 42, locality 18), revealed dark bluish grey mudstone interbedded with brownish grey, fissile, shelly mudstone (oil shale), with small (0.4 m diameter) cementstone nodules. Fossils from the shelly mudstone include common *Amoeboceras (Nannocardioceras)*, together with *Aulacostephanus*; these indicate the Eudoxus, or possibly low Autissiodorensis Zone.

There are intermittent exposures of dark grey and brownish grey shelly mudstones (Plate 5, 13) and oil shales, locally with cementstone nodules, along the Darknoll Brook [8031 1123 to 8048 1276], between Okeford Fitzpaine and Fiddleford (Figure 42, locality 19). The strata range from the Mutabilis Zone in the north to the Elegans and Scitulus zones in the south. The exposed sequence is described by Bristow and Cox (1991), and is summarised here.

The Nannocardioceras Cementstone (Plate 5, 11), with *Lingula, Isocyprina minuscula, Protocardia* and *A. (Nannocardioceras)* in solid preservation, was found at one point [8066 1169]. Farther upstream, a bed of septarian nodules, overlain by a 'cemented' oil shale, undulates above and below stream level over a distance of 40 m [8036 1130 to 8031 1123]. At the eastern end, the oil shale yielded the gastropod *Semisolarium hallami* and small '*Astarte*'. From the succeeding brownish grey oil shale, small trochiform gastropods, '*Astarte*' and *Nanogyra virgula* were collected. A similar sequence and fauna with, in addition, *Pectinatites* was found at the western end of the section. The cementstone band corresponds to the Yellow Ledge Stone Band (basal Scitulus Zone and base KC 37).

A well at Okeford Fitzpaine [c.8110 1097] (Figure 42, locality 20) proved 5.1 m of Kimmeridge Clay, beneath 0.9 to 1.8 m of made ground (White, 1923, p.40). The Kimmeridge Clay is described as dark blue-grey shale with seams of shells, and a bed of black and white septaria 3.7 m below the top of the well. The fossils include *Protocardia*, common tiny '*Astarte*' and iridescent ammonite fragments showing bifurcate ribs (*Pectinatites?*); they possibly indicate KC 35 (Autissiodorensis Zone).

A river cliff [8261 1251] (Figure 42, locality 21) on the east bank of the River Stour near Child Okeford exposes about 2.5 m of brownish grey, fissile shelly mudstone (oil shale) with small bivalves including *Isocyprina minuscula*, small *Liostrea, Protocardia, Aulacostephanus* ex gr. *mammatus, A.* ex gr. *pseudomutabilis – fallax*, 'perisphinctid' fragment, fish scales and shell fragments, debris and spat. The fauna and lithology indicate the Autissiodorensis Zone (possibly KC 35) (Bristow and Cox, 1991).

Upper Kimmeridge Clay

There are several small exposures [7982 1076; 7986 1082; 7985 1084] in dark grey, shelly mudstones in the stream bed south of Etheridge Farm, Okeford Fitzpaine. At the first locality (Figure 42, locality 22), the strata appear to be affected by a north-west-trending fault and dip 60°SW. A slab of shelly oil shale from this locality was covered with plasters of *Isocyprina minuscula, Semisolarium hallami*, small pyritised *Camptonectes* and poorly preserved *Pectinatites*. The fauna and lithology suggest the Scitulus Zone (possibly KC 37). A similar horizon was recorded in the Darknoll Brook [8036 1130 to 8031 1123].

Samples of shelly and very shelly mudstone obtained from beneath the Lower Greensand between 9.1 and 11.3 m depth in the Child Okeford Borehole [8358 1330] (Figure 42, locality 23) yielded '*Procerithium*', '*Discinisca*' [= *Pseudorhytidopilus*], *Corbulomima suprajurensis* (common), *Isocyprina minuscula* (common), *Nanogyra virgula* (rare), *Protocardia morinica, Pectinatites* fragment and a fish scale. This assemblage suggests a level low in the Upper Kimmeridge Clay, possibly KC 36 (Elegans Zone).

In the Bedchester Borehole [8567 1792] (Figure 42, locality 24), samples from beneath the Lower Greensand between 7.0 and 9.95 m depth showed medium to pale grey, calcareous, silty, sparsely to moderately shelly mudstone with medium-sized infaunal bivalves including an arcid and lucinid, together with *Corbulomima?*, a nuculoid and poorly preserved *Pectinatites*. The basal 0.45 m showed much less silty, smooth-textured, sparsely shelly or barren mudstone with occasional dull pyritic trails or shell fragments. This assemblage is suggestive of a high level in the Upper Kimmeridge Clay, probably the Pectinatus Zone.

An exposure in the bed of the Twyford Brook [8571 1900] (Figure 42, locality 25) reveals fissile, medium grey, weathering brown, moderately shelly mudstone and quasi-oil shale, with common '*Discinisca*' [= *Pseudorhytidopilus*] *latissima*, several *Quadrinervus ornatus*, bivalves including *Isocyprina minuscula, Mesomiltha concinna?* and *Protocardia morinica*, indeterminate ammonite fragments (presumed pectinatitid), fish scales and fragments, shell fragments/debris/spat and foram-spotting. The fauna and lithology indicate the Upper Kimmeridge Clay, probably Wheatleyensis Zone at about the level of KC 42.

An excavation [8485 2028] at Guy's Marsh sewerage farm, Shaftesbury (Figure 42, locality 26), showed:

	Thickness m
KIMMERIDGE CLAY	
Clay, mottled orange and grey	2.00
Mudstone, greyish brown, shelly, bituminous [oil shale]	2.00
Cementstone doggers, septarian (?dicey weathering)	0.15
Mudstone, calcareous, pale grey, shelly	1.00

Material from the spoil heaps, mainly from the cementstone doggers, yielded whorl fragments of large, iridescent, thick-shelled pectinatitid ammonites. A sample from the lowest clay was devoid of calcareous microfossils, but yielded fairly common cephalopod hooks, together with abundant framboidal pyrite, rare chips of black wood and fish debris. This assemblage, without calcareous microfossils, suggests a disaerobic facies approaching an oil shale and could be accommodated in the lower part of the Upper Kimmeridge Clay (Elegans, Scitulus or lower Wheatleyensis zones). The cementstone doggers could be the Grey Ledge Stone Band (in KC 40) of the Dorset coastal sections (Figure 43).

A section [8619 2112] in a bank at Gear's Mill to the south of Shaftesbury (Figure 42, locality 27) exposes a total of about 2.2 m of pale to medium grey shelly mudstone with thin interbeds of brown bituminous mudstone. At the base of the section is a 0.5 m thick bed of soft coccolith-rich limestone with a pale weathering patina. The high coccolith content has been confirmed by examination using a scanning electron microscope (Plate 5, 12). The coccolith-rich limestone yielded *Protocardia*, oyster? and ammonite fragments. It is almost certainly the White Stone Band (basal Pectinatus Zone; basal KC 46).

A stream section east of Holm and Ivy Farm to the south of Shaftesbury has poor exposures in medium and dark grey shelly mudstones [8577 2144 to 8597 2162] (Figure 42, locality 28). Brown bituminous mudstones are locally present [8595 2161]. Impressions of relatively finely ribbed, indeterminate*Pectinatites* fragments were found in some of the mudstones [8586 2162].

The Church Farm No. 2 Borehole [8555 2223], Shaftesbury (Figure 42, locality 29), penetrated c.7 m of Upper Kimmeridge Clay beneath a phosphatic nodule bed at the base of the Lower Greensand at 27.22 m. Specimens of *Pectinatites* (*Pectinatites*) at 27.40 m and 34.25 m strongly suggest that the Kimmeridge Clay lies within the Pectinatus Zone. As well as sand laminae and burrowfills, the mudstones include a weak oil shale parting and a cementstone; other macrofauna includes *Lingula, Pseudorhytidopilus* and *Protocardia*. Foraminifera, including flood proportions of *Trochammina globigeriniformis* (at 31.50 to 33.50 m) and *Ammobaculites agglutinans* (at 27.50 to 28.50 m), also indicate the Pectinatus Zone with the highest beds no younger than the early Pallasioides Zone.

FIVE

Lower Cretaceous

Cretaceous rocks crop out principally in the eastern part of the district, but there is a small outcrop in the extreme south-west; these rocks cover almost half the map area. The Lower Cretaceous rocks present in the district comprise, in ascending sequence, the Lower Greensand, Gault and Upper Greensand formations.

LOWER GREENSAND

The Lower Greensand was laid down during a series of marine transgressions that successively overstepped westwards across the eroded top of the Jurassic formations (Figure 44).

The main outcrop of the Lower Greensand is between Child Okeford and Cann, just south of Shaftesbury. North of Cann, c.4 m of Lower Greensand has been proved in a borehole [8555 2223] at Shaftesbury, but it has not been possible to trace the outcrop beneath landslipped ground. West of the River Stour, only small scattered pockets of Lower Greensand are preserved beneath the Gault.

The Lower Greensand is divided into mappable members: a lower unit, the Child Okeford Sands, of fine- to very fine-grained, poorly sorted, glauconitic sand, with some beds of medium-grained sand, clayey beds and thin, ferruginously cemented beds; and an upper unit, the Bedchester Sands, consisting of very clayey, fine-grained, poorly to very poorly sorted, glauconitic sand, or of very fine-grained sandy clay. Locally, such as south-west of Child Okeford, the Bedchester Sands overstep the Child Okeford Sands to rest in pockets on the Kimmeridge Clay. In turn, in the same general area, the Gault oversteps the Bedchester Sands to rest on Kimmeridge Clay. The base of the Gault is usually marked by a pebble bed; where this is absent or has not been recognised, it is probable that the Bedchester Sands have been included with the Gault on the geological map.

The maximum thickness of the Lower Greensand is estimated to be about 10 m in the tract north-east of Child Okeford. Near Bedchester [855 175], the thickness probably does not exceed 8m; between Twyford and Cann, it is about 5 m thick. In the scattered outliers south-west of Child Okeford, the maximum thickness is about 3 m.

The age of the Lower Greensand in the district is uncertain. Three samples from the Bedchester Sands yielded a sparse fauna of the agglutinating foraminifera *Trochammina concava* (abundant), *Reophax minuta* (rare), *Rhizammina* cf. *dichotoma, Glomospira gaultina, Haplophragmoides chapmani* and *Textularia* sp., typical of the latest Aptian (*jacobi* Zone) and early Albian (*tardefurcata* and *mammillatum* zones). The absence of *Arenobulimina*, which first appears in the *D. mammillatum* Zone, however, suggests that none of the assemblages is younger than the *tardefurcata* Zone. The fauna is similar to that described from the foraminiferal assemblage zone 1 (*tardefurcata* Zone) of Price (1977).

The biostratigraphy of the Lower Greensand and basal Gault in the Okeford Fitzpaine brickpit [815 109] was reviewed by Owen (1971). He concluded that the 0.45 m of brown sandy ironstone beneath the pebbly basal bed of the Gault, which by its description resembles the Bedchester Sands, was of Lower Albian, *kitchini* Subzone age. The 0.45 m of 'clean' green sand beneath probably corresponds to the Child Okeford Sands. A section at Dinton [c.SU 010 318], Wiltshire, in the Salisbury district, appears to have exposed a similar sequence to that in the Shaftesbury area (Casey, 1956). There, 0.76 m of bluish grey, sandy, ferruginous, fossiliferous clay (?Bedchester Sands) separates the Gault from the Kimmeridge Clay. The fauna from the ?Bedchester Sands indicates the *kitchini* Subzone.

The Child Okeford Sands have not been dated. It is assumed that they are of Aptian age.

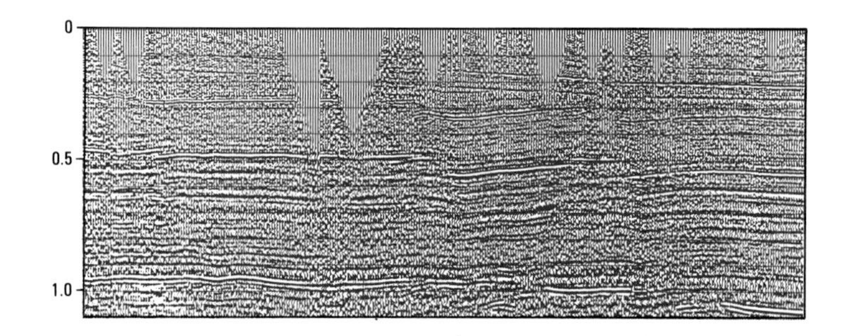

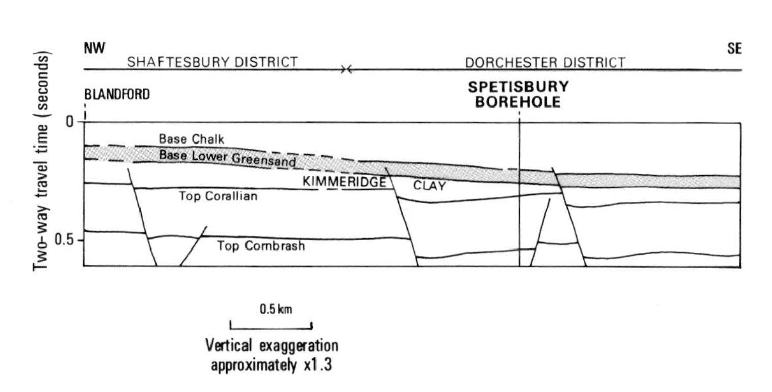

Figure 44 Seismic section showing truncation of Jurassic strata beneath the sub-Cretaceous unconformity (see Figure 4 for location).

DETAILS

Higher Ansty and Stoke Wake

Auger holes locally proved the Bedchester Sands, but it was not possible to map the member. One hole [7653 0491] near Higher Ansty showed 0.2 m of greenish grey to organic brown, micaceous, clayey, fine-grained sand, below 2.1 m of drift and Gault.

The Bedchester Sands may also occur near Stoke Wake, where an auger hole [7620 0556] proved 2.3 m of glauconitic, extremely sandy clay to clayey fine-grained sand; however, because the characteristic basal Gault grit was not encountered, these fine-grained sands could belong to the Gault.

Belchalwell–Okeford Fitzpaine area

An auger hole [8024 1141] 150 m east-north-east of Etheridge Farm, Okeford Fitzpaine, proved the following sequence:

	Thickness m
HEAD	
Sand, fine-grained, clayey, with scattered flints	0.6
LOWER GREENSAND (Bedchester Sands)	
Clay, sandy, fine-grained	1.4
Clay, micaceous, dark grey, fine-grained, sandy, becoming less sandy downwards	2.8

A sample from the basal clay yielded the foraminifera *Trochammina concava* (common), *Glomospira gaultina, Reophax minuta, Haplophragmoides chapmani* and rare *Textularia* sp., indicative of the latest Aptian (*jacobi* Zone) or early Albian (*tardefurcata* Zone).

Conygar Coppice [813 120], north-east of Okeford Fitzpaine, is formed of an outlier of Gault, with pockets of Bedchester Sands beneath it resting on an irregular surface of Kimmeridge Clay. Three out of five boreholes proved up to 3.1 m of Lower Greensand beneath Gault. The first [8114 1205] passed out of the Gault at a depth of 7.2 m, and proved 2.2 m of stiff bluish grey clay with sand partings, on 0.9 m of soft, bluish grey, silty, clayey sand, above Kimmeridge Clay. Lower Greensand was absent in a second borehole [8112 1198] 70 m to the south. Another borehole [8125 1195] passed through 3 m of Gault with a 'gravelly' base into 2.7 m of greyish brown silty clay with pockets of brown sandy clay, above Kimmeridge Clay.

Between 0.3 and 0.9 m of 'clean' green sand (Child Okeford Sands), beneath 0.46 m of brown sandy ironstone (Bedchester Sands) was recorded between the Gault and Kimmeridge Clay in the Okeford Fitzpaine brickyard [815 109] (White, 1923).

Child Okeford–Farrington

Representative auger holes in the Lower Greensand [8309 1337] in the tract between Child Okeford and Farrington are given below:

	Depth m
LOWER GREENSAND	
Bedchester Sands	
Sand, fine- to medium-grained, orange-brown	0.60
Sand, very clayey, ferruginous, orange-brown, with scattered well-rounded pebbles up to 1 cm	1.65
Clay, fine-grained sandy, micaceous, laminated, brown	2.25
Child Okeford Sands	
Sand, fine-grained, glauconitic	2.35

A sample from the clay yielded rare specimens of the foraminifera *Glomospira gaultina.*

A borehole [8358 1330], 0.5 km to the east, proved:

	Thickness m	*Depth* m
GAULT		
Clay, sandy, orange-grey; coarse sand grains and rounded pebbles up to 5 mm across at 1.45 m, and up to 10 mm at 2 m	2.00	2.00
LOWER GREENSAND		
Bedchester Sands		
Sand, very fine-grained, glauconitic, clayey	0.10	2.10
Sand, fine-grained, very clayey, glauconitic, becoming less clayey at 3.3 m; very clayey at 3.9 to 4.2 m	2.30	4.40
Child Okeford Sands		
Sandstone, hard	0.30	4.70
Sand, soft, clayey, glauconitic, waterlogged	3.90	8.60
KIMMERIDGE CLAY		
Mudstone, dark grey, shelly	0.50	9.10

An auger hole [8385 1488] 1.5 km to the north proved 1.1 m of brownish grey sand, finely interlaminated with very fine-grained sand, above Kimmeridge Clay. Foraminifera from these laminated strata included abundant *Trochammina concava*, rare *Reophax minuta* and very rare *Rhizammina* cf. *dichotoma*, indicative of the latest Aptian *jacobi* Zone or early Albian *tardefurcata* Zone.

Farrington–Cann

Jukes-Browne (1891) had a trench [c.8568 1702] dug in the Lower Greensand near Piper's Mill which proved:

	Thickness m
Brown loamy soil	0.30–0.61
GAULT	
Clay, mottled brown and grey containing, in the lower part, pebbles of vein quartz and lydianite as large a beans	0.61
LOWER GREENSAND	
Bedchester Sands	
Sand, greenish brown with clay mottles	0.61
Clay, soft, purple-brown (0.3 m) passing into dark green sandy clay with patches of greenish brown sand, and finally into mottled brown, yellow and green sand	1.52
Clay and sand, laminated, purple-black	0.30
Clay, greenish black, full of glauconite grains	0.76
Sandstone, hard, brown, cemented with iron oxide	0.15
Child Okeford Sands	
Sand, rather coarse-grained, yellowish brown	0.46
Sand, fine-grained, greenish grey	0.61

Ferruginous fine-grained sandstone floors the Fontmell Brook [8585 1703], just east of Piper's Mill, and its tributary [8570 1720; 8566 1732] to the north of the mill. To the south of Piper's Mill, a small exposure [8555 1685] showed glauconitic, clayey, fine-grained sand with ferruginous boxstones.

North of Piper's Mill, Jukes-Browne (1891) saw the following section in the Child Okeford Sands in a sandpit [8558 1763]: ferruginous and greenish brown sand, with two layers of coarse-grained yellow sand (no thickness given), on 1.82 m of fine-grained, greenish brown sand, on 1.52 to 1.82 m of dark green glauconitic sand. Bedchester Sands crop out north-west of the pit. A borehole [8567 1792] in this area proved:

	Thickness m
HEAD	
Clay, sandy, stony, orange-brown	1.60
LOWER GREENSAND	
Bedchester Sands	
Sand, clayey, glauconitic	2.30
Child Okeford Sands	
Sand, glauconitic, dark grey	6.60
KIMMERIDGE CLAY	
Mudstone, dark grey, stiff	9.95

Near Twyford, the outcrop of the Lower Greensand probably corresponds only to the Child Okeford Sands, the Bedchester Sands being thin, absent or included with the Gault. The formation maintains a fairly constant thickness of about 5 m. An auger hole [8576 1890] proved, beneath 1.1 m of Gault, 0.2 m of fine- to medium-grained sand, on 0.1 m of coarse-grained glauconitic sand, on more than 0.9 m of clayey, greyish brown, glauconitic sand.

At Hartgrove, there is a small outlier of Gault capping a thin bed of Lower Greensand. A pit [8389 1819] at Hartgrove Farm encountered:

	Thickness m
Topsoil	0.3
GAULT	
Clay, silty, mottled orange and grey, passing down into medium grey clay	1.2
LOWER GREENSAND (Bedchester Sands)	
Sandstone, fine-grained, ferruginous, nodular	0.3
Sand, mostly fine-grained, but with some lenses of poorly sorted, coarse-grained sand	0.3
Sand, clayey, fine-grained, glauconitic, with purplish brown thin clay beds and lenses	0.3
Sand, fine-grained, clayey, bioturbated, glauconitic	0.7
KIMMERIDGE CLAY	
Clay, dark greyish brown, brecciated	0.3
Clay, dark grey, finely bedded	0.7

Cann–Shaftesbury area

Because of landslips, there is little undisturbed outcrop of the Lower Greensand in this tract. The Church Farm Borehole (Figure 45) [8555 2223], Shaftesbury, proved 4.36 m of fine-grained, glauconitic, burrowed and bioturbated sand; towards the base, there is a 0.99 m thick bed of micaceous, glauconitic, fine-grained sandy silt. A sample from this bed yielded the foraminifer *Trochammina* aff. *wetteri* (a species which occurs throughout much of the overlying Gault). The basal bed consisted of a phosphatic nodule bed resting on the burrowed top of the Kimmeridge Clay.

GAULT

The Gault was laid down during the Albian transgression, which swept across the district from the east. The deposits were laid down on a surface of tilted and eroded Upper Jurassic strata, such that the base rests on progressively older strata as the outcrop is followed south-westwards. In the north-east of the district, the Gault rests on Upper Kimmeridge Clay, but gradually cuts down onto the Lower Kimmeridge Clay near Okeford Fitzpaine, to rest on Corallian strata in the south-west near Lyon's Gate.

The base of the Gault is marked by a pebbly gritty sandstone, up to about 2.5 m thick in the central tract near Okeford Fitzpaine, but only 0.1 m thick over much of the outcrop. The pebbles consist of small (up to 1 cm), well-rounded clasts of quartz, quartzite and lydite. They may be scattered, or concentrated into a bed up to 10 cm thick; the latter type is well developed around Okeford Fitzpaine. At the old Okeford Fitzpaine brickpit [815 109], the basal bed comprises c.1.5 m of brown, yellowish and bluish grey, micaceous sand, partly ferruginous and oolitic, and containing many scattered small siliceous pebbles (White, 1923). South-west of Okeford Fitzpaine, the basal pebble bed passes into a coarse-grained gritty sand.

The bulk of the Gault is made up of clayey, fine-grained micaceous, patchily glauconitic, sandy silts, and clayey fine-grained sands and sandstone. Locally, the sands and sandstones are thick enough to be mapped and named separately, e.g. the Fontmell Magna Sand (Bristow and Owen, 1991). Work by Bloodworth (1990) on the clay mineralogy of the Gault in the Church Farm Borehole [8555 2223] at Shaftesbury shows that the clay fraction is dominated by smectite (5 to 38 per cent, but averaging about 22 per cent), with subordinate mica and minor kaolinite (Figure 45). The proportion of kaolinite relative to smectite and mica increases with depth; quartz, feldspar and cristobalite were detected in all the samples examined. The quartz-feldspar-smectite-mica-kaolinite assemblage is typical of the Gault and overlying Upper Greensand (Perrin, 1971). Increasing levels of smectite towards the top of the Gault have been noted elsewhere in southern England, but not from the western part of the Wessex Basin (Jeans et al., 1982, fig.14). In this latter area, the clay mineralogy of the upper part of the Gault is dominated by mica-kaolinite. The smectite is thought to result from an increase in volcanogenic input (Jeans et al., 1982).

The formation is thinnest in the south-west between Lyon's Gate and Buckland Newton, where it is 12 to 14 m thick. Between Stoke Wake and Okeford Fitzpaine, the maximum thickness is about 15 m, although at Conygar Coppice [813 118], north of Okeford Fitzpaine, it is at least 20 m thick. It continues to thicken north-north-eastwards; at Fontmell Magna, a borehole [8672 1657] proved 23 m of Gault, and at Pen Hill [851 167], it is estimated at 25 m. Northwards, the thickness decreases, and only 17 m were proved in the Church Farm Borehole, Shaftesbury [8555 2223] (Figure 45).

The age of the formation is poorly documented in the district (Figure 45). At Okeford Fitzpaine, a brick pit [815 109] formerly exposed 8.7 m of the Lower Gault. The fauna was re-examined by Owen (1971), who concluded that a specimen of *Douvilleiceras inaequinodum* (= *Acanthoceras mammillatum* of Newton

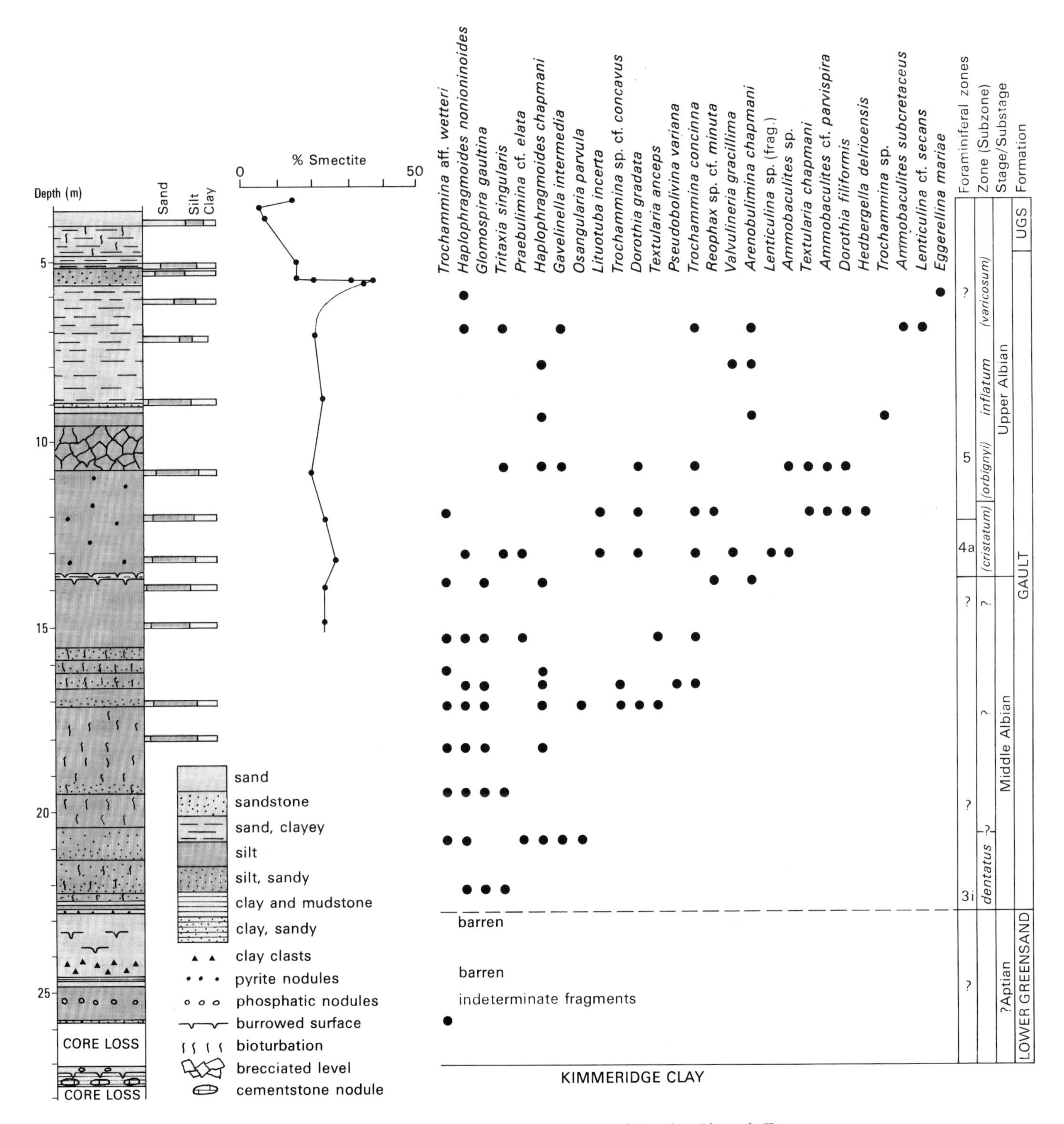

Figure 45 Stratigraphy, clay mineralogy and foraminifera of the Gault in the Church Farm Borehole, Shaftesbury.

(1897), as redetermined by Spath (1925, p.73)) came from the basal bed of the Gault as here defined (see p.94). Beds 4 m above the base were identified as of *lyelli* Subzone age, and the succeeding strata, which yielded no fauna, were regarded as falling in the *spathi* Subzone (Owen, 1971).

The Gault in the Church Farm Borehole [8555 2223] yielded a sparse foraminiferal fauna. The oldest sample dated, 0.7 m above the base, is no older that the *dentatus* Zone. The Middle/Upper Albian boundary can be drawn c.9 m above the base of the Gault; it is based on the presence of *Arenobulimina chapmani*, which first appears 8.76 m above the base of the Gault in the borehole (Figure 45). *A. chapmani* is unknown below the base of the *cristatum* Subzone. The sample came from just below a prominent erosion surface, with fine-grained, sand-filled burrows of the overlying bed piped down into the clayey silt of the sample. It seems probable that the erosion surface coincides with the Middle/Upper Albian boundary, and that the specimen of *A. chapmani* was brought down into its present position by burrowing or bioturbation. Samples c.12.8 m above the base indicate the *orbignyi* Subzone or younger; there is no evidence of the *dispar* Zone.

At Fontmell Magna, a temporary exposure [8670 1708] in clayey, fine-grained, micaceous sand and sandstone of the Fontmell Magna Sand, yielded a fauna of *inflatum* Zone (late *orbignyi* to early *varicosum* Subzone) age (Bristow and Owen, 1991).

North-east of the district, at Fovant, a section [SU002 290] showed 3.6 m of 'silty rock' forming a transition from the Gault to the Upper Greensand (Mottram, 1957). Fossils collected from the uppermost Gault at this locality indicate the *inflatum* Zone, *varicosum* Subzone. The lower part of the Upper Greensand at Winterborne Kingston has yielded ammonites indicative of the same subzone (Morter, 1982), and hence the Gault/Upper Greensand junction falls within it.

Details

Buckland Newton–Higher Ansty

West of Buckland Newton, much of the Gault is obscured by landslip. North of Buckland Newton, an auger hole [6895 1083] proved, beneath 1.2 m of gravelly sandy clay, brownish and greenish grey, becoming dark grey to olive-grey, silty, micaceous, clayey, fine-grained sandy to extremely sandy clay down to a depth of 4.3 m.

An auger hole [7649 0435] north-west of Higher Ansty showed the following section:

	Depth m
GAULT	
Clay, greenish and yellowish grey, very sandy	1.90
Grit, quartzose, ferruginous; grains to 5 mm in size	2.00
LOWER GREENSAND	
Clay, very sandy, brownish grey, micaceous	3.00
Sand, fine-grained, clayey, lilac-brown, micaceous	3.30

Farther north, an auger hole [7653 0491] showed 1 m of gravelly clay on 2 m of brownish olive-grey, very sandy, micaceous clay, on 0.1 m of clayey sandy grit with rounded quartz grains, on clayey silty sand of the Lower Greensand.

Stoke Wake–Belchalwell

In an auger hole [7702 0692], east of Stoke Wake, the basal quartz grit was overlain by silty, yellowish brown clay.

Around Belchalwell, most auger holes proved grey, silty to very silty, rather micaceous clay with some clayey silt [e.g. 7920 0996]:

	Depth m
GAULT	
Clay, brownish grey, slightly sandy	2.2
Grit, clayey, quartz grains	2.3
LOWER GREENSAND	
Clay, dark grey, fine-grained sandy, with ferruginous layers; becomes blocky fractured towards base	3.2
KIMMERIDGE CLAY	
Clay, brown, carbonaceous	3.8
Shale, black to very dark grey	4.0

Grit was seen in debris from a pond [7957 0972]; sieving showed the mean grain size to be 0.37ø, with a sorting of 1.22ø.

Two auger holes [7951 1063; 7963 1062] proved essentially similar sequences in the basal beds of the Gault and the top of the Lower Greensand on the outlier north-east of Stroud Farm, Belchalwell (Bristow, 1989b). In the first, the following succession was proved:

	Depth m
GAULT	
Sand, fine-grained, orange-brown, with scattered small pebbles	0.50
Sand, clayey, with grit	0.80
Grit, clayey, slightly micaceous, with scattered rounded pebbles of quartzite up to 2 cm across at base	1.20
LOWER GREENSAND	
Bedchester Sands	
Sand, fine-grained, glauconitic, clayey, micaceous, ferruginous	1.30
Clay, sandy, fine-grained, medium grey, slightly micaceous, laminated	1.80
Clay, sandy, gritty, orange-brown	1.85
Clay, stiff, dark grey, finely sandy, mottled orange-buff	2.40
Clay, dark grey, finely sandy, micaceous, bioturbated	3.15

There are exposures of the basal gritty bed in the stream bed to the east of the above auger holes [e.g. 7987 1085].

Okeford Fitzpaine

Near Etheridge Farm, an auger hole [7999 1133] proved 1 m of brown sandy clay, resting on 0.1 m of coarse- to very coarse-grained clayey sand, resting on Kimmeridge Clay. About 100 m south, another auger hole [7998 1122] proved 1 m of coarse- to very coarse-grained, weakly glauconitic sand at the base of the Gault, on orange-buff, clayey, fine-grained sand. To the south-east, a section [8018 1109] showed 0.4 m of glauconitic, fine-grained, sandy clay, with a 0.2 m basal bed of fine-grained ferruginous sandstone, resting on Kimmeridge Clay.

In the bank [8035 1133] of the Darknoll Brook, there is a loose 0.15 m thick block of the basal bed of the Gault, which consists of a coarse-grained, pebbly, ferruginous sandstone.

A section at the former Okeford Fitzpaine brickyard [815 109] was described by Newton (1897), Jukes-Browne and Hill (1900), White (1923) and Owen (1971). The following is a composite section based on the above authors:

	Thickness m
SUBSOIL	0.3
HEAD	
Clay, yellow, with flint and chert pebbles, and chalk rubble	0.9
GAULT	
Clay, brown	1.5
Clay, sandy, micaceous, dark grey, with phosphatic nodules	3.3 to 4.5

	Thickness m
Clay, silty, dark grey, shelly, glauconitic	1.2
Sandstone, ferruginous, dark brown, fossiliferous in upper part	1.2
Loam, micaceous, clayey, brown, greenish grey and yellow, with partly phosphatised fossils; small pebbles, particularly in lower part	1.5
LOWER GREENSAND	
Bedchester Sands	
Ironstone, brown sandy, ?oolitic	0.46
Child Okeford Sands	
Sand, green, clean	0.3 to 0.9
KIMMERIDGE CLAY	
Clay, stiff, blue, with a bed of hard sandy rock	2.4 or 3.0

The pebbly, clayey, micaceous sand was originally assigned to the *mammillatum* Zone on the strength of the zone fossil, but Spath (1925) redetermined the '*Acanthoceras mammillatum*' as *Douvilleiceras inaequinodum* [= *eodentatus* Subzone (Owen, 1971)]. Work by Owen (1971) established the presence of the succeeding *lyelli* Subzone in the dark grey, glauconitic, shelly clay; the higher *spathi* Subzone was inferred for the succeeding strata.

There is an outlier of Gault at Conygar Coppice [812 119], to the north-east of Okeford Fitzpaine. The basal grit bed was found at a number of localities, particularly on the western and northern sides of the outlier. Five boreholes on this outlier commenced in the Gault and terminated in Kimmeridge Clay. The basal pebbly bed was described by the drillers as 0.2 to 0.9 m of silty clay with some fine-grained sand and 'gravel'.

Child Okeford–Fontmell Magna

A representative auger hole [8421 1450] in the Gault and Lower Greensand in the tract north-east of Child Okeford is given below:

	Depth m
HEAD	
Sand, fine-grained, orange	0.7
Sand, clayey, orange-brown	1.0
Sand, clayey, with small well-rounded pebbles	1.1
GAULT	
Clay, very sandy, orange-brown, passing down into greyish brown, laminated, very sandy clay; small well-rounded pebbles at 1.8 m	1.8
LOWER GREENSAND	
Bedchester Sands	
Sand, fine-grained, very clayey, brownish grey	2.1

The old brickpit [8540 1635], now known as the Homestead, to the north-west of Sutton Waldron, was described by Jukes-Browne and Hill (1900) (here metricated):

	Thickness m
Clay, grey and brown, sandy, micaceous	1.52
Ferruginous rubble, loose, red and brown	0.10
Clay, mottled grey and light brown, with patches of yellow micaceous sand and glauconitic sand	0.61
Clay, dark reddish brown, micaceous and sandy, with many phosphatic septaria	0.46
Clay, sandy, dark greenish grey	0.61
Clay, brown, tough	0.22
Sandy loam, dark grey	0.76

White (1923) recorded a glauconitic sandy loam and ferruginous concretions in a shallow cutting in this pit. One of the concretions contained a cast of '*Inoceramus* cf. *anglicus*'.

Farther north, White (1923) noted the occurrence of '*Syncyclonema orbicularis*' in an ironstone with phosphatic nodules in the basal Gault on the west side of the Fontmell Brook [c.856 171].

Jukes-Browne and Hill (1900) saw a temporary section [8669 1705] in the Fontmell Magna Sands at the northern end of Fontmell Magna, but erroneously thought that the deposits were part of the Upper Greensand. A nearby temporary section [8670 1708] in 1989 revealed poor sections, up to 1 m high, in bluish grey, clayey, fine-grained, micaceous sand and very sandy clay. The fauna includes *Birostrina sulcata, Entolium orbiculare, Neithea syriaca, Limaria gaultina, Leptosolen* sp. and *Jurassiphorus granosus,* indicative of a late *orbignyi* Subzone age. On the surface of the site were several blocks of yellowish brown, fine-grained shelly sandstone, and some harder nodules of grey sandstone. The exact stratigraphical position of these blocks was not determined, but their fauna suggests that they had come from slightly higher excavations, above the bluish grey sand and sandy clay. The fauna includes *Birostrina subsulcata, ?Oxytoma* or *Meleagrinella* sp., *Entolium orbiculare, Limaria gaultina, Plagiostoma globosum* and *Hysteroceras binum,* indicative of the earliest part of the *varicosum* Subzone (Bristow and Owen, 1991).

Fontmell Magna–Shaftesbury

There are several exposures [8694 1759; 8696 1740; 8690 1727; 1785 1717] along the Fontmell Brook in clayey, thinly bedded, fine-grained, locally shelly sandstone.

The pebbly basal bed of the Gault can be traced across the fields to the west of Sutton Waldron [8567 1752 to 8571 1794]. The pebble bed was also encountered in auger holes [8568 1948; 8543 1928] at Twyford, and on the surface of the field [8542 2004], 500 m west of Allan's Farm.

In the outcrop just south of Gear's Mill [861 208], ferruginous sandy clay at the base of the Gault was commonly augered; one auger hole [8567 2012] encountered:

	Thickness m
Sand, fine-grained, clayey, orange-brown, with one well-rounded pebble	0.5
?LOWER GREENSAND	
?Bedchester Sands	
Clay, sandy, glauconitic, mottled orange and grey	1.0
Clay, sandy, ferruginous, micaceous, glauconitic	0.2
?Child Okeford Sands	
Sand, fine-grained, clayey, micaceous	0.2

The log of the Church Farm Borehole [8555 2223], Shaftesbury, is shown in Figure 45. Nearby [8570 2176], stiff pinkish brown clay was encountered.

Outliers west of Shaftesbury

Some 300 m west of Hawker's Farm, Stour Row, an outlier [816 219] of Gault consisting of glauconitic mottled orange and grey clay, with a sandy and pebbly base, rests on Kimmeridge Clay.

There are three outliers of glauconitic, mottled orange and grey, sandy clay on the east side of East Stour Common [around 817 233], 500 m west of Lower Duncliffe Farm [822 235] and south of Lower Duncliffe Farm [828 233].

UPPER GREENSAND

Jukes-Browne and Hill (1900, pp.157, 159) made a fivefold subdivision of the Upper Greensand in the northeast of the Shaftesbury district and gave broad geographical distributions for the five units. Their sequence (a–e) and the terminology adopted in this account is given in Table 9 and Figure 46.

Bed a was divided into a lower, c.6 m thick, tough, grey, micaceous sandstone, dark and argillaceous near the base, and an upper, c.6 m thick, micaceous sand and sandstone.

The evolution of the terminology of the various members, together with their lithology, thickness and fauna is given below under each member.

The full Upper Greensand sequence is present only in the north of the district, from Compton Abbas northwards (Figure 47). To the south, the chert beds are absent, and in the tract as far as Iwerne Courtney, the Melbury Sandstone rests on the Shaftesbury Sandstone. West of the River Stour, the Melbury Sandstone is represented only by a pebble bed with a phosphatised remanié fauna, which rests on the eroded top of the Shaftesbury Sandstone or the Cann Sand. The pebble bed, with a slightly chalky matrix, persists south-westwards to beyond the sheet margin. It is named the Bookham Conglomerate in this account because it is particularly well developed and exposed at Bookham Farm [7065 0414]. In this south-western tract, a calcareously cemented, glauconitic sandstone underlies the Bookham Conglomerate and forms a marked bench feature. It is regarded as the lateral equivalent of part of the Shaftesbury Sandstone.

The thickness of the Upper Greensand near Shaftesbury was thought to be about 42 to 45 m, whilst on Duncliffe Hill, 3 km west of Shaftesbury, some 60 m was assigned to this formation (White, 1923). It is now known that on Duncliffe Hill, only about 6 m of Upper Greensand are preserved; the gross difference in the quoted thickness is because White did not recognise the extensive landslips that ring the hill. A well [8675 2325] at Shaftesbury, which commenced in the Boyne Hollow Chert, proved 55.47 m of Upper Greensand. To this can be added about 5 m of sand and chert, which crop out between the borehole and the base of the Chalk, to give a total thickness of about 60 m for the Upper Greensand at Shaftesbury.

The Upper Greensand around Shaftesbury is probably close to its maximum development in the region. All the members thin south-westwards, and some, notably the Boyne Hollow Chert, disappear within 5 km of Shaftesbury (Figure 47). Drummond (1970, p.683) thought that the absence of the Boyne Hollow Chert from south of Shaftesbury indicated either non-deposition and/or erosion across the Mid-Dorset Swell.

Facies change, associated with overall thinning, condensation and presumed higher energy environments, could also explain the absence of certain Upper Greensand rocks over the swell. The absence of chert pebbles in the basal Chalk supports the idea of facies change, since chert clasts would be an expected component of the basal bed of the Chalk if the absence of the chert beds is due to erosion. However, Tresise (1960) attributed the absence of chert pebbles in the intraformational pebble beds in the Upper Greensand to the chert still being a silica gel, which disintegrated on exhumation. This could account for their absence from the base of the Melbury Sandstone and from the Bookham Conglomerate.

Following Drummond's interpretion of non-deposition and/or erosion, the effect of the north-eastern margin of the Mid-Dorset Swell on the Upper Greensand between Compton Abbas, the most southerly locality at which the chert beds have been recognised, and Stour Bank is as follows. At Clayesmore School [8642 1466], Iwerne Minster, and Manor Farm [8565 1296], Iwerne Courtney, a pebble bed of *carcitanense* Subzone age at the base of the Melbury Sandstone rests on the ragstone at the top of the Shaftesbury Sandstone. At Stour Bank [846 107] and at Okeford Fitzpaine, the Bookham Conglomerate rests disconformably on the eroded top of the ragstone. At Bookham Farm [7065 0414], the Bookham Conglomerate rests at a lower level on the eroded top of

Table 9 Classification of the Upper Greensand in the north-east of the district.

Present survey		Jukes-Browne and Hill (1900)	
Melbury Sandstone	Thickness (m) 0–6	e. Greensand, fine-grained, glauconitic	Thickness (m) c.3
Boyne Hollow Chert	4–15	d. Chert beds (or Beds) and sands and sandstone	c.6
Shaftesbury Sandstone (Ragstone at top)	10–25	c. Sandstone, fine- to medium-grained, glauconitic; uppermost beds hard and shelly, lower beds only weakly cemented b. Sands, buff and grey, soft	c.3 c.21
Cann Sand	25–30	a. Sandstone and sand, micaceous	c.12

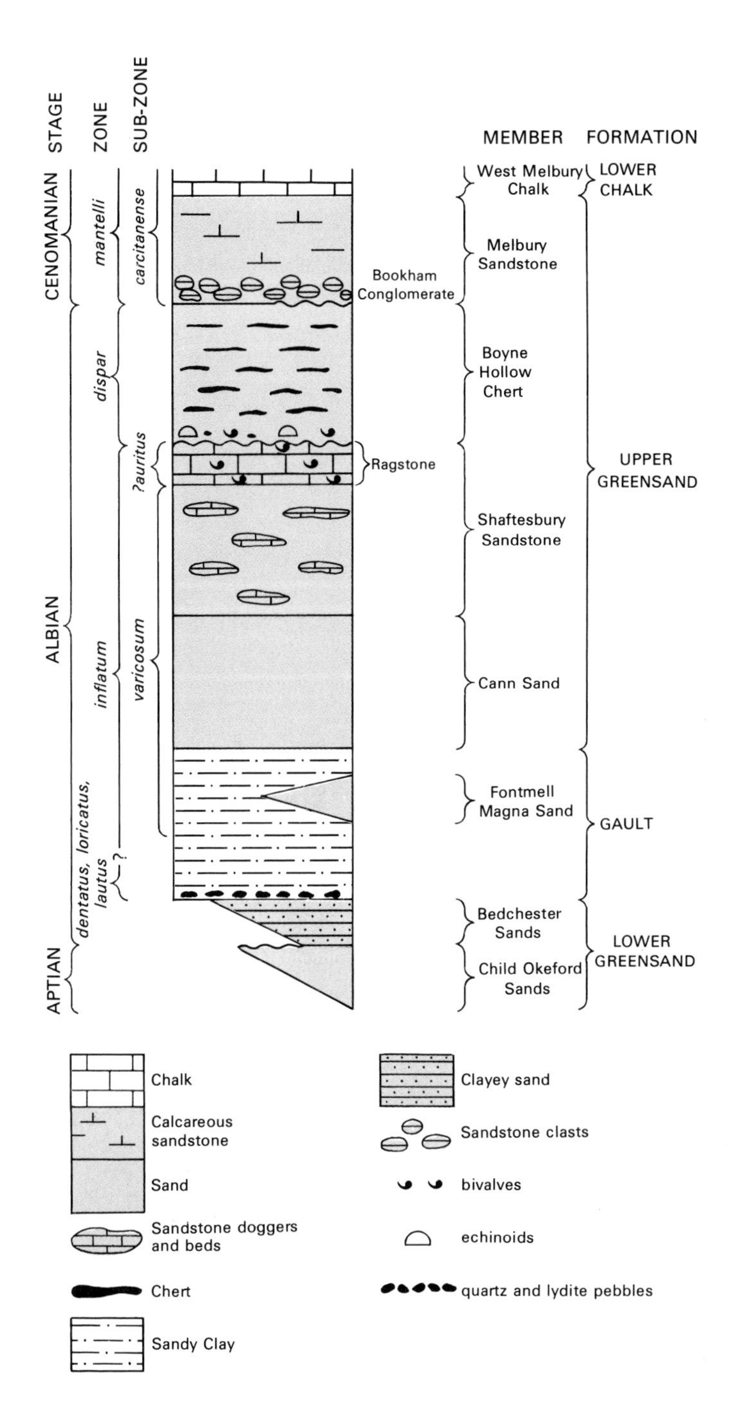

Figure 46 Generalised stratigraphy of the Lower Greensand, Gault and Upper Greensand in the north-eastern part of the Shaftesbury district (not to scale).

the Shaftesbury Sandstone. If, however, the absence of the chert beds is due to facies change, then the sandstone beneath the Melbury Sandstone at Clayesmore School and Manor Farm, and beneath the Bookham Conglomerate at Stour Bank, Okeford Fitzpaine, Bookham Farm etc., would be younger than the ragstone and be of equivalent age (*dispar* Zone) to the chert beds.

The lithology of the Upper Greensand of Wessex has been described by Tresise (1960, 1961); he included few observations from the Shaftesbury district, but his general conclusions are relevant. Tresise (1960, p.237) recognised five types of glauconite within the Upper Greensand:

1. Dark green opaque grains of rounded or botryoidal form.
2. Glassy grains of varying shades of green, fine-grained, commonly of silt grade.
3. Casts of organic remains.
4. Glauconite replacing mineral grains.
5. Pigmentary glauconite staining pebble surfaces.

The dark green opaque grains are the commonest type and characterise glauconite-rich horizons. They usually occur in fine- to medium-grained sands as rounded, ovoid or botryoidal grains. Abrasion is thought to result in the more rounded grains characteristic of the finer grades.

The glassy type occurs as silt or very fine-grained sand grains, which are angular or irregularly rounded, and translucent to transparent. They are characteristic of very fine-grained, silty or only sparsely glauconitic rocks. The glauconite of the Boyne Hollow Chert is of this type.

Glauconitic internal moulds of foraminifera occur, but they are not common. Glauconite also fills pore spaces of echinoids and holothurian fragments, and the axial canals of sponge spicules. The glauconite filling may be of either the glassy or the opaque type.

In the fourth type, glauconite coats various mineral grains such as corroded quartz grains. Feldspars may have slight glauconitisation along cleavage planes. Some minerals, possibly including potash feldspar, may be completely replaced.

Some abraded limestone pebbles have a marginal rim, up to 6 mm thick, stained by glauconite. The impregnating glauconite appears to have replaced calcareous mud, since it occurs as dendritic patches within the groundmass and also infills the chambers of foraminifera. The pebble bed at the base of the Melbury Sandstone is of this type.

Phosphatic nodules, which have a more sporadic distribution in the Upper Greensand than the glauconite, are particularly developed in those beds rich in dark green opaque glauconite. Near Shaftesbury, they have been seen only locally in sands above the Ragstone and within the Boyne Hollow Chert (Jukes-Browne and Hill, 1900). The nodules, usually dark brown or black, but locally pale brown or cream, are rounded, ovoid or slab-like in shape; they are thought to have formed by phosphatisation of calcareous sandstone or of fossils. They commonly have a pitted surface and include pockets of the surrounding sediment (Tresise, 1960, p.330).

In the Upper Greensand of the district, chert is mostly confined to the Boyne Hollow Chert, but scattered small chert nodules occur in the Cann Sand at Bookham Farm in the south-west. The chert nodules of the Boyne Hollow Chert usually consist of a brown, black or grey, massive flinty core surrounded by up to 25 mm of more porous, white, yellow or brown siliceous material; they are up to 0.6 m thick. In places, the massive core is lacking. The beds associated with the chert consist of silt and fine-grained sand, richly calcareous, but non-ferrug-

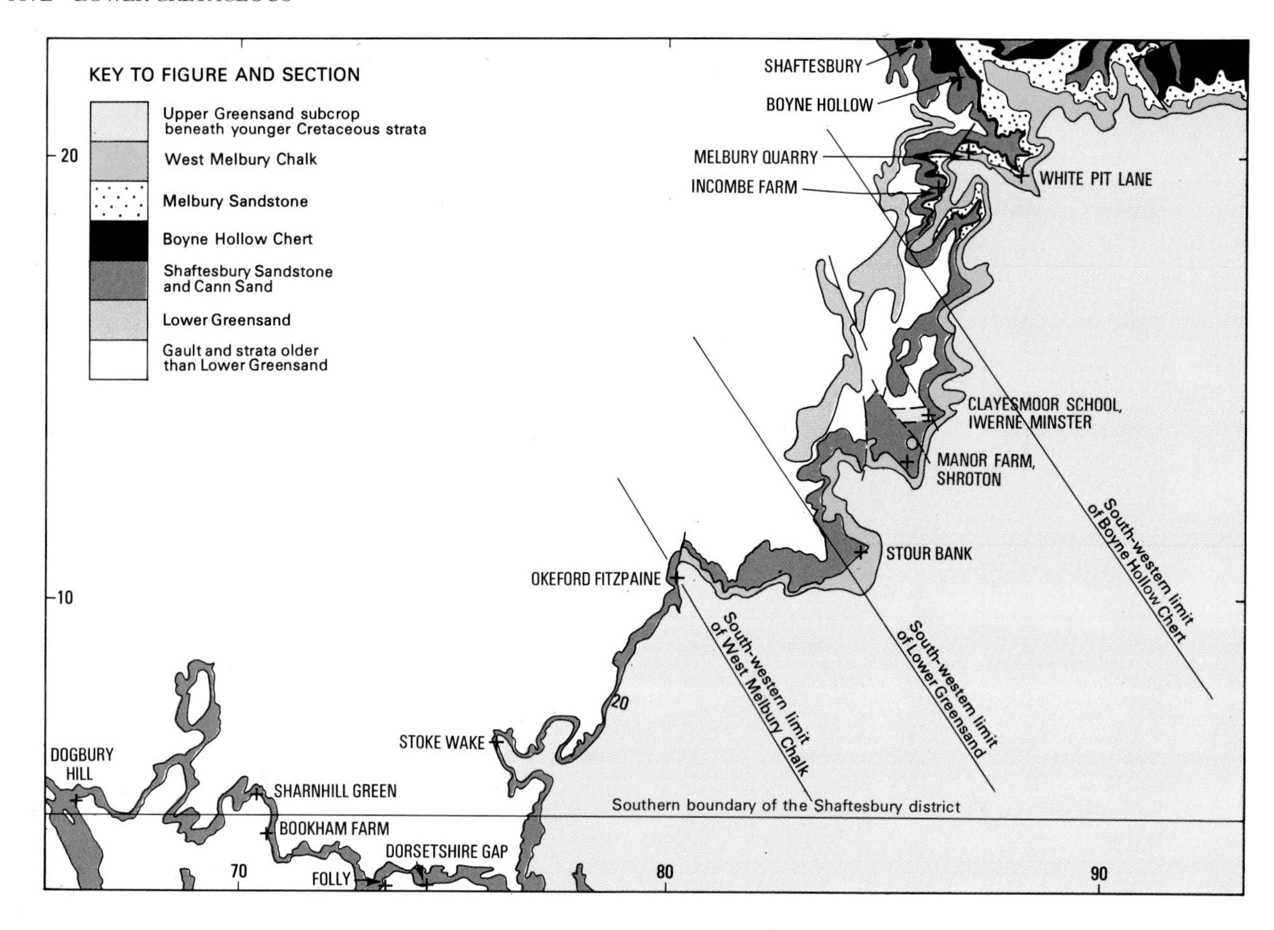

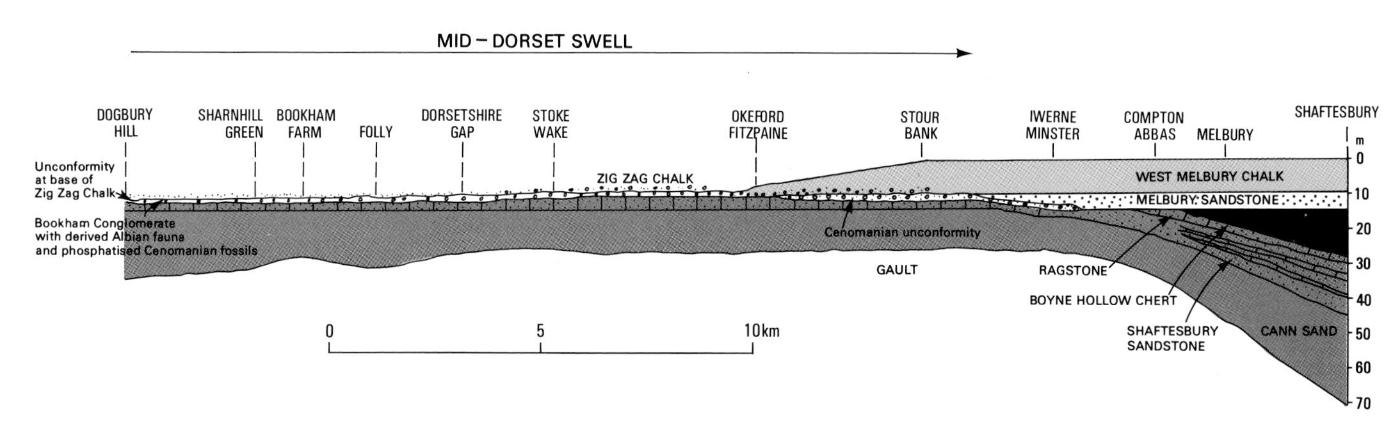

Figure 47 Sketch map and schematic strike section of the outcrop of the Upper Greensand showing the principal localities mentioned in the text.

inous and only sparsely glauconitic. The nature and origin of the Upper Greensand chert in the Wessex Basin is discussed in detail by Tresise (1961).

Within the district, there is little faunal evidence for the age of most of the Upper Greensand, except for part of the Melbury Sandstone. Few ammonites have been found and, for some of them, the exact stratigraphical provenance is uncertain. However, by combining this sparse ammonite evidence with data from western Dorset, where the succession is similar to that around Shaftesbury, and from outcrops farther north, a broad chronostratigraphical framework can be established. The Upper Greensand ranges from the Middle Albian *varicosum* Subzone, to the Lower Cenomanian *mantelli* Zone, *carcitanense* Subzone. The faunal evidence is discussed under each member, and in the regional correlation.

Cann Sand

The term Cann Sand was introduced by Bristow (1989a) for the fine-grained, in part micaceous sand and very weakly cemented sandstone, up to 30 m thick, that occupies a prominent shelf-like feature below the Upper Greensand scarp formed by the Shaftesbury Sandstone and Boyne Hollow Chert members. There is no permanent section in the sands, but there is an undisturbed outcrop (the sands are commonly involved with landslips in the underlying Gault) at the village of Cann [872 213], with the prominent feature of the Shaftesbury Sandstone rising to the north of the village.

A borehole [8853 2032] near Melbury proved 9.4 m of pale to dark grey, fine-grained, micaceous, patchily clayey, glauconitic sandstone with scattered bivalves in the lower part of the member. Its junction with the Gault is sharp, with fine-grained clayey sand resting on a 1.3 m thick bed of dark grey, very sandy, micaceous clay. In the field, the junction is commonly marked by springs. Glauconitic, fine-grained sand is consistently encountered in auger holes. The upper boundary of the Cann Sand is nowhere exposed, but is taken at a sharp feature break at the foot of the Shaftesbury Sandstone.

In the tract south-west of Belchalwell Street, most of the Upper Greensand, which is 16 to 20 m thick, consists of slightly clayey, fine-grained, glauconitic sand. The basal beds are the most clayey, and there is a gradual upward decrease in clay content to give a clay-free sand in the upper part.

No fossil has been identified from the Cann Sand. At Fovant [SU 002 290], north-east of the district, 'transition beds' from the Gault to the Upper Greensand yielded a fauna of *varicosum* Subzone age (Mottram, 1957; Bristow and Owen, 1991), which establishes a maximum age for the base of the Cann Sands. Evidence from the Shaftesbury Sandstone (see below) suggests that the upper part of the Cann Sand also falls in the *varicosum* Subzone.

A fauna from the 'basal beds of the Upper Greensand' near Mosterton [4748 0569] on the Yeovil Sheet, consisting of 'fossiliferous ?cowstone and grey micaceous clayey sand' yielded a fauna which included the ammonite *Callihoplites* aff. *auritus* (Wilson et al., 1958, pp.148, 155) indicative of the *auritus* Subzone. However, the fauna on which this date is based was collected from a gully within a slipped mass of Upper Greensand.

SHAFTESBURY SANDSTONE

The term Shaftesbury Sandstone was introduced by Bristow (1989a) for the fine-grained sands and sandstone, capped by a hard well-cemented sandstone, that overlie the Cann Sand. The member, capped by the Boyne Hollow Chert, forms a prominent escarpment along most of the Upper Greensand outcrop near Shaftesbury. In the past, the Shaftesbury Sandstone has been referred to the Ragstone Beds (White, 1923, p.46), the Ragstone and Freestone Beds (White, 1923, p.51) and the Ragstone (Drummond, 1970, p.687), although the ragstone lithology is confined to the uppermost beds of the member. Drummond (1970, fig.2) correlated the Shaftesbury Sandstone with the Exogyra Sandstone of south-west Dorset and the Exogyra Rock of south Dorset.

The Shaftesbury Sandstone was formerly worked for building stone in quarries around Shaftesbury, but there is now no permanent exposure. One of these quarries [8737 2227], originally recorded by Jukes-Browne and Hill (1900, p.159), is designated the type section, and is described on p.108.

The member consists of alternating beds of glauconitic sand and weakly calcite-cemented, glauconitic sandstone in the lower part, and hard, shelly, calcite-cemented, glauconitic sandstone (ragstone) in the upper part. The base is nowhere exposed, but the sharp feature break below the escarpment occurs at the junction with the underlying uncemented or only weakly cemented Cann Sand. The upper boundary is taken at the top of the Ragstone.

On the south-western margin of the district, the Shaftesbury Sandstone crops out between Dogbury Gate [656 053] and Woolland [776 070], although it cannot be mapped separately from the Cann Sand over most of that area, nor between Minterne Magna [659 043] and Cerne Abbas [665 011] in the Dorchester district, where it forms marked bench features below the Chalk.

The sandstone is fine- to medium-grained and very glauconitic, and has a variable calcareous cement, which locally gives rise to a very doggery or nodular, rubbly sand. Fossils in the upper beds include the oyster *'Exogyra columba'* sensu Woods *non* Lamarck and, less commonly, *Amphidonte obliquatum*, which may indicate the *auritus* Subzone.

In the type area, the thickness of the Shaftesbury Sandstone is probably about 20 to 25 m; elsewhere, in the eastern part of the district, it varies from 10 to 25 m. In the south-west of the district, the maximum thickness is 2 to 3 m.

Age-diagnostic fossils are rare in the Shaftesbury Sandstone. The stratigraphical position of a specimen of *Prohysteroceras (Goodhallites) applanata* in Gillingham Museum, collected from 'Boyne Hollow', is not known; it is of *varicosum* Subzone age. Both the Shaftesbury Sandstone and the Boyne Hollow Chert were formerly exposed in this pit [8737 2227]. However, since there is evidence to suggest that the basal part of the Boyne Hollow Chert is of *dispar* Zone age (see below), the *Prohysteroceras* almost certainly came from the Shaftesbury Sandstone. The matrix of the specimen suggests that it may have come from the 'firm, compact glauconitic sandstone' (see p.108), i.e. the freestone beneath the Ragstone. The Ragstone could therefore also be of *varicosum* Subzone age, or younger. A fragment of cf. *Mortoniceras (Mortoniceras)* sp., moulded onto an oyster attachment area, was found in brash on the Shaftesbury Sandstone outcrop at East Compton [8770 1892]. The inner whorls are not tri- or quadrituberculate, which indicates an age not later than early *auritus* Subzone, and probably *varicosum* Subzone. Specimens of *Mortoniceras (Mortoniceras) cunningtoni*, *Anahoplites picteti* and *Idiohamites* sp., indicative of a late *varicosum* Subzone or early *auritus* Subzone age, were found in loose fragments of a calcareously cemented, glauconitic sandstone at Hill Farm [7725 0678], near Woolland. Because of faulting, the exact stratigraphical position of this occurrence is uncertain. The above three occurrences suggest an age no older than the *varicosum* Subzone for most of the Shaftesbury Sandstone, although the uppermost part could be of *auritus* Subzone age. *Pycnodonte (Phygraea)? vesiculosum,* usually fragmented, together with other oysters, is locally common in the ragstone at the top of the member.

Boyne Hollow Chert

The term Boyne Hollow Chert was introduced by Bristow (1989a) for the chert-bearing beds that overlie the Shaftesbury Sandstone and cap the Upper Greensand escarpment. Despite its feature-forming nature, and the fact that the chert beds were extensively quarried for road metal, there is no good exposure in this member. It was formerly well exposed in a quarry [8737 2227] in Boyne Hollow (Jukes-Browne and Hill, 1900, p.159) (see p.108).

The member consists of up to 15 m of glauconitic quartz sand and weakly glauconitic sandstone with cherty and siliceous concretions and, in places, beds of chert up to 0.6 m thick. The basal bed consists of up to 1 m of fossiliferous glauconitic sand and weakly cemented sandstone with phosphatic nodules that rests, possibly disconformably, on the ragstone at the top of the Shaftesbury Sandstone. For mapping purposes, the top of the member is taken above the highest chert bed or nodule, but up to 2 m of chert-free glauconitic sandstone intervene between the highest chert bed and the fossiliferous phosphatic nodule bed at the base of the Melbury Sandstone.

Jukes-Browne and Hill (1900) obtained a diverse fauna from the 0.9 m thick glauconitic sand with phosphatic pebbles at the base of the Boyne Hollow Chert in a quarry [8690 1935] west of Melbury Hill. The fauna (see p.108) includes brachiopods, bivalves, echinoids and the ammonite *Idiohamites* aff. *elegantulus*. Exposures at the Boyne Hollow pit (Jukes-Browne and Hill, 1900) and on the northern outskirts [8630 2335] of Shaftesbury, revealed unfossiliferous glauconitic sand with phosphatic nodules, resting on Shaftesbury Sandstone. A similar sequence was noted in the Bridport district, north of Beaminster [4839 0317] (Wilson et al., 1958, p.156). This phosphatic nodule bed probably represents an erosive event and subsequent transgression that can be recognised throughout southern and eastern England. Comparable beds are the Horish Wood Greensand in Kent (Owen, 1976) and Bed XII at Folkestone (Owen, 1976) which, until recently, have been regarded as falling in the top part of the *auritus* Subzone. However, there is evidence that the fossils of *auritus* subzone age in Bed XII are derived and incorporated in *dispar* Zone rocks (Morter and Wood, 1983). Support for this hypothesis in the Shaftesbury area is provided by the occurrence of *Merklinia aspera,* which appears to be unknown below the higher part of the *auritus* Subzone (post-Potterne Rock), the occurrence of *Idiohamites elegantulus* which ranges from the Albian *dispar* Zone, to the Cenomanian *mantelli* Zone, and the probable presence within the chert beds of *Catopygus columbarius* and *Holaster geinitzi* (see below), which also range from the *dispar* Zone into the Cenomanian.

In the district, the only fossils that have been found unequivocably within the cherty sequence are from a pit [8690 1935] near Melbury. Material in the Geological Survey collections from this quarry, but not referred to by Jukes-Browne and Hill (1900), includes *Merklinia aspera triplicata.* Brash overlying the Boyne Hollow Chert near East Compton [8774 1892] yielded the echinoids *Catopygus columbarius, Epiaster lorioli* and *Holaster geinitzi* (common). The Boyne Hollow Chert is probably of the same *dispar* Zone age as the chert beds in the Bridport district (Spath, 1933, p.423).

Sandstone at the top of the chert beds at West Melbury yielded (suggested modern names are given in square brackets) *'Aequipecten asper'* [*Merklinia aspera*], *'Exogyra conica'* [*Amphidonte obliquatum*] and *'Discoidea subucula'* [*Discoides subuculus*] (White, 1923, p.58; Mottram, 1957, p.164). However, in view of the extensive homeomorphy found in late Albian exogyrine oysters, the record of *Exogyra conica* cannot be accepted at face value.

Melbury Sandstone

The term Melbury Sandstone was used by Drummond (1970, pp.695, 698) for the strata between the Boyne Hollow Chert and the base of the Chalk. Earlier workers had referred to these beds as the Warminster Greensand (Edmunds, 1938, p.185), Passage Beds (Mottram, 1957, p.164) and 'Cenomanian sands' (Kennedy, 1970, p.622).

The type section was that formerly exposed in the Melbury quarry [8753 2015] (see p.110). Jukes-Browne and Hill (1900) commented on the difficulty of drawing the boundary between the Upper Greensand and Lower Chalk, and initially placed it at the base of Bed 3; they subsequently (1903, p.104) placed it at the top of Bed 3. White (1923, p.58), however, included beds 2 and 3 in the Lower Chalk.

In this account, we follow the later interpretation of Jukes-Browne and Hill (1903) and place the base of the Melbury Sandstone at the base of the pebbly, shelly, phosphatic bed (Bed 2). In the field, in the absence of sections, it is not always possible to distinguish the pebbly basal bed from the underlying glauconitic sandstone, locally up to 2 m thick, at the top of the Boyne Hollow Chert. The upper boundary in the field is taken at the base of the incoming of marl.

Southwards from Melbury, the stratigraphical break at the base of the Melbury Sandstone becomes more pronounced. At Clayesmore School [8642 1487], a well-developed conglomerate, 0.3 m thick, with clasts of the underlying bed, rests on the eroded top of the Shaftesbury Sandstone. It is overlain by at least 1.4 m of shelly calcareous sandstone and clayey sand (see p.110). A similar sequence was noted at Manor Farm, Iwerne Courtney [8565 1296] (p.109). South-west of Iwerne Courtney, the Melbury Sandstone is condensed and is represented only by a conglomerate, the Bookham Conglomerate. Locally, the conglomerate is richly fossiliferous, with a mixture of well-preserved indigenous fossils and a rolled and abraded, phosphatised fauna.

In the absence of good exposure, or borehole control, it is difficult to give a precise figure for the thickness of the Melbury Sandstone, but it probably varies from 0 to 6 m.

The Melbury type section has yielded a rich fauna of brachiopods, bivalves, ammonites and echinoids. The ammonites include *Hyphoplites curvatus arausionensis, Mantelliceras* cf. *cantianum, M. lymense* and *Mariella* sp., indicative of an earliest Cenomanian, *carcitanense* Subzone age (see details). The many sections between Melbury and Manor Farm, Iwerne Courtney, and the condensed sequence at Stour Bank [c.8465 1070] (Kennedy, 1970), have all yielded *carcitanense* Subzone faunas.

Bookham Conglomerate

The Bookham Conglomerate, which occurs in pockets up to 1 m thick in the top of the Shaftesbury Sandstone, is best developed in the western part of the district between Dogbury [657 052] and Buckland Newton [684 046]. It persists beyond the latter to Okeford Fitzpaine [803 103] and Stour Bank [846 107], and from there, extends northwards as the basal bed of the Melbury Sandstone. Small pockets of the conglomerate are probably present in the Woolland to Stoke Wake [759 065] area. The type locality lies just south of the district, at Bookham Farm [7065 0414] (Figure 48). Because it is only about 1 m thick at most, and impersistent, it has not been mapped separately; in the western part of the district, it is generally visible only as fragments mixed with Shaftesbury Sandstone brash.

The bed consists of clasts of very glauconitic shelly sandstone up to cobble size, commonly with a phosphatic rind, together with phosphatised shells, mainly bivalves and ammonites, set in a matrix of sandy glauconitic chalk; locally, it may comprise a weakly cemented, coarse-grained, glauconitic sand, as at Knole Quarry [7030 0519], Sharnhill Green. The base is irregular and at Bookham Farm it descends into clefts in the eroded top of the Shaftesbury Sandstone.

The fauna includes reworked and phosphatised shells (typically *Glycymeris sublaevis, Fenestricardita?, Cucullaea mailleana, Ostlingoceras*), silicified bivalves (*Merklinia* ex gr. *aspera, Neithea* spp. and *Pycnodonte (Phygraea)* aff. *vesiculosum*) which are also probably derived from the highest (unnamed) subzone of the *dispar* Zone. An indigenous Cenomanian fauna, including *Mantelliceras mantelli, Mariella* sp? and *Ostlingoceras* sp. (see p.109; Smart, 1955; Kennedy, 1970) occurs in the matrix.

Regional correlations

The name 'Exogyra Rock' or 'Sandstone' has been applied by Drummond (1967; 1970, p.690) and Kennedy (1970) to shelly sandstones developed at more than one level at or near the top of the Upper Greensand in the Shaftesbury and Bridport districts. The term Exogyra Rock, has also been applied to deposits in south Dorset of *dispar* Zone age (Wright *in* Arkell, 1947, pp.185–192). Deposits named the Exogyra Sandstone, which occur beneath the Chert Beds in west Dorset, have been assigned, with no supporting evidence, to the *inflatum* Zone, *auritus* Subzone by Wilson et al. (1958, p.141).

Drummond's 'Exogyra Rock', on the west side of Melbury Hill [8690 1935], is the ragstone at the top of the Shaftesbury Sandstone; at other localities in the Shaftesbury district (e.g. Stour Bank, Okeford Fitzpaine and Bookham Farm), his 'Exogyra Rock' refers to a cemented sandstone developed at the eroded top of the Shaftesbury Sandstone, beneath the unconformable Bookham Conglomerate. Where the Boyne Hollow Chert is absent, the stratigraphical position of this sandstone is uncertain. Between Okeford Fitzpaine and Compton Abbas, where the chert beds are present, it is most probably the southern continuation of the ragstone of the Shaftesbury area. South-west of Okeford Fitzpaine, the cemented sandstone is probably developed at a lower stratigraphical level within the Shaftesbury Sandstone.

In the Bridport district, the Chert Beds are of *dispar* Zone age (Spath, 1933, p.423), although this date was questioned by Carter and Hart (1977, p.105, fig.45, 46) who, on foraminiferal evidence, date the Chert Beds as Cenomanian in age. Thus the chert beds in the Bridport and Shaftesbury districts, on either side of the Mid-Dorset Swell, appear to be roughly contemporaneous.

The Melbury Sandstone appears to correlate with the upper part of the Eggardon Grit of the Bridport district (Wilson et al., 1958; Kennedy, 1970).

DETAILS

Cann Sand

Dogbury Gate area

Jukes-Browne and Hill (1900) described beds below the Shaftesbury Sandstone near Dogbury Gate [?6550 0525] as green sand with concretions of calcareous stone, overlying 1m of sand full of broken shells, in turn overlying 1m of irregular, lumpy, calcareous sandstone, with '*Ammonites rostratus, Trigonia aliformis, T. meyeri, Cardium hillanum, Lucina* sp., *Exogyra conica, Cyprina cuneata, C. rostrata, Pecten (Neithea) quadricostatus* and *Cucullaea glabra*'. Another pit [probably 6570 1030] showed a lower section in 4 m of yellowish green, slightly micaceous, well-sorted glauconitic quartz sand (Jukes-Browne and Hill, 1900). A band of loamy sand full of '*Exogyra conica* and broken *Pecten quadricostata*' occurred in the middle of the section.

Dungeon Hill–Buckland Newton

At Bookham Farm [7065 0412], just south of the district, some 0.65 m of poorly cemented, bioturbated, glauconitic, shelly sandstone with a few cherty sandstone lenses is exposed beneath the Shaftesbury Sandstone (Figure 48). It yielded *Amphidonte obliquatum* [abundant], *Entolium* sp.juv., *Neithea gibbosa, Mimachlamys* ex gr. *robinaldina* and *Lima subovalis*? Beneath this sandstone, a further 1m of soft, well-burrowed glauconitic sand containing *Amphidonte obliquatum* occurs. Between the sandstone and the sand there is an impersistent seam, up to about 20 mm thick, of sandy pale brown clay.

Stoke Wake–Belchalwell Street–Okeford Fitzpaine–Shillingstone

Around Ibberton, springs are common at the base of the Upper Greensand. A cutting [7992 0914] near Belchalwell, near the top of the Cann Sand, showed 2.5 m of well-sorted, burrowed, very fine-grained, silty sand, with many bivalves and gastropods, and an impersistent calcareous cemented layer.

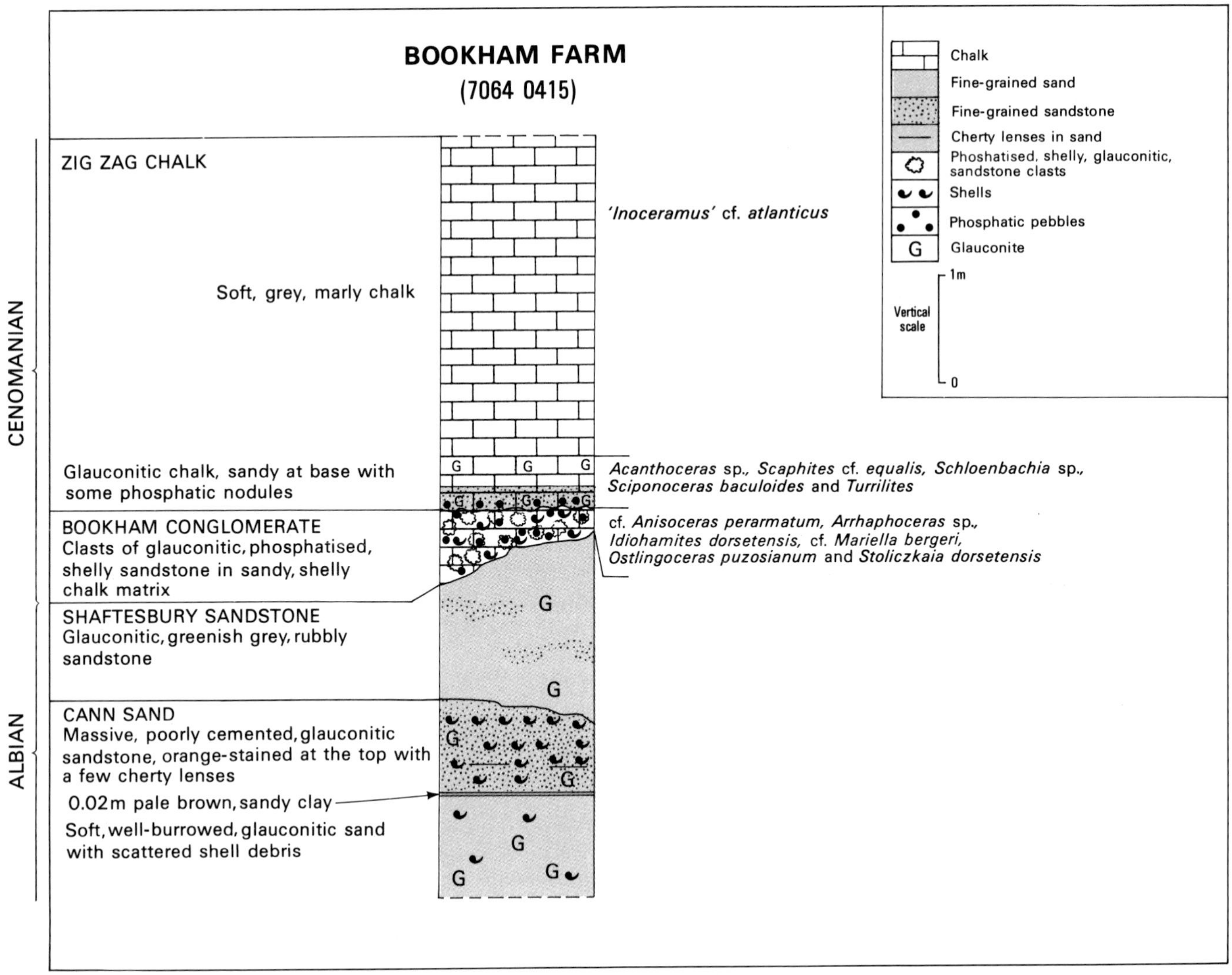

Figure 48 Section in the basal Chalk and top Upper Greensand at Bookham Farm.

The Upper Greensand is about 24 m thick near Okeford Fitzpaine. From the Cann Sand at Hartcliff Farm [8037 1019], Jukes-Browne and Hill (1900, 1903) obtained '*Exogyra conica*, *Terebratella pectita* and *Serpula concava*'.

At Stour Bank, sandstone brash [8467 1070] includes *Pycnodonte (Phygraea)? vesiculosum*, cf. *Gryphaeostrea canaliculata*, *Mimachlamys* ex gr. *robinaldina* and *Neithea gibbosa*.

The east bank of the River Stour at Hanford [8397 1089] is a 4 m high cliff; buff, glauconitic fine-grained sand is exposed at river level. Springs are common at the base of the cliff [8397 1133; 8385 1149 to 8382 1152]. Around Little Hanford [840 115], the Upper Greensand varies from 14 to 20 m in thickness. A well [8386 1192] near Little Hanford, proved 1.5 m of head deposits, on 13.7 m of soft, very fine-grained green sand, patchily cemented into sandstone, on 3 m of blue sandy clay of the Gault.

Shaftesbury area

North of Compton Abbas, the Cann Sand is about 10 m thick near Twyford and 20 m near Melbury Abbas. The base is commonly marked by springs.

Fine-grained micaceous sandstone crops out in the road cutting [8823 2037] at Melbury. Farther east, a borehole [8853 2032] in the bottom of the valley proved, beneath 5.6 m of head and alluvium, 9.4 m of grey, fine-grained, micaceous, shelly, glauconitic sand and weakly cemented sandstone, resting on dark grey, very sandy micaceous clay of the Gault.

At Cann, the type area, the sand is about 30 m thick. A borehole [8737 2153], which commenced about 5 m below the top of the member, proved 9.2 m of loose, dark green, clayey, silty fine-grained sand. West of Cann, the outcrop is either cambered [8697 2107] or broken by landslips. A piston sampler hole [8667 2147] on the east side of French Mill Lane, in an area where the sand is thought to be in situ, proved:

	Depth m
Sand, stony, brown	1.20
Sand, fine- to medium-grained, glauconitic, soft; water table at 1.9 m; bright green sand 2.5 to 3.1 m	c.5.00
Sand, fine- to medium-grained, firm	6.25
Sand, fine-grained, glauconitic	6.95
GAULT	
Clay, sandy, micaceous, dark grey	7.05

At Shaftesbury, the Cann Sand forms a well-defined flat or gently sloping area on three sides of the town. In the St James district [860 225] on the south side, there are large camber flaps of Cann Sand [e.g. 8535 2218, 8555 2176, 8598 2168]; the last rests on Kimmeridge Clay at its foot.

Duncliffe Hill

Duncliffe Hill is capped by about 5 m of buff, shelly, glauconitic, fine-grained sand and weakly cemented sandstone. Springs commonly issue from the base of the formation, particularly on the north side of the hill.

Shaftesbury Sandstone

Lyon's Gate–Dungeon Hill–Buckland Newton

Glauconitic sandstone occurs at the roadside [6549 0526], just below the base of the Chalk, and beside a track [6577 0519] near Lyon's Gate. An exposure [?6550 0525] west of the main road showed Shaftesbury Sandstone passing down into 'greenish sand with concretions of hard calcareous stone' (Jukes-Browne and Hill, 1900). The Shaftesbury Sandstone is exposed close to the Lower Chalk boundary at Dogbury [6627 0529], where over 1 m of glauconitic sandstone occurs. To the south-west, there are small (up to 0.3 m) exposures [6588 0494; 6594 0487; 6620 0457] on the side of the valley draining to Cerne Abbas.

At Sharnhill Quarry, Knole Hill, Kennedy (1970, p.627) reported 1 m of hard, nodular, rubbly, glauconitic sandstone with a rich fauna of silicified bivalves: '*Chlamys aspera, Exogyra obliquata, Entolium orbiculare* and *Neithea gibbosa* and occasional echinoids: *Discoides subuculus, Catopygus columbarius* and *Holaster laevis*'. This is overlain by the Bookham Conglomerate and underlain by loose glauconitic sand.

The Shaftesbury Sandstone is exposed at Bookham Farm [7065 0414; 7065 0412] where it consists of 1.2 to 1.5 m of glauconitic, rubbly, greenish grey sandstone with *Proliserpula* sp?, *'Rotularia' concava* s.l., *Cyclothyris* cf. *punfieldensis, Amphidonte obliquatum, Ceratostreon? undata, Entolium orbiculare, Neithea gibbosa* and *Discoides* cf. *subuculus.* It is overlain disconformably by the Bookham Conglomerate, and underlain by poorly cemented, bioturbated, glauconitic, fine-grained sandstone with *Amphidonte obliquatum.*

Bulbarrow Hill–Woolland

Beside a spring [7725 0678] at Hill Farm, close to the faulted junction of the Upper Greensand and Chalk, debris of weakly cemented glauconitic shelly sandstone was found. The fauna includes *Rotularia* sp., *Callistina plana, Calva (Chimela) caperata, 'Cucullaea'* aff. *carinata, 'C'. (Ideonarca) glabra, 'C'.* aff. *fibrosa, Entolium orbiculare, Gryphaeostrea* aff. *canaliculata, Mimachlamys* sp., *Modiolus reversus, ?Nannonavis* sp., *Neithea* sp., *Panopea mandibula, Protocardia* sp., *Pterotrigonia?* cf. *aliformis, Anahoplites picteti, Idiohamites* sp., *Mortoniceras (Mortoniceras) cunningtoni* and *Scaphites* aff. *simplex,* indicative of a late *varicosum,* possibly early *auritus,* Subzone age.

Stoke Wake–Woolland

Rubbly glauconitic sandstone was reported below glauconitic marly chalk at Stoke Wake (Kennedy, 1970); this is probably the Shaftesbury Sandstone.

Stour Bank–Compton Abbas

At Stour Bank [8467 1070], brash of glauconitic sandstone from beneath the Bookham Conglomerate (see p.109) includes cf. *Gryphaeostrea canaliculata?, Mimachlamys* ex gr. *robinaldina, Neithea gibbosa* and *Pycnodonte (Phygraea)? vesiculosum.*

Between Stour Bank and Compton Abbas, the ragstone at the top of the Shaftesbury Sandstone has been recognised in several sections, but the member does not form a mappable unit until north of Compton Abbas. From there northwards, it forms a well-defined feature and thickens to about 5 m near Twyford, 15 m near Melbury Abbas and 25 m near Shaftesbury.

The Shaftesbury Sandstone was worked in several pits on the north side of Hambledon Hill [8420 1334; 8425 1340; 8437 1337]. A coarse-grained or gritty, glauconitic sandstone, overlying 4.5 m of soft green sand, occurs at the top [8420 1330] of the first pit, and in a nearby track [c.8420 1323] (White, 1923, p.53). On the north-east side of Hambledon Hill [8473 1304], *Entolium orbiculare, Parsimonia antiquata?, Mimachlamys* ex gr. *robinaldina, Merklinia aspera, Neithea gibbosa* and *Pycnodonte (Phygraea)? vesiculosum* occurred in a shelly glauconitic sandstone brash. Similar brash beneath the Melbury Sandstone on the hill [8590 1572] south-west of Sutton Waldron included *Neithea gibbosa* and abundant *Pycnodonte (Phygraea)? vesiculosum.*

At Manor Farm [8565 1296], Iwerne Courtney, spoil from a trench included a block of hard, glauconitic sandstone with a large nautiloid fragment; nearby [8551 1307], echinoids and *Neithea gibbosa* were collected.

At Clayesmore School [8642 1466], Iwerne Minster, the top Shaftesbury Sandstone yielded: *Rotularia* aff. *polygonalis, Ceratostreon, Entolium orbiculare* (Plate 8m), *Pycnodonte (Phygraea)? vesiculosum* (Plate 8m) in rock-building quantities, *Lima subovalis?, Neithea gibbosa* (Plate 8m) and *Pseudolimea composita* (Plate 8m).

In the sunken lane [8695 1850] south-east of Compton Abbas church, and also below a feature break near East Compton [8760 1847], there are beds of glauconitic sandstone (ragstone), up to 0.4 m thick. In the field [8770 1892] north of Gourd's Farm, East Compton, a sharp feature break [8774 1892] is also probably formed by the ragstone. In brash west of the feature, and slightly higher stratigraphically, the following fauna in a matrix of glauconitic sandstone was found: *Entolium orbiculare, Pycnodonte (Phygraea)? vesiculosum, Mimachlamys* ex gr. *robinaldina* and *Neithea gibbosa.*

North-west of Whitehall, 1.8 m of firm sandrock is overlain by a 0.9 m thick massive bed of glauconitic, calcareous sandstone (the Ragstone of Drummond, 1970), overlain by the Boyne Hollow Chert (see p.108) (Jukes-Browne and Hill, 1900) [8690 1935]; the beds dip at 5°ESE. From the ragstone, were obtained serpulid-like bryozoan, a small terebratulid, *Amphidonte obliquatum, Entolium orbiculare, Gryphaeostrea canaliculata* and *Neithea gibbosa.*

Exposures [8653 1973 to 8663 1978] in the scarp face 700 m farther north-west reveal the following composite section in the upper part of the Shaftesbury Sandstone and basal Boyne Hollow Chert.

	Thickness m
BOYNE HOLLOW CHERT	
Chert in discontinuous lenses in a sandy matrix	0.20
Sand, unconsolidated	0.06
SHAFTESBURY SANDSTONE	
Sandstone, fine-grained, massive, glauconitic, shelly (ragstone)	0.84
Sand, fine-grained, glauconitic	0.30
Sandstone, fine-grained, glauconitic, friable	0.50
Sand, fine-grained, glauconitic	0.30

The basal sand yielded *Ditrupa* sp., *Glomerula gordialis* and cf. *Amphidonte obliquatum.* From the lower part of the overlying sandstone, *Ditrupa* sp., *Glomerula gordialis*, abundant *Amphidonte obliquatum*, (Plate 8k) *'Exogyra columba'* Woods non Lamarck, *Entolium orbiculare*, cf. *Gryphaeostrea canaliculata* and *Neithea gibbosa* (Plate 8a) were collected. A similar fauna, but with only two *Amphidonte obliquatum* and, in addition, *Rotularia*, *Mimachlamys* ex gr. *robinaldina* and a possible *Pterotrigonia aliformis* occurred in the upper part. *Amphidonte*, *Entolium orbiculare*, *Ceratostreon? undata*, *Gryphaeostrea canaliculata*, *Lima subovalis*, *Mimachlamys* ex gr. *robinaldina*, *Neithea gibbosa*, *Pseudoptera* cf. *haldonensis*, *Pycnodonte (Phygraea)? vesiculosum* and *Syncyclonema* were collected from the ragstone.

Melbury–Shaftesbury–Ludwell

The Shaftesbury Sandstone is about 10 m thick near Melbury. Ragstone was formerly exposed in the road cutting [8823 2005] below Melbury Abbas church (Jukes-Browne and Hill, 1900), and it was from here that Pulteney (1813) collected the type specimens (present whereabouts unknown) of *Neithea gibbosa* and *Amphidonte obliquatum* from a friable, greyish green, quartzose sandstone. The precise level in the sequence from which they came is unclear; they were probably collected from the base of ragstone, or lower in the sequence.

At East Melbury, ragstone, consisting of c.2 m of massive, fine- to medium-grained sandstone, in beds 1 m thick, occurs at the side of a track [8898 2020]. In Dinah's Hollow, West Melbury, glauconitic shelly sandstone, in beds up to 0.5 m thick, is exposed at the top of the cutting [8829 2056].

The old pit [8737 2227] at Boyne Hollow formerly exposed 3.9 m of Shaftesbury Sandstone, beneath the Boyne Hollow Chert (see p.108). From its preservation, it is thought that the magnificent specimen of *Prohysteroceras (Goodhallites) applanata* in Gillingham Museum from 'Boyne Hollow' came from the Shaftesbury Sandstone, possibly from the 'freestone' near the base of the section. South of Shaftesbury, firm, compact, glauconitic sandstone with a calcite cement ('freestone') was worked in the large quarry known as the Wilderness [866 225 to 8660 2205] (White, 1923, p.51).

Around Shaftesbury, the Shaftesbury Sandstone is about 25 m thick. Ragstone, overlain by Boyne Hollow Chert (see p.108), was noted by Dr H G Owen on the northern outskirts [c.8630 2337] of the town.

Along Coombe Valley, east of Shaftesbury, the member is about 10 to 15 m thick; this increases to 15 to 20 m between Coombe and the northern margin of the district. On the south side of Long Bottom, a cave [8834 2268] has been dug into the ragstone. Close by, there are many large (1.5 × 0.8 × 0.5 m) blocks of shelly calcareous sandstone. The ragstone was worked in a quarry [8832 2251] 150 m south of the cave, but no section remains. Farther down the valley, a 3 m high section [8887 2267] exposes weakly cemented, glauconitic, fine-grained sand in the middle of the Shaftesbury Sandstone.

Around Ludwell, the Shaftesbury Sandstone is only about 10 m thick. Ragstone, 3 m thick, is exposed in the cutting [9048 2251] south-west of the village. The more easterly exposure [9049 2250] dips at about 40°S and is presumably cambered. Farther up the Ludwell Valley [9080 2215], at the watercress beds, there is a 1.5 m thick massive glauconitic sandstone at the base of a 5 m high scarp. There, and also in the tributary valley to the west, powerful springs issue from the base of the sandstone. North-east of Ludwell, the Shaftesbury Sandstone is mostly between 10 and 15 m thick.

Boyne Hollow Chert

The most southerly locality where chert fragments were found is north-east of Manor Farm [8717 1795], Compton Abbas. West of Compton Abbas, the chert beds are about 2 to 3 m thick.

Above a feature break formed by the ragstone, and amongst chert debris, *Catopygus columbarius*, *Epiaster lorioli* and several *Holaster geinitzi* were collected in a field [877 189] at East Compton. These echinoids probably come from the basal bed of the Boyne Hollow Chert, as at Whitehall (see below).

In a pit [8690 1935] north-west of Whitehall, above the ragstone (see p.106), Jukes-Browne and Hill (1900, p.160) saw a shelly phosphatic nodule bed with a sandy matrix, 0.2 to 0.3 m thick, passing up into a bed of soft, grey, glauconitic, shelly sand with scattered phosphatic nodules, which in turn passed up to sandstone with chert. From the basal 0.3 m, *Cyclothyris* sp. nov. (Plate 8l), *Amphidonte obliquatum*, *Entolium orbiculare*, *Merklinia aspera*, *Neithea gibbosa*, *Trigonia carinata*, *Nautilus laevigatus* [not seen] and *Discoides subuculus* were collected. The uppermost 0.6 m yielded *Smilotrochus?*, *Rotularia* aff. *umbonata*, *Grasirhynchia grasiana*, *Ovatathyris?*, *Avellana* sp., *Gryphaeostrea canaliculata*, *Merklinia aspera* or *M.* ex gr. *aspera*, *Neithea gibbosa*, *N. (Neithella?) notabilis*, *Neohibolites* sp., *Idiohamites* aff. *elegantulus* [det. H G Owen], *Nautilus laevigatus*, *Discoides subuculus*, *Hyposalenia clathrata*, *Polydiadema* sp. and *Cretoxyrhina mantelli?*

The basal sand in a nearby section [8663 1978] (see p.106), probably the same as that above, yielded bryozoa, *Parsimonia* cf. *antiquata*, *'Rotularia'* cf. *concava*, *Amphidonte obliquatum*, *Entolium orbiculare*, *Gryphaeostrea canaliculata*, cf. *Lima subovalis* juv?, *Mimachlamys* ex gr. *robinaldina*, *Neithea gibbosa*, *Pycnodonte (Phygraea)?* aff. *vesiculosum* and *Discoides* sp.

The Boyne Hollow Chert has a wide outcrop south-west [8650 1965] and west [8620 2015] of West Melbury, where it is about 5 m thick.

Plate 8 Cretaceous fossils.

a. *Neithea gibbosa* (Pulteney); Upper Greensand, Cornhills Farm, near West Melbury (MW 232)
b. cf. *Cretoxyrhina mantelli* (Agassiz); Upper Greensand (Melbury Sandstone), Clayesmore School, Iwerne Minster (BRI 3732).
c. *Merklinia aspera* (Lamarck); Upper Greensand (Melbury Sandstone), Whitehall, Compton Abbas (BRI 3786).
d.(a,b) *Schloenbachia varians* (J Sowerby); Lower Chalk, Shroton Lines, Shroton (BRI 3905).
e. *Micraster cortestudinarium* (Goldfuss), Upper Chalk, Lewes Chalk, Shillingstone Chalk Pit, Shillingstone (VRB 18).
f. *Margaritella? tiara* (J de C Sowerby), Upper Greensand, Melbury Sandstone, Clayesmore School, Iwerne Minster (BRI 3761).
g. *Phalacrocidaris merceyi* (Cotteau), Upper Chalk, Lewes Chalk, Hatts Barn, near Ashmore (BRI 4260).
h. *Mytiloides subhercynicus* (Seitz), Middle Chalk, New Pit Chalk, Melbury Abbas (BRI 4183)
i. *Echinocorys* sp., Upper Chalk, Sutton Waldron (BRI4232).
j.(a,b) *Mantelliceras cantianum* Spath, Upper Greensand (Melbury Sandstone), Manor Farm, Shroton (TNN 1434).
k. *Amphidonte obliquatum* (Pulteney), Upper Greensand (Shaftesbury Sandstone), Cornhills Farm, near West Melbury (MW 188).
l. *Cyclothyris* sp.nov?, Upper Greensand (Boyne Hollow Chert), Incombe Farm, Compton Abbas (ZAZ 848).
m.(a) *Pycnodonte (Phygraea)? vesiculosum* (J de C Sowerby). (b) *Neithea gibbosa* (Pulteney). (c) *Entolium orbiculare* (J Sowerby). (d) *Pseudolimea composita* (J de C Sowerby); Upper Greensand (Shaftesbury Sandstone), Clayesmore School, Iwerne Minster (BRI 3783).

a
MW 232
b
c
e
f
d(a)
d(b)
h
i
g
j(a)
j(b)
k
l
d
a
b
c
m

In a borehole near East Melbury [8789 2014], beneath the Chalk, were proved 6 m of fine-grained, glauconitic sandstone [Melbury Sandstone], on 2 m of fine- to medium-grained, glauconitic 'limestone', with chert nodules in the lower 1 m, above 6.5 m of fine-grained, calcareous, glauconitic, shelly sandstone [Shaftesbury Sandstone]. North-west of East Melbury, large fragments of dark grey chert are common in brash just above the feature break which marks the base of the member [8840 2035; 8848 2053].

Boyne Hollow, near Mayo Farm, is the type section [8737 2227] of the Boyne Hollow Chert (Jukes-Browne and Hill, 1900, p.160) (here metricated):

	Thickness m
BOYNE HOLLOW CHERT	
Soil and rubble	0.46
Rather soft, pale grey, sandy stone with siliceous cherty concretions	1.07
Soft grey glauconitic sand with smaller (siliceous) concretions	1.37
Layer of brown siliceo-phosphatic masses	0.23
Firm grey, glauconitic, silty sand	0.61
Fine greyish sandstone, firm, but not hard, with grey cherty concretions	1.07
Pale whitish grey powdery sandstone, full of hard calcareo-siliceous concretions, some of which have centres of blue-grey chert	1.22
Massive layer of blue-grey chert, with thick whitish rind	0.61
Firm grey glauconitic sand with a layer of small brownish concretions at the base	0.38
Greenish grey, very glauconitic, sandy rock with a few brown phosphates	0.61
SHAFTESBURY SANDSTONE	
Very hard, glauconitic, semi-crystalline (i.e. calcitic) sandstone 'Rag'; many fossils at the top	0.91
Softer, but firm compact glauconitic sandstone, with calcite cement (freestone)	1.37
Soft, greenish grey, glauconitic sand, seen for	0.61

Much of the older part of Shaftesbury is underlain by the Boyne Hollow Chert (White, 1923, p.52). Beds and nodules of chert were seen in several excavations on the east side of the town. The most northerly of these [8702 2318] revealed about 1.5 m of shelly glauconitic sand with common pectinids, and chert nodules up to 8 cm thick. Farther south [8705 2308], 1.6 m of shelly, calcareously cemented, glauconitic sandstone with buff phosphatic nodules up to 5 by 10 cm and thin (up to 4 cm) chert nodules were seen. Large blocks of chert up to 0.2 m thick occur on the surface at the site entrance [8693 2304].

Sections [c.8630 2337] seen by Dr H G Owen on the northern outskirts of Shaftesbury in 1972 showed;

	Thickness m
BOYNE HOLLOW CHERT	
Sandstone, cherty	0.45
Sand, glauconitic, loamy, with lines of small chert nodules	1.50
Sandstone, cherty	0.30
Sand, pale buff, ferruginous-streaked, with nodules of chert	0.75
Sandstone, white	0.40
Nodular cherty sandstone in greyish white soft sand	1.50–1.80
Sandstone, highly glauconitic, with pale buff scattered phosphates	0.45
SHAFTESBURY SANDSTONE	
Sandstone, glauconitic, tough, massive bedded, shelly	2.00
Sand, glauconitic, some white (?phosphatic) patches	3.00

South of Coombe, there are few arable fields and the boundary with the overlying Melbury Sandstone is conjectural [882 224 to 900 227]. Scattered chert fragments occur in the the field [around 895 225] south-east of Coombe.

Around Ludwell, chert debris occurs in the roadbank [9050 2309], as brash north-east of the village [around 906 228], in a temporary exposure [9061 2262] east of the village, and in the cutting [9045 2251] of the main road. A little farther west [9038 2252], silicified sandstone and chert crop out in the road bank.

East of the Ludwell Brook, 3 m of chert beds just north of Manor Farm [around 9093 2190], thicken to about 12 m north of Birdbush [around 913 234]. Within this tract, grey and brown chert fragments are common in the brash. A quarry [9170 2336] near Brook Waters exposes about 1.5 m of fine- and very fine-grained glauconitic sand with irregular layers of glauconitic sandstone and chert nodules up to 0.4 m across.

In the most westerly [922 215 to 917 223] valley south of the Ferne Park Fault, the chert beds are very thin and only a few chips and scattered larger fragments of chert occur in the soil. In a second valley [9285 2150 to 925 227], chert debris is more common [around 9265 2210]. A giant form of *Birostrina?* preserved in glauconitic sandstone was found [9203 2246] south-west of Rowberry House. In the valley that runs through Ferne [9325 9170 to 932 228], grey-centred chert is common in the upper part of the valley; lower down, large grey chert nodules, interbedded with glauconitic sandstone, protrude through the turf [9320 2238]. Pits [9320 2226; 9322 2245] on either side of the above outcrop show poor exposures in glauconitic sand, sandstone and chert. Fragments of grey chert, silicified and also glauconitic sandstone are common north of the road leading to Berwick St John [around 9355 2275].

Melbury Sandstone and Bookham Conglomerate

Although members of the Upper Greensand can be mapped only locally south of Compton Abbas, it is possible to recognise the Melbury Sandstone in brash and sections as far south as Iwerne Courtney. From Hambledon Hill south-westwards, the Melbury Sandstone is represented only by the Bookham Conglomerate.

Lyon's Gate area–Buckland Newton–Bookham Farm

Jukes-Browne and Hill (1900) recorded green nodules and phosphatised *'Ammonites studeri, Turrilites bergeri, Actaeonella* sp., *Natica* sp., *Holaster laevis, Catopygus carinatus, Rhynchonella latissima* and *Pecten asper'.* in the top 15 cm of a sandstone in a quarry [6579 0520] near Dogbury Gate.

Recent (1990) excavations [7064 0415] at Bookham Farm, just south of the district, revealed a section across the Upper Greensand–Lower Chalk boundary (Figure 48). The section has previously been described by Jukes-Browne and Hill (1900; 1903), Smart (1955) and Kennedy (1970). The conglomerate consists mainly of sub-rounded clasts of very glauconitic sandstone up to cobble size in a matrix of sandy glauconitic chalk. Some of the clasts, commonly shelly, have phosphatised rims; others are phosphatised fossils. The sandy matrix contains angular to sub-rounded quartz grains up to 800 μm in size. The base of the conglomerate is irregular and fills clefts in the Shaftesbury Sandstone. The orange-stained top of the bed is planar and sharply defined.

A diverse reworked and strongly phosphatised fauna mainly of bivalves and ammonites, of *perinflatum* Subzone age, has been collected from this bed. Smart (1955) recorded '*Aucellina gryphaeoides, Chlamys aspera, Cucullaea mailleana, Neithea quadricostata, N. quinquicostata, Anisoceras* sp., *Arrhaphoceras studeri, Pleurohoplites* sp. group of *renauxianus* and *Turrilites* sp. aff. *wiesti*'. During the present survey, the following additional fauna was collected: *Dereta* juv?, *Cucullaea* or *Isoarca, Entolium orbiculare, Fenestricardita* or *Ludbrookia* cf. *cottaldina, Mimachlamys* ex gr. *robinaldina, Neithea gibbosa, Pycnodonte (Phygraea)? vesiculosum*, cf. *Anisoceras perarmatum, Arrhaphoceras* sp., *Idiohamites dorsetensis*, cf. *Mariella bergeri, Ostlingoceras puzosianum, Stoliczkaia dorsetensis* and *Discoides subuculus*.

At Knole Quarry [7030 0519], Sharnhill Green, the Bookham Conglomerate, 0.15 to 0.23 m thick, consists of clasts of highly glauconitic, grey, fossiliferous (including *perinflatum* Subzone ammonites (Kennedy, 1970)) sandstone with thin phosphatic rinds, in a matrix of weakly cemented, coarse-grained sand, with many molluscs, including very large *Neithea*, nautiloids and partly phosphatised, indigenous, Cenomanian ammonites.

Woolland–Okeford Fitzpaine area

A pit [8037 1019] at Okeford Fitzpaine formerly exposed the base of the Chalk and the upper part of the Upper Greensand (Jukes-Browne and Hill, 1900, 1903; White, 1923; Kennedy, 1970; Drummond, 1967, 1970; Carter and Hart, 1977, fig.21). The following section is based on Jukes-Browne and Hill (1900, 1903):

	Thickness m
LOWER CHALK: WEST MELBURY CHALK	
Marl, chalky, glauconitic, with phosphatised nodules and fossils	0.46
BOOKHAM CONGLOMERATE	
Nodular concretions, closely packed in a matrix of sandy glauconitic chalk	0.61
CANN SAND	
Sand, shelly, dark green, glauconitic; varies in thickness from 0.2 to 0.6 m, average	0.40
Sandstone, nodular, glauconitic	0.75
Sand, dark green, glauconitic	3.66

The strata dip 5° east. The Bookham Conglomerate was originally referred to as 'Cornstone' (Jukes-Browne and Hill, 1900, p.164) and later as the 'Rye Hill Sand' (Jukes-Browne and Hill, 1903). Fossils collected from the Bookham Conglomerate by Jukes-Browne and Hill (1900) and Mr J Pringle in 1922 have been redetermined. They include bryozoa, *Glomerula gordialis, Rotularia* sp., *Micrabacia coronula, Boubeithyris diploplicata, Grasirhynchia grasiana, Aucellina* ex gr. *gryphaeoides, Cucullaea mailleana, Glycymeris sublaevis, Gryphaeostrea canaliculata, Mimachlamys* ex gr. *robinaldina, Pseudolimea composita*, cf. *Rhynchostreon plicatulum, Mantelliceras mantelli, Mariella* sp?, *Ostlingoceras* sp.juv., *Schloenbachia varians, Catopygus columbarius, Ochetes* sp., *Hyposalenia clathrata* and *Discoides subuculus*. The fauna is an earliest Cenomanian assemblage, comparable with that from the basal bed of the Melbury Sandstone at Melbury Quarry (see p.110). *Glycymeris sublaevis* and *Ostlingoceras* sp., which are heavily phosphatised, are reworked elements of the late *dispar* Zone assemblage in the Bookham Conglomerate. Kennedy (1970, p.624) found the ammonites *Ostlingoceras puzosianum* and *Durnovarites* sp., indicative of the *perinflatum* Subzone, presumably in the Bookham Conglomerate.

Stour Bank

Phosphatised fossils from probable Bookham Conglomerate brash [8390 1023] north-east of Shillingstone Hill include ?*Stoliczkaia dorsetensis* and *Ostlingoceras puzosianum*.

At Stour Bank [c.8465 1070], the junction between the Lower Chalk and Upper Greensand was formerly exposed (Mottram, 1957; Drummond, 1967; Kennedy, 1970). Blocks from the top of the Upper Greensand in the Natural History Museum resemble the ragstone, which appears to be overlain by the Bookham Conglomerate, consisting of nodular and rubbly, glauconitic, calcareous sandstone, with hard cobbles set in a softer greensand. The top is glauconitised and rust-stained, and there is some piping of glauconitic chalk into it. Some of the cobbles are glauconitised and phosphatised; silicified fossils and some small brown phosphates also occur (Kennedy, 1970). Kennedy recorded *Ostlingoceras puzosianum, Neithea gibbosa* and other bivalves from this section, but did not specify whether they came from the cobbles or the matrix; they are presumed to be derived.

Iwerne Courtney–Iwerne Minster

Brash from an excavation [8565 1296] at Manor Farm, 650 m north-west of the Iwerne Courtney Church, included a pebbly, shelly, glauconitic, calcareous, fossiliferous sandstone of the Melbury Sandstone. Fossils include: *Margaritella? tiara* ['*Solarium bicarinatum*'] (Plate 8f), *Periaulax?, Cucullaea mailleana*, cf. '*Cyprina*' *quadrata*, (possibly derived from the Shaftesbury Sandstone), '*Inoceramus*' *crippsii*, '*I*.' sp., *Linotrigonia vicaryana, Ludbrookia, Merklinia aspera, Mimachlamys* ex gr. *robinaldina, Neithea* cf. *gibbosa, Plicatula inflata, Pycnodonte (Phygraea)? vesiculosum, Trigonarca, Eutrephoceras* sp., *Anisoceras* sp., *Hyphoplites, Idiohamites* cf. *elegantulus, Mantelliceras cantianum* (Plate 8j), *M. mantelli, M.* cf. *picteti, M.* cf. *saxbii, M.* sp., *Mariella dorsetensis, Neostlingoceras* aff. *carcitanense, Schloenbachia varians* and *Sciponoceras*. The rich, indigenous ammonite fauna clearly indicates the *carcitanense* Subzone.

At Clayesmore School, Iwerne Minster, the following temporary section [8642 1466] was seen:

	Thickness m
Topsoil	0.20
Chalky gravel	0.15
MELBURY SANDSTONE	
Sand, fine-grained, clayey	0.90
Sandstone, calcareous, flaggy, with scattered fossils including *Avellana* sp., *Periaulax* sp., *Mantelliceras, Puzosia (P.) mayoriana* juv. and *Schloenbachia varians*	0.50
Pebble bed, with indurated pebbles of calcareous, glauconitic sandstone up to 15 cm across; fossils very common in matrix, and particularly on top (see below)	0.30
SHAFTESBURY SANDSTONE	
Sandstone, glauconitic, shelly (see p.105 for fauna)	0.50

The rich fauna from the basal pebble bed of the Melbury Sandstone includes: *Boubeithyris diploplicata, 'Jurassiphorus'* sp., *Margaritella? tiara* ['*Solarium bicarinatum*'], *Cucullaea, Entolium*, juv., *Plicatula inflata, Trigonarca* cf. *passyana* juv., *Eutrephoceras ex-*

pansum, Hyphoplites curvatus pseudofalcatus, Hypoturrilites, Idiohamites alternatus vectensis, Mantelliceras, Mariella, Neostlingoceras carcitanense, Puzosia (P.) mayoriana juv., *Schloenbachia varians, Catopygus columbarius, Discoides subuculus, Tiaromma michelini* and cf. *Cretoxyrhina mantelli* (Plate 8b). Additional fossils collected from the spoil heap and believed to have come from this same bed include *Oncotrochus* sp., cf. *Periaulax* sp., *'Inoceramus'* ex gr. *crippsii*, cf. *Cymatoceras deslongchampsianum* and *Hyphoplites* cf. *curvatus.*

Sutton Waldron–Compton Abbas–Melbury Abbas

Faunas found at two places on the hill south-west of Sutton Waldron included *Grasirhynchia grasiana, Terebratulina protostriatula, Pycnodonte (Phygraea)? vesiculosum, Schloenbachia varians, Catopygus columbarius* and *Holaster* cf. *geinitzi* [from 8590 1572], and *Grasirhynchia grasiana, Cucullaea* sp., *Eutrephoceras, Hyphoplites* cf. *curvatus arausionensis, H.* cf. *c. pseudofalcatus* and *Schloenbachia varians* [from 8580 1537].

A scrape [8626 1809] in calcareous, fine-grained sandstone on the west side of Elbury Hill yielded *Avellana* sp., *Cucullaea mailleana, Mantelliceras, Schloenbachia varians* and *Discoides subuculus.*

Jukes-Browne and Hill (1900) saw an exposure [8689 1969] on the west side of Melbury Hill of soft greenish grey sandstone with *'Pecten orbicularis, P. (Neithea) quadricostatus, Exogyra conica, E. columba* and *Trigonia carinata'*.

Temporary exposures [around 8869 1960] on the south side of White Pit Lane, Melbury Abbas, revealed about 1 m of fine-grained, glauconitic, weakly cemented, calcareous sandstone, out of a total thickness of 2 m of Melbury Sandstone. The fauna includes: *Porosphaerella, Grasirhynchia grasiana, Ovatathyris ovata, Avellana, 'Jurassiphorus', Periaulax?, Anomia* sp., *Gryphaeostrea canaliculata, 'Inoceramus'* ex gr. *crippsii, Merklinia aspera* (Plate 8c), *Mimachlamys* ex gr. *robinaldina, Neithea* aff. *quinquecostata, Plagiostoma* cf. *semiornata, Rhychostreon* sp. juv., *Hyphoplites campichei?, H. curvatus* cf. *pseudofalcatus, Hypoturrilites, Mantelliceras mantelli, Mariella* sp., *Schloenbachia varians, Catopygus columbarius* and *Discoides subuculus.*

One of the best sections in the Melbury Sandstone was that formerly exposed in the Melbury quarry [8753 2015]. The record below, here metricated, is from White (1923, p.58); it is slightly modified from the section of Jukes-Browne and Hill (1900, pp.160–161):

	Thickness m
LOWER CHALK: WEST MELBURY CHALK	
4. Firm marly chalk with grains of quartz and glauconite; partly shaly at base; *Grasirhynchia grasiana, Avellana* sp., *Lima* aff. *aspera* and *Mantelliceras* cf. *dixoni;* passing down into	0.76
MELBURY SANDSTONE	
3. Glauconitic sandy marl or marly sandstone, drying to light greenish grey; many fossils including *'Stauronema carteri', Avellana* sp., *Margaritella? tiara, Entolium orbiculare, 'Inoceramus* ex gr. *crippsii, Linotrigonia (Oistotrigonia) vicaryana, Mimachlamys* ex gr. *robinaldina, Plicatula inflata, Pseudolimea composita, Trigonarca* cf. *passyana, Cymatoceras elegans,* cf. *Pseudocenoceras largilliertianum, Hyphoplites curvatus arausionensis, Mantelliceras lymense, Scaphites* cf. *obliquus, Schloenbachia varians, Mariella* sp., *Catopygus columbarius, Discoides subuculus* and *Holaster* sp.	1.22
2. Rough glauconitic sandstone with hard brown material like impure chert and phosphatic concretions; many fossils including: *'Rotularia'* sp., cf. *Boubeithyris diploplicata, Cyclothyris difformis, Ovatathyris ovata, Terebratulina protostriatula, Entolium orbiculare, 'Inoceramus'* ex gr. *crippsii, Merklinia aspera, Mimachlamys* ex gr. *robinaldina, Neithea* sp., *Plagiostoma meyeri, Plicatula inflata, Mantelliceras* cf. *cantianum, Mariella* sp., *Schloenbachia varians, Stomohamites duplicatus, Holaster* sp. and *Catopygus columbarius*	1.22
BOYNE HOLLOW CHERT	
1. Softer, fine-grained, glauconitic sandstone, with few fossils: *Aequipecten asper, [Exogyra conica]*	1.83

The various stratigraphical classifications applied to the above sequence are discussed on p.102.

Cann–Charlton–Shaftesbury

On Cann Common, brown calcareous sandstone with *Catopygus columbarius* was formerly exposed [8856 2095] (White, 1923, p.57). It is probable that 'ragstone' recorded by Jukes-Browne (MS, BGS) in a quarry [8815 2108] on the west side of Cann Common, and 'Warminster Greensand' in a quarry [8665 2008] at West Melbury (B H Mottram, MS, BGS) are both part of the Melbury Sandstone.

A pit [8986 2156] near Charlton formerly exposed fossiliferous, marly, glauconitic sand (Jukes-Browne and Hill, 1903) with *'Rotularia'* sp., *Limaria* sp.nov., *Merklinia aspera, Mimachlamys* aff. *robinaldina,* cf. *Rhynchostreon plicatulum* and *Catopygus columbarius,* overlain by 'Glauconitic Marl' (?Bed 3 of the Melbury quarry) with *Margaritella? tiara* [*'Solarium bicarinatum'*], *Pycnodonte vesicularis, Hyphoplites curvatus arausionensis, Schloenbachia varians, Scaphites* sp., *Mariella* sp. and *Discoides subuculus.*

Calcareous glauconitic sandstone from a excavation [8717 2308] on the east side of Shaftesbury included *Schloenbachia varians* and *Mantelliceras* sp. A fragment of chert in the spoil suggests that the Boyne Hollow Chert was exposed at the base of the section at a depth of about 1.5 m. West Melbury Chalk occurs at the top of the exposure. Thus the Melbury Sandstone is here not more than 1.5 m thick. Farther north, the proximity of the Lower Chalk to temporary exposures [around 8702 2318] revealing chert, suggests that the Melbury Sandstone is absent or very thin.

SIX

Upper Cretaceous

CHALK

The Chalk crops out over about 180 km^2 in the eastern half of the district and forms a prominent escarpment, which rises to 274 m above OD at Bulbarrow Hill [778 056] in the south, extends north-eastwards to Melbury Hill [873 197] (263 m above OD) just south of Shaftesbury, and then swings eastwards through Win Green [925 206] (277 m above OD) to the margin of the district. The escarpment is breached by the River Stour near Shillingstone.

The ground falls gradually south-eastwards from the crest of the escarpment to about 35 m above OD in the valley of the River Tarrant in the extreme south-east. The dip of the Chalk over this tract is slightly steeper than the topographical fall, so that successively higher units crop out in a south-easterly direction.

The Chalk throughout much of southern Britain has been divided into Lower, Middle and Upper formations, the base of the Middle and the Upper Chalk being taken at the Melbourn Rock and Chalk Rock respectively, both of which, generally, are laterally persistent markers. However, in the district it was found that the Chalk Rock was not everywhere identifiable, because chalk-rock lithologies occur at several levels in the Middle and Upper Chalk and cannot readily be distinguished from one another.

By using a combination of feature mapping, lithological variations, macro- and micropalaeontology, geophysics and aerial photography (including Landsat imagery), each of varying importance throughout the sequence, it was found that the Chalk could be divided into nine mappable units, all of which are laterally persistent in the Shaftesbury and adjacent districts. All but the highest occur in the district (Table 10). In south-eastern England, different names have been used for the Chalk of the North Downs (Robinson, 1986) and South Downs (Mortimore, 1986). Some of Mortimore's nomenclature is applicable, with minor modification, to the Shaftesbury district and is used here. New names have been introduced for the Lower Chalk (not subdivided by Mortimore or Robinson) and for much of the Upper Chalk. The revised nomenclature is shown in Table 10.

The use of the terms Lower, Middle and Upper Chalk is retained in this account for ease of description and to provide continuity with traditional nomenclature. The traditional Middle/Upper Chalk boundary in the district falls in the lower part of the Lewes Chalk.

The lack of cored boreholes or good continuous exposure means that the base of some units is not precisely defined lithologically; this does not diminish their usefulness, because they allow over 300 m of Upper Chalk to be divided into five mappable units. The use of these informal units in the Shaftesbury and adjacent districts enables complex fault patterns to be recognised in the Upper Chalk for the first time; they can also be used to define the outcrops of commercially important (e.g high-brightness) chalk.

Lower Chalk

The Lower Chalk, as mapped in the Shaftesbury district, corresponds to the Lower Chalk of other areas in southern England. The junction with the Upper Greensand may be transitional over a metre or so and is marked by a fairly abrupt incoming of marl and a rapid decrease in sand content.

The Lower Chalk is divided into the West Melbury Chalk, overlain by the Zig Zag Chalk (Bristow, 1989a). The West Melbury Chalk consists of up to 16 m of soft, off-white, creamy and buff marly chalk, which is glauconitic, sandy and marly in the basal part; there are a few thin harder beds of chalk. The Zig Zag Chalk comprises 20 to 50 m of firm, off-white chalk. The boundary between the two members is usually well marked by a sharp feature break, with the West Melbury Chalk forming a low shelf, and the Zig Zag Chalk rising steeply from it. There is a distinct difference in gamma-ray signature between the two units (Figure 49). The West Melbury Chalk is equivalent to the lower part of the Chalk Marl of the South Downs succession, and the Zig Zag Chalk to the upper part of the Chalk Marl, together with the overlying Grey Chalk and Plenus Marls.

West Melbury Chalk

The West Melbury Chalk extends from beyond the northern boundary of the district to just west of Shillingstone [824 098]. Although it is about 10 m thick in the type area, the outcrop width is only about 70 to 80 m. The outcrop widens considerably to the north-east, where it forms a shelf up to 0.5 km wide [e.g. around 882 220 and 905 210] at the foot of the chalk escarpment. South and south-westwards, the outcrop varies from a few tens of metres to almost a kilometre in width south-east of Shillingstone [845 102]. West of Shillingstone, the member disappears [around 824 098], but reappears about 1 km west of the Shillingstone pit and persists as far as Okeford Fitzpaine [804 103] before dying out against the Mid-Dorset Swell (Drummond, 1970; Kennedy, 1970). West Melbury Chalk may, however, be locally present north of Ansty [7719 0477] (Jukes-Browne and Hill, 1903), but it has not been possible to map it. The progressive cutting out of the member onto the Mid-Dorset Swell can be demonstrated by a number of scattered localities.

In the subcrop, the West Melbury Chalk has been recognised from geophysical logs in the Quarleston (Figure 49), Spetisbury and Shapwick boreholes; it is absent from the Winterborne Kingston Borehole on the Mid-Dorset Swell.

Table 10 Chrono- and lithostratigraphy of the Chalk.

Stages		Benthonic foraminiferal zones			Macrofossil Zones	Macrofossil Subzones	Traditional southern England Chalk subdivisions		Mortimore's (in Lake et al., 1987) Chalk subdivisions		Subdivisions in the Shaftesbury area	
CAMPANIAN		Swiecicki (1980)	B2	iii / ii / i	*Gonioteuthis quadrata*		Upper		Culver		Upper Chalk	Spetisbury / Tarrant
CAMPANIAN		Swiecicki (1980)	B1	iii / ii	*Offaster pilula*		Upper		Newhaven		Upper Chalk	Blandford
SANTONIAN		Swiecicki (1980)	B1	i	*Marsupites testudinarius*		Upper		Newhaven		Upper Chalk	Blandford
SANTONIAN					*Uintacrinus socialis*		Upper		Newhaven		Upper Chalk	Blandford
SANTONIAN / CONIACIAN					*Micraster coranguinum*		Upper		Seaford		Upper Chalk	Blandford / Lewes
CONIACIAN					*Micraster cortestudinarium*		Upper	Top Rock	Lewes		Upper Chalk	Lewes
TURONIAN					*Sternotaxis plana*		Upper	Chalk Rock	Lewes		Upper Chalk	Lewes
TURONIAN					*Terebratulina lata*		Middle	Spurious Chalk Rock	Ranscombe	New Pit Beds	Middle Chalk	New Pit
TURONIAN					*Mytiloides labiatus* s.l. (*pars*)		Middle		Ranscombe	Holywell Beds U / M / L	Middle Chalk	Holywell
CENOMANIAN	U				*Neocardioceras juddii*		Middle	Melbourn Rock	Ranscombe	Melbourn Rock	Middle Chalk	Holywell
CENOMANIAN	U	Carter ad Hart (1977)	14		*Metoicoceras geslinianum*		Lower	Plenus Marls	Lower Chalk		Lower Chalk	(Plenus Marls)
CENOMANIAN	M	Carter ad Hart (1977)	13		*Calycoceras guerangeri*		Lower	Grey Chalk	Lower Chalk		Lower Chalk	Zig Zag
CENOMANIAN	M	Carter ad Hart (1977)	12		*Acanthoceras jukesbrownei*		Lower	Grey Chalk	Lower Chalk		Lower Chalk	Zig Zag
CENOMANIAN	M	Carter ad Hart (1977)	11ii		*Acanthoceras rhotomagense*	*Turrilites acutus*	Lower	Grey Chalk / Chalk Marl	Lower Chalk		Lower Chalk	Zig Zag
CENOMANIAN	M	Carter ad Hart (1977)	11i		*Acanthoceras rhotomagense*	*Turrilites costatus*	Lower	Chalk Marl	Lower Chalk		Lower Chalk	West Melbury
CENOMANIAN	L	Carter ad Hart (1977)	10		*Mantelliceras dixoni*		Lower	Chalk Marl	Lower Chalk		Lower Chalk	West Melbury
CENOMANIAN	L	Carter ad Hart (1977)	9		*Mantelliceras mantelli*	*Mantelliceras saxbii*	Lower	Chalk Marl	Lower Chalk		Lower Chalk	West Melbury
CENOMANIAN	L	Carter ad Hart (1977)	8		*Mantelliceras mantelli*	*Neostlingoceras carcitanense*	Lower	Glauconitic Marl	Lower Chalk	Glauconitic Marl	Up. G'sand	Melbury Sandstone
CENOMANIAN	L	Carter ad Hart (1977)	7		*Mantelliceras mantelli*	*Neostlingoceras carcitanense*	Lower	Warminster Greensand	Lower Chalk	Glauconitic Marl	Up. G'sand	Melbury Sandstone
ALBIAN			6		*Stoliczkaia dispar*	*Durnovarites perinflatum*	UPPER GREENSAND (pars)		Upper Greensand		Up. G'sand	Boyne Hollow Chert
ALBIAN			6		*Stoliczkaia dispar*	*Mortoniceras (M.) rostratum*	UPPER GREENSAND (pars)		Upper Greensand		Up. G'sand	Boyne Hollow Chert

No one locality exposes a complete sequence of the West Melbury Chalk. A representative section in the basal beds, here metricated, is that in the old Melbury Quarry [8753 2015] (Jukes-Browne and Hill, 1900, p.190).

	Thickness m
5. Greyish marly chalk, including a rough hard bed near the top, and becoming sandy in lower part with *Grasirhynchia grasiana*, *Mantelliceras* sp., *Schloenbachia* spp., *Turrilites scheuchzerianus?* and *T.* cf. *costatus*	3.20
4. Firm marly chalk with grains of quartz and glauconite; partly shaly at base, but passes into the next bed below; *Grasirhynchia grasiana*, *Avellana* sp., *Lima* aff. *aspera* and *Mantelliceras* cf. *dixoni*	0.76

MELBURY SANDSTONE (see p.110)

Fossils are common and include '*Inoceramus*' *crippsi*, *I.* ex gr. *virgatus*, *Mantelliceras dixoni*, *M. saxbii* and *Schloen-*

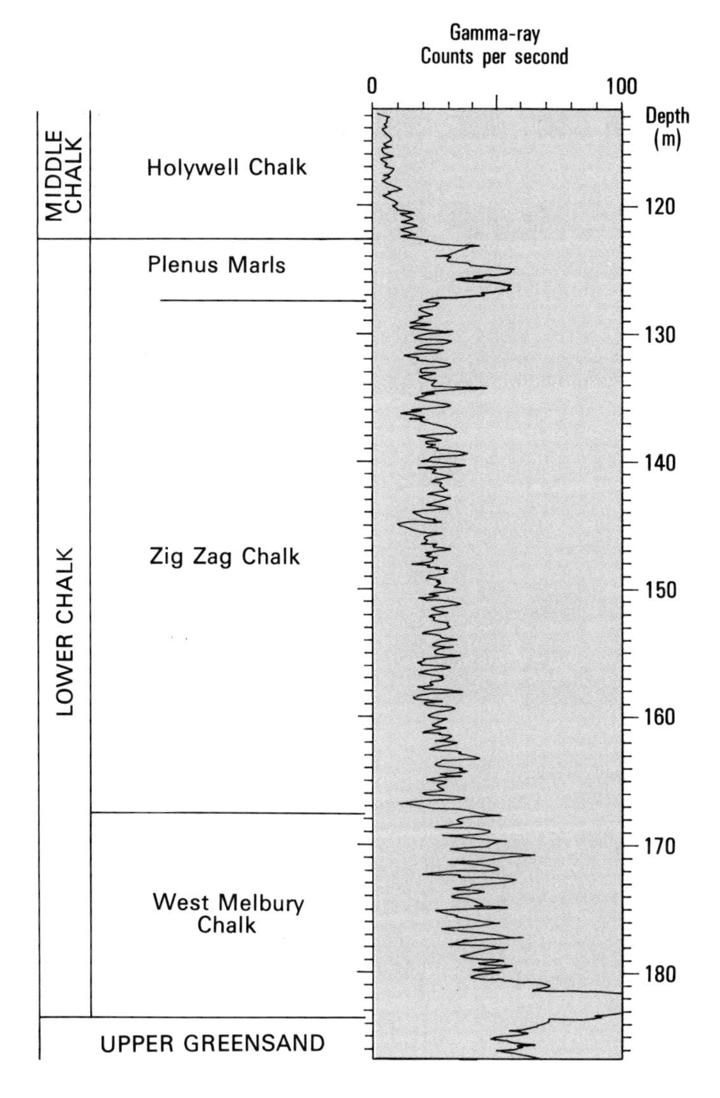

Figure 49 Gamma-ray log of the Lower Chalk of the Quarleston Borehole [8594 0565].

bachia varians (commonly abundant), together with rarer specimens of *Acompsoceras inconstans* and *Hypho-plites falcatus interpolatus*. The fauna is indicative of the *saxbii* Subzone, the *dixoni* Zone and basal part of the *costatus* Subzone.

The type section at West Melbury shows a transition from Melbury Sandstone into West Melbury Chalk. At Stour Bank [8465 1070] and Okeford Fitzpaine [803 102], the West Melbury Chalk, with a basal pebble bed with a derived *carcitanense* Subzone fauna, rests on the Bookham Conglomerate (Kennedy, 1970; Jukes-Browne and Hill 1900, 1903).

Zig Zag Chalk

The term Zig Zag Chalk was introduced by Bristow (1989a) for the 20 to 50 m of firm white chalk that overlies the West Melbury Chalk or, in its absence, the Upper Greensand. The base is taken at a marked feature break, which appears to correspond to the incoming of thick beds of firm to hard chalk above the gently sloping ground developed on the more marly West Melbury Chalk.

The member has a fragmented outcrop in the south-west around Dogbury, Ridge Hill, Dungeon Hill and near Buckland Newton. East of these occurrences, the Zig Zag Chalk has an unbroken outcrop from Moots Copse [772 048] in the south to Monk's Down [942 208] in the north-east.

There is no one section within the Shaftesbury district which exposes the whole of the Zig Zag Chalk, although pits at Shillingstone [8235 0980], on the north side of Hambledon Hill [8416 1304] (Jukes-Browne and Hill, 1903; Kennedy, 1970), on the west side of Melbury Hill [8697 1973] (Jukes-Browne and Hill, 1903; White, 1923) and at the foot of White Sheet Hill [9350 2395], just beyond the northern margin of the district (Jukes-Browne and Hill, 1903; Bristow, 1990b), must formerly have exposed good sections. However, there are sufficient sections in the Shaftesbury area to establish that the major elements of the Middle Cenomanian stratigraphy established in Sussex and Kent can be recognised in the present district (Figure 50).

Recent research on Middle Cenomanian sections at Southerham near Lewes, and on the coast between Folkestone and Dover (Gale, 1989, 1990, in preparation) has revealed a consistent stratigraphy in the lower part of the *rhotomagense* Zone. Two massive, prominent limestones (Gale, 1989, fig.3) are of particular use for long-range correlation. The lower limestone overlies a dark grey, highly fossiliferous, silty bed characterised by *'Aequipecten' arlesiensis, Inoceramus schoendorfi, Oxytoma seminudum* and *Cunningtoniceras*. The higher limestone, which marks the entry of the ammonite genus *Acanthoceras*, contains numerous large *Inoceramus tenuis*, associated with *Turrilites costatus*, including a distinctive variant. This 'Tenuis Limestone' (not named in Gale, 1989, fig.3) is overlain by the 'Cast Bed', a bed of brown, silty chalk with a rich fauna of poorly preserved composite moulds of aragonite-shelled molluscs, notably gastropods. The 'Cast Bed', which equates with the Totternhoe Stone of the East Anglia to Yorkshire outcrop, is characterised by common *Oxytoma seminudum*, and marks the entry of a distinctive and diverse assemblage of small brachiopods including *Modestella geinitzi*. The Cast Bed is overlain by an alternation of marls and marly limestones with common *Orbirhynchia mantelliana*, comprising the highest of the three mantelliana bands of southern England (the Upper mantelliana Band of Sussex (Lake et al., 1987)). The sequence is capped by a limestone with abundant *Sciponoceras baculoides*, the top of which marks the position of the mid-Cenomanian non-sequence (Carter and Hart, 1977) and the base of the *acutus* Subzone. A rhythmic alternation of marls and marly limestones continues for 1 to 2 m above the Upper mantelliana Band, above which there is a change to more massively bedded chalk with thin marl bands, constituting the Grey Chalk.

In the Shaftesbury district, the 'Tenuis Limestone' has been recognised on Hambledon Hill [8439 1312], north-east of Ferne [9368 2290], and at the foot of Zig Zag Hill [8911 2071]. The only locality where the 'Cast Bed' has

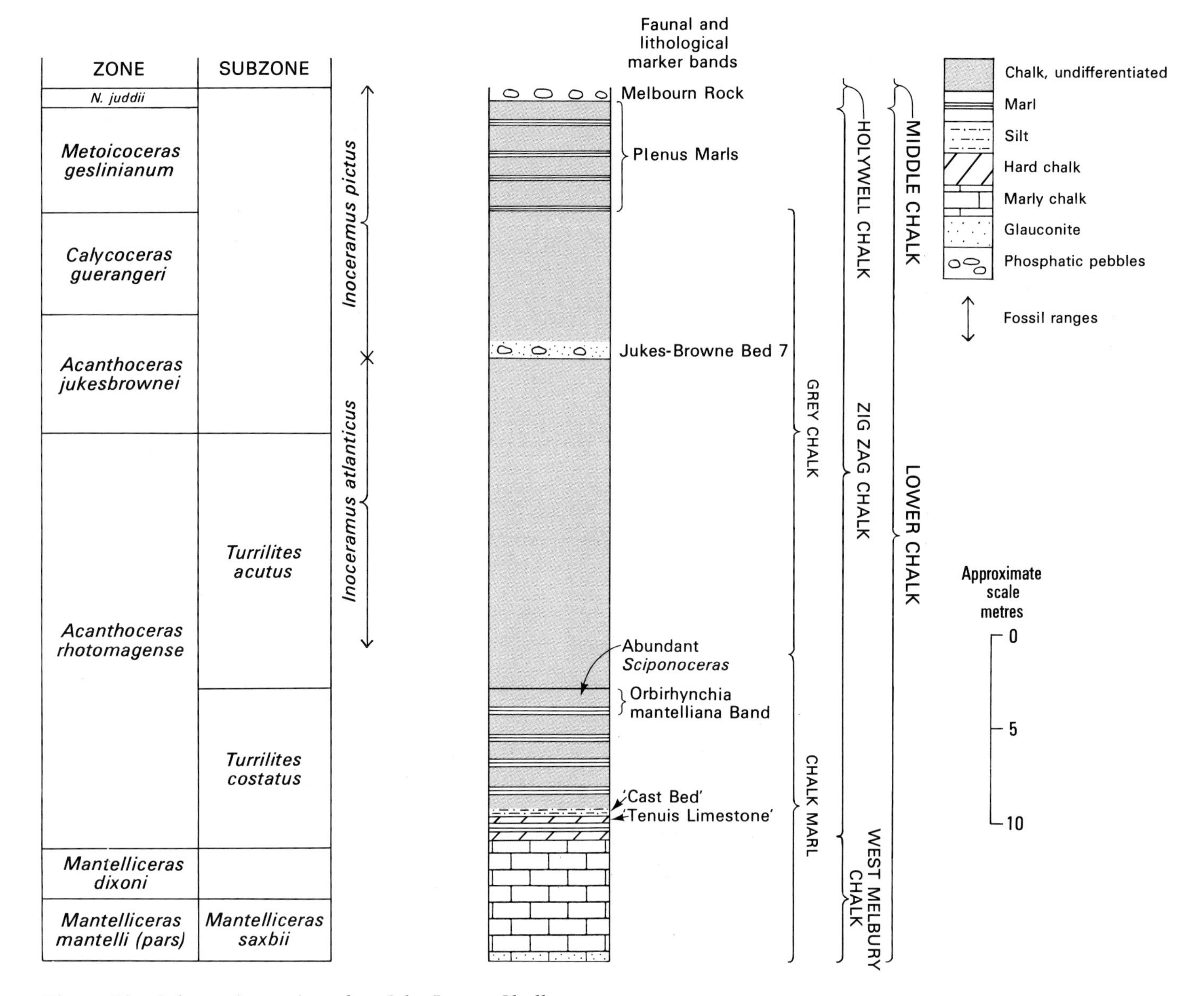

Figure 50 Schematic stratigraphy of the Lower Chalk.

been found is in the pit [8417 1312] on the north side of Hambledon Hill, where the sequence from the 'Cast Bed' to the basal part of the traditional Grey Chalk was exposed. Above the 'Cast Bed', occurred c.9 m of alternating marls and harder beds, including the O. mantelliana Band, overlain by 6 m of massive grey chalk (Jukes-Browne and Hill, 1903; Kennedy, 1970)

The upper part (Grey Chalk eqivalent) of the Zig Zag Chalk consists of firm white chalk, divisible into a lower unit, up to the level of 'Jukes-Browne Bed 7' (see Figure 50), with common *'Inoceramus' atlanticus*, and an upper unit with *I. pictus*. The range of *'I'. atlanticus* is from the base of the *jukesbrownei* Zone to the base of Jukes-Browne Bed 7; *I. pictus* s.l. ranges from within the *jukesbrownei* Zone to the top of the Plenus Marls, with related species extending up into the Holywell Chalk.

The higher part of the Zig Zag Chalk is only poorly exposed, although fossiliferous brash with either *'Inoceramus' atlanticus* or *I. pictus* is common at several localities. The highest c.15 m of the sequence, from the probable Jukes-Browne Bed 7 to the top of the Plenus Marls, was exposed in the quarry [8133 0995] and track [8130 0992] at the foot of Okeford Hill (Figure 51).

The Plenus Marls are included within the top part of the Zig Zag Chalk, because it is not possible to map them separately from the beds below. Although rarely well exposed, they are present along the whole outcrop of the Lower Chalk of the district. Surprisingly, Jukes-Browne and Hill (1903) do not refer to the Plenus Marls by name, although marl seams just below the Melbourn Rock were recorded by them in the district. This may account for their absence in this area from Jefferies' (1963, fig.6) map, despite several references to the Plenus Marls by White (1923, pp.60, 61, 62, 64). The Plenus Marls give a good gamma-ray signal and are readily recognisable in geophysical logs.

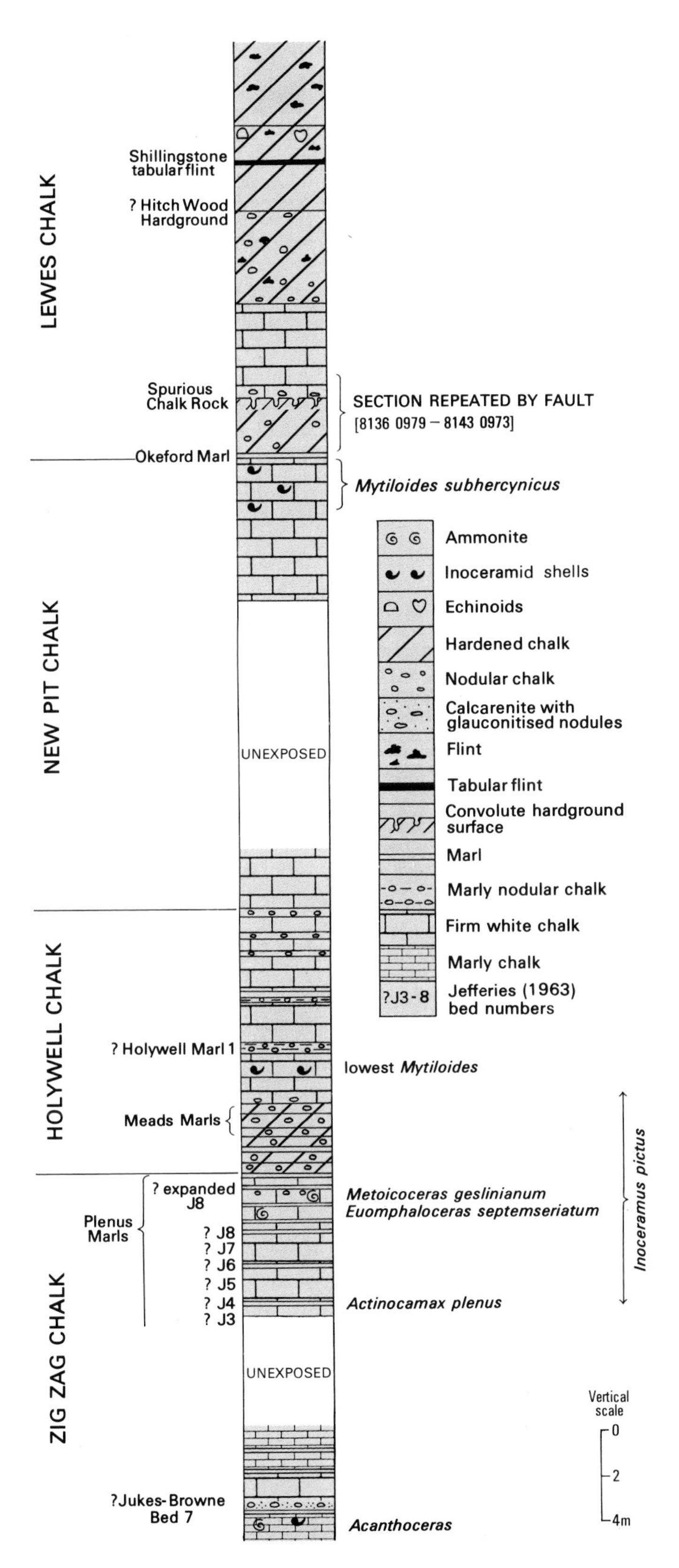

Figure 51 Section in Chalk on track [8133 0995 to 8164 0975] up Okeford Hill.

In a detailed study of the Plenus Marls in the Anglo-Paris Basin, Jefferies (1963) divided them into eight beds of alternating marl and chalk or chalky marl, each with a distinctive fauna, which he recognised throughout southern England. At the base of each marl seam, there is an erosion surface, the most strongly developed, Jefferies' sub-plenus erosion surface, being the basal one. The Plenus Marls are coextensive with the greater part of the *geslinianum* Zone. They contain common *Inoceramus pictus,* but few specimens of the zone fossil, *Actinocamax plenus,* have been found in the district.

The Plenus Marls in the district consists of an alternating sequence of blocky, white chalk in beds up to 1.2 m thick, and beds of medium grey silty marl, mostly between 1 and 10 cm thick, though the highest (Jefferies' Bed 8) can be up to 50 cm thick. Smooth-bedded chalk occurs between the highest marl and nodular chalk at the base of the Holywell Chalk. This is 0.60 m thick in the track section [8130 0994] on Okeford Hill, 0.76 m in a pit [8104 1026] on Okeford Hill, 0.6 m thick at Sharnhill Green [7026 0515] (White, 1923) and 1.67 m thick at Maiden Newton [SY 585 980] (Jukes-Browne and Hill, 1903), just beyond the district. Because of its smooth texture, this unit is included in the Zig Zag Chalk, but it may not be a correlative of the Plenus Marls of the standard succession.

In the type area, at Merstham in Surrey, the Plenus Marls are 2.13 m thick. They have their maximum development in England at Eastbourne, where they are 11 m thick (Lake et al., 1987). In the two of Jefferies' sections nearest to the district, at Lulworth and Winchester, the thicknesses are 3.3 and 2.6 m respectively. On Okeford Hill [8130 0994], where the Plenus Marls are about 6 m thick, it is evident that the 'Plenus Marl' noted by White (1923) is only the uppermost marl (Bed 8) of Jefferies (1963).

The younging of the base of the chalk in a south-westerly direction in the district, evident in the West Melbury Chalk, continues in the Zig Zag Chalk. The Chalk Marl component of the Zig Zag Chalk is present in a pit south of Belchalwell [7955 0870] (White, 1923; Freshney, 1990), but is absent farther south-west in sections at Bookham Farm [7066 0415], Dorsetshire Gap [7420 0320] and Sharnhill Green [703 052] (White, 1923; Jukes-Browne and Hill, 1903; Smart, 1955; Kennedy, 1970); it may, however, be locally present at Stoke Wake [771 048] and near Armswell [730 035] on the northern edge of the Dorchester Sheet (Jukes-Browne and Hill, 1903). As the outcrop is followed south-westwards, the beds with '*Inoceramus*' *atlanticus* (see below), which occur at an estimated 25 m above the base of the Zig Zag Chalk at White Sheet Hill in the north, are less than 5m above the base near Cerne Abbas, south of the district. They may even be absent in this latter area.

Middle Chalk

The Middle Chalk is defined in East Anglia as the beds between the base of the Melbourn Rock and the base of the Chalk Rock. Recent work (see for example Mortimore, 1986; Mortimore and Wood, 1986; Gale et al.,

1987) shows that the Chalk Rock cannot be used as a precise lithological marker in southern England. Other nodular and glauconitised horizons, and chalkstones, span the Middle/Upper Chalk boundary. These beds were named the Lewes Chalk by Mortimore (1986). The term Sussex White Chalk Formation was introduced for the chalk above the Lower Chalk, and the division into Middle and Upper Chalk was discontinued (Mortimore, 1986). However, to preserve continuity, the terms Middle and Upper Chalk are here retained for descriptive purposes.

A sequence of nodular chalks, 9 to 20 m thick, at the base of the Middle Chalk forms a readily identifiable and mappable unit across the district. Those at the base of the sequence probably equate with the Melbourn Rock of the type locality in Cambridgeshire, but the higher beds of nodular chalk in the district include strata stratigraphically higher than the Melbourn Rock.

In Sussex, the Ranscombe Chalk (Mortimore, 1986), which is equivalent to most of the Middle Chalk, comprises the Melbourn Rock, succeeded by the Holywell Beds, divisible into lower, middle and upper units, and the New Pit Beds (Mortimore, 1986). There, nodular chalks extend from the base of the Melbourn Rock to the top of the middle Holywell Beds.

In the Shaftesbury district, following discussions with Professor Mortimore, a modified name, the Holywell Chalk, has been introduced to embrace the nodular chalks of the Melbourn Rock and the lower and middle Holywell Beds; the smooth chalks of the upper Holywell Beds and New Pit Beds have been renamed the New Pit Chalk (Bristow, 1989a).

HOLYWELL CHALK

The Holywell Chalk consists dominantly of nodular chalks, with some weak chalkstones, thin marl seams and smooth-textured chalks. *Mytiloides* is common to abundant in the higher beds. The upper limit of the Holywell Chalk corresponds to the top of the shell-detrital nodular chalk. Good sections occur in Shillingstone pit [8235 0990], where the member is 14 m thick, and a poorer section can be seen in the track [8130 0995] up Okeford Hill, where it is about 10 m thick (see Figure 51). From Shillingstone to the north-east of the district, the thickness probably does not exceed 10 m; in the north-east, it is about 15 m.

The Holywell Chalk has a broken outcrop in the south-west of the district around Dogbury [660 051], Ridge Hill [685 055], Dungeon Hill [690 073] and near Buckland Newton [690 049]. The main outcrop extends from just south of Bulbarrow Hill [around 775 055] to the north-eastern margin of the district. Locally, such as at the top of Zig Zag Hill [895 210], there is a marked feature break at the top of the Zig Zag Chalk, with the Holywell Chalk cropping out over a wide area on the dip slope. Elsewhere, on the scarp face, the Holywell Chalk forms a weak to good feature. Along the feature, nodular chalk, commonly *Mytiloides*-rich, occurs in brash and minor exposures.

The fauna of the Holywell Chalk, particularly the inoceramids, shows a rapid succession of species. At the base, and continuing up from the Zig Zag Chalk to the top of the Cenomanian part of the Holywell Chalk, is *Inoceramus pictus*. The Turonian part is characterised by a succession of *Mytiloides* species, and terminates in beds with abundant *Mytiloides mytiloides*.

NEW PIT CHALK

The New Pit Chalk, 10 to 20 m thick, as used in this account (Bristow, 1989a), comprises the upper Holywell Beds and New Pit Beds of Mortimore (1986). The base is taken at the incoming of smooth, firm, white chalks with conspicuous marl seams. Towards the top of the New Pit Chalk, the lowest of the New Pit Marls, a widespread marker, has been recognised in the Shillingstone Pit by Mortimore (1987, fig.2).

In, and just beyond, the north-eastern margin of the district, grey flints occur in the upper 10 m of the New Pit Chalk [e.g. 9484 2475]. They have been recorded at two localities [8962 2078; 8738 1633] in the district. Chalk at the first locality yielded *Mytiloides subhercynicus*, *Collignoniceras woollgari* and *Lewesiceras peramplum*. At the second locality, on Sutton Hill, flints were recorded by White (1923, p.63), but are not common; only one was found during the present survey. In the Chalk there, *Collignoniceras* and *Lewesiceras* are fairly common; rare *Romaniceras?* also occur. In the Shillingstone Pit [824 098], on the track up Okeford Hill [8134 0484 to 8149 0476] and in other sections farther south-west, no flint has been seen in the upper part of the New Pit Chalk, but elements of the above fauna, particularly *M. subhercynicus*, have been recognised at several localities. The New Pit Chalk spans the *labiatus/lata* zonal boundary.

Upper Chalk

Although the Chalk Rock, the base of which defines the base of the Upper Chalk, falls within the lower part of the Lewes Chalk, for descriptive purposes, the Lewes Chalk is here included within the Upper Chalk.

The Lewes Chalk comprises a succession of hard, locally nodular or porcellanous chalks, and is succeeded by a thick sequence of firm white chalk with scattered flints and thin marl seams. Dip and scarp features within the higher succession, first identified by Bristow (1987) in the Bournemouth district, have now been recognised in the Shaftesbury and surrounding districts. Because of poor exposure, the features have not yet been directly related to lithological units, but their persistence and apparently constant stratigraphical position allow four informally named units bounded by these features to be mapped. The lowest of these, the Blandford, Tarrant and basal part of the Spetisbury Chalk, crop out in the Shaftesbury district.

LEWES CHALK

The Lewes Chalk (Mortimore, 1986) in the district comprises fossiliferous, hard, commonly yellow-stained and glauconitised chalkstones, and nodular, in places glauconitised, chalks. There is a complex internal stratigraphy (Figures 52 and 53). The Lewes Chalk is about 30 to 40 m thick over much of the northern part of the district, but south of Iwerne Minster [880 150] the thickness increases to 50 m.

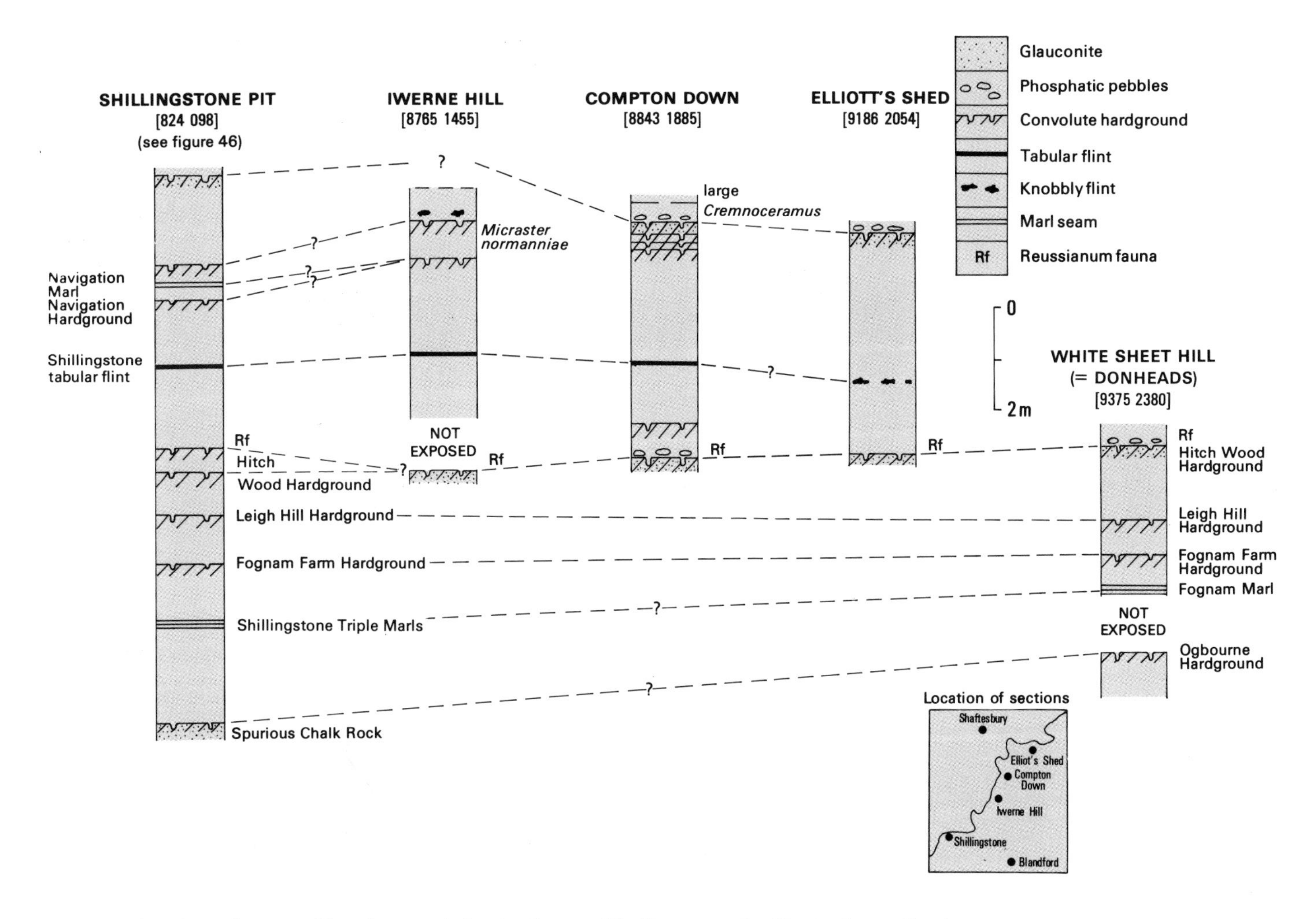

Figure 52 Correlation of hardgrounds in the Lewes Chalk across the Shaftesbury district.

The base of the Lewes Chalk, readily identifiable on sonic logs, is taken at a persistent marl seam, above which occur nodular chalks and chalkstones. This marl has been correlated with the Fognam Marl (Bromley and Gale, 1982), Glynde Marl 1 (Mortimore, 1987; Wood et al., 1982) and one of the New Pit Marls (Wray and Gale, 1992). Because of uncertain long-distance correlations, a local name, Okeford Marl, named from the section [8133 0995 to 8164 0975] up Okeford Hill (Figure 51) is introduced in this account.

About 2 m above the Okeford Marl occurs the Spurious Chalk Rock, a distinctive bed of hard, unfossiliferous, glauconitised chalk and pebble bed. It has been recorded at several localities between Piddletrenthide [7043 0093] and the Winterborne Kingston Borehole [SY8470 9796] in the Dorchester district, north-eastwards as far as Shillingstone [824 098], Stourpaine [865 086] and Ranston Hill [8676 1194] in the present district. Glauconitised Spurious Chalk Rock has not been identified north of Ranston Hill, but it could be present as far north as Sutton Hill [8753 1639]. From Sutton Hill southwards, the 'Lower' Rock of the older authors (e.g. White, 1923) refers to the Spurious Chalk Rock, but north of Sutton Hill, the 'Lower Rock' refers to a higher glauconitised hardground, the Hitch Wood Hardground (see below).

The Spurious Chalk Rock is generally regarded as the lowest hardground of the Chalk Rock Formation (Bromley and Gale, 1982) or Member (Gale et al., 1987), although the correlation is uncertain (see below). The Chalk Rock in the type area of Ogbourne Maizey, Wiltshire, comprises five principal hardgrounds developed on chalkstones and separated by weakly to strongly developed nodular chalks. The two lowest, the Ogbourne and Pewsey hardgrounds, are overlain by a well-developed marl seam, the Fognam Marl. Above the Fognam Marl, Bromley and Gale recognised three major hardgrounds, the Fognam Farm, Leigh Hill and Hitch Wood in ascending order. The Hitch Wood Hardground is characterised by the so-called 'reussianum fauna', comprising phosphatised moulds of molluscs and ammonites, including the eponymous heteromorph *Hyphantoceras reussianum.*

The three closely spaced marl seams which occur 1.5 m above the Spurious Chalk Rock in the Shillingstone pit have been correlated with the Fognam Marl (Bromley and Gale, 1982, fig.10), the Latimer Marl (Bromley and Gale, 1982, p.294) and Southerham Marl 1 (Mortimore, 1987; Wray and Gale, 1992). Additionally, Wray and Gale

Figure 53 Section in Shillingstone Chalk Pit (including work by Mortimore (1987; Ms) and Bromley and Gale (1982; Ms)).

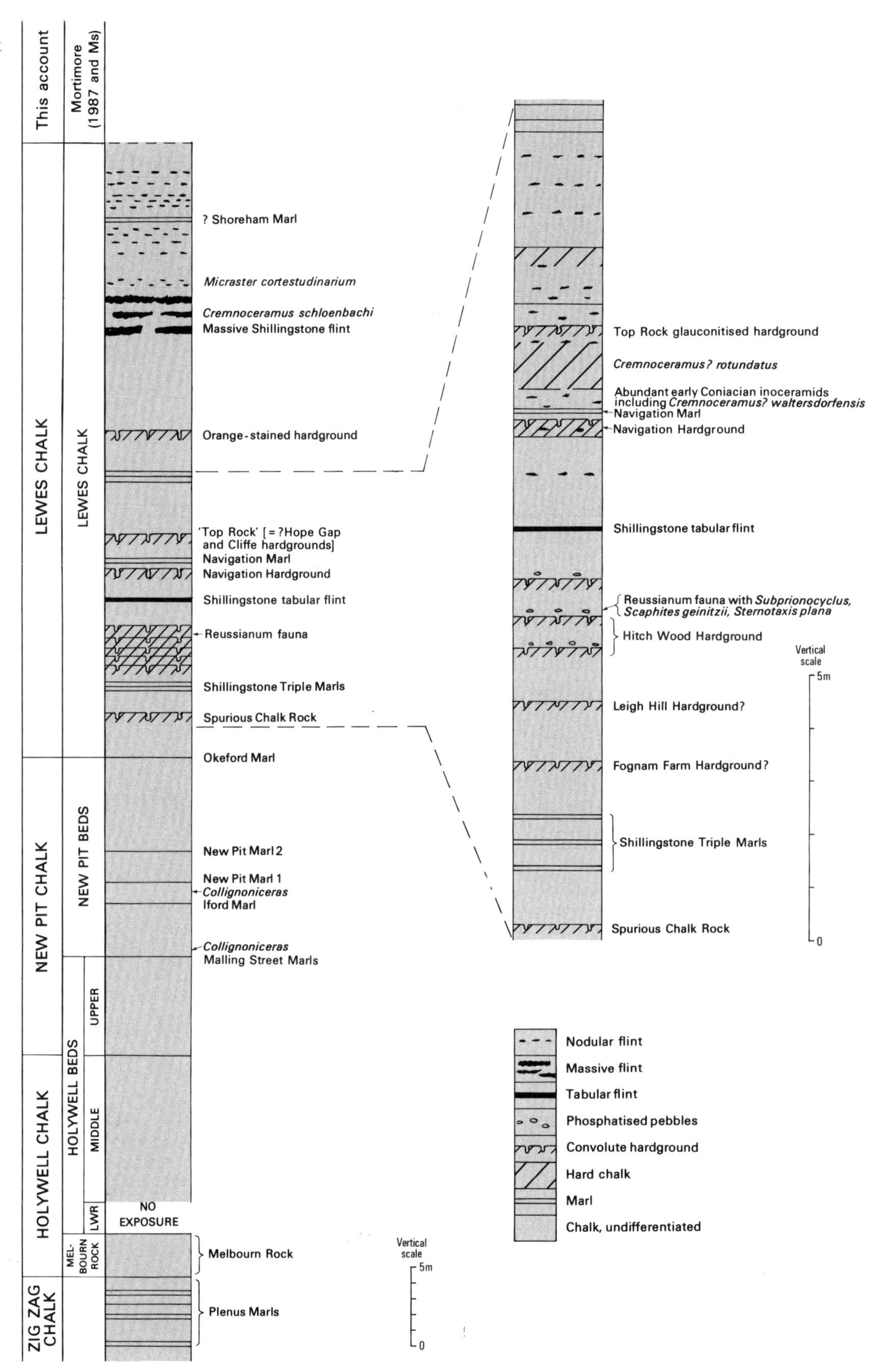

(1992) suggest that the Fognam Marl of Wiltshire might be a correlative of the Glynde Marl of Sussex. As with the marl beneath the Spurious Chalk Rock, because of the uncertain correlation, a local name, the Shillingstone Triple Marls, named from the Shillingstone pit [823 099] (Figure 53), is introduced in this account.

In the district, only the Hitch Wood Hardground of the type area in Wiltshire has been identified with certainty. The correlation of the lower hardgrounds, especially the Spurious Chalk Rock, and the interbedded marl seams with those of Wiltshire and Sussex, is speculative and unresolved. The Okeford Marl beneath the Spurious Chalk Rock is discussed above. The Spurious Chalk Rock could correlate with the either the Ogbourne Hardground or with the Fognam Farm Hardground (Bromley and Gale, 1982, pp.293, 294).

The pit [9375 2382] on White Sheet Hill (the Donheads locality of Bromley and Gale, 1982) formerly exposed a 5 m section, with the supposed Fognam Marl in the floor of the quarry and the inferred Fognam Farm, Leigh Hill and the richly fossiliferous Hitch Wood hardgrounds in the quarry face. Only the last of these is still exposed. At an unknown distance below the marl, a nearby exposure showed a glauconitised hardground, which Bromley and Gale (1982, fig.10) tentatively correlated with the Ogbourne Hardground, but which (in field notes) they regarded as the Spurious Chalk Rock.

In the north of the district, the Hitch Wood Hardground with its 'reussianum fauna' has been recognised with certainty in the floor of the quarry [919 206] at Elliott's Shed (Bromley and Gale, 1982), at the base of the pit [8845 1885] on Compton Down, and on Iwerne Hill [8765 1461], its most southerly occurrence as a glauconitised hardground. At Compton Down, the Hitch Wood Hardground is composite, comprising a glauconitised hardground with green pebbles, overlain by 0.65 m of nodular limestone with scattered phosphatic intraclasts and abundant glauconite grains, terminating ina phosphatised surface, on and below which the 'reussianum fauna' is concentrated. Small carious finger flints rest on the surface and penetrate 0.2 m into the underlying nodular chalk. Traced southwards, the bed loses its distinctive glauconitic staining and much of its associated fauna. At Shillingstone [824 098], the equivalent hardground is phosphatised, and both phosphatic intraclasts and glauconite grains are absent from the overlying nodular limestone. The latter is penetrated by a more extensive network of finger flints than at Compton Down and terminates in a phosphatised hummocky surface overlain by a gritty chalk with phosphatised reussianum faunal elements.

In the north of the district, the Hitch Wood Hardground is overlain by up to 10 m of hard nodular chalks, terminating in a complex of glauconitised, pebble-strewn hardgrounds (the 'Upper Rock' of White, 1923), which was correlated with the Top Rock by Bromley and Gale (1982). In the more basinal succession at Shillingstone [824 098], a conspicuous semicontinuous tabular flint 1.5 m above the equivalent of the Hitch Wood Hardground, appears to correlate with a tabular flint developed high in the *plana* Zone at Mupe Bay and White Nothe on the Dorset coast (Mortimore and Pomerol, 1987). This flint also occurs on the track up Okeford Hill (Figure 51), and is presumably represented by the thin flint near the base of the exposed section on Iwerne Hill (Figure 54). It is possibly identi-

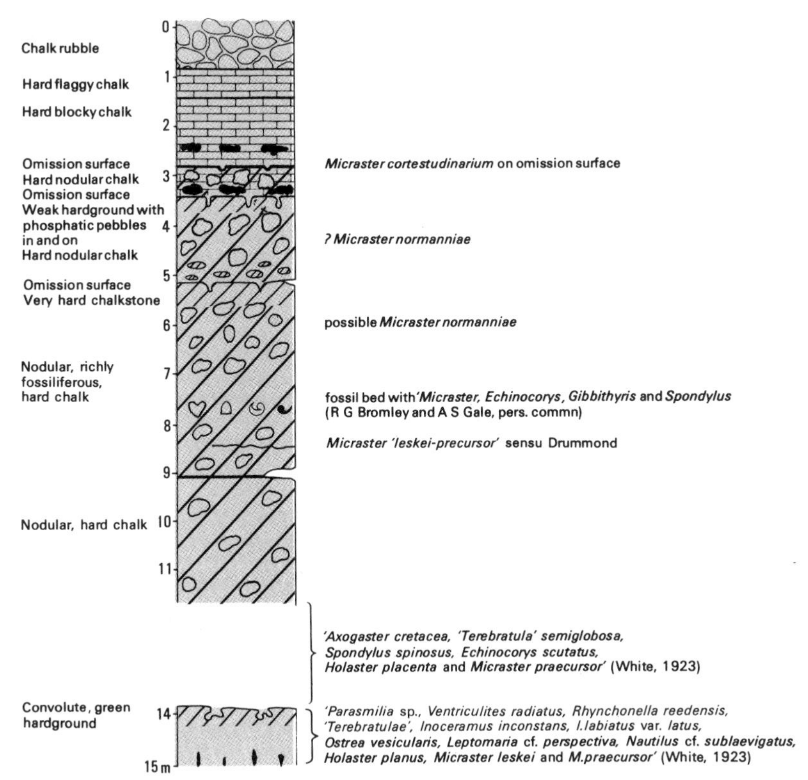

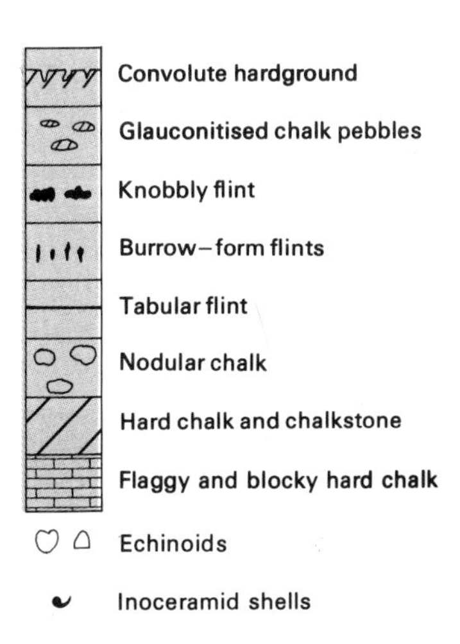

Figure 54 Section in the Lewes Chalk, Iwerne Hill.

fiable at Compton Down (Figure 52) and, tenuously, at Elliott's Shed.

Above this tabular flint, an iron-stained hardground and an overlying marl seam at Shillingstone have been correlated respectively with the Navigation Hardground and one of the Navigation Marls of the Sussex succession. This correlation of the boundary successions between the *plana* and *cortestudinarium* zones has been confirmed during the recent survey by the finding of a shell bed about 1m above the marl, which is rich in the basal Coniacian inoceramids *Cremnoceramus? waltersdorfensis* and *C? rotundatus*; a comparable shell bed is known from the base of the *cortestudinarium* Zone at Dover, Kent and in Sussex. A glauconitised hardground at 5.5 m above the Hitch Wood Hardground, and 1.2 m above a tabular flint at the top of the shell bed, was correlated by Mortimore (1987, fig.5) with the Top Rock and equated with the Hope Gap Hardground of the Sussex succession.

The Top Rock/Hope Gap Hardground is succeeded by about 30 m of hard, nodular, flinty chalk and chalkstone, locally porcellanous, but with some interbeds of firm white chalk. There are discontinuous exposures of this succession above the main faces at Shillingstone. The biostratigraphy is poorly known, but there are manuscript records (R N Mortimore and A S Gale) of the high *cortestudinarium* Zone inoceramid *Cremnoceramus schloenbachi* from a level of three closely spaced, thick tabular flints, about 5 m above a hitherto unrecognised orange hardground, and c.11 m above the Hitch Wood Hardground. During this survey, inoceramids, possibly including *I. kleini*, were collected at this level, but no evidence of *C. schloenbachi* was found. The thickest (lowest) of these tabular flints was incorrectly identified by Mortimore and Pomerol (1987, fig.9) with the Seven Sisters Flint in the lower part of the *coranguinum* Zone. Current work suggests that the entire succession (c.12 m) at Shillingstone above the Top Rock, up to the level of a marl seam near the top of the section, belongs to the *cortestudinarium* Zone; the marl seam is tentatively correlated with the higher of the Shoreham Marls at the base of the *coranguinum* Zone (Mortimore, 1986).

Fossils, dominantly brachiopods and echinoids, are locally common in the Lewes Chalk and indicate that the lower part of the Chalk Rock falls within the *lata* Zone (Table 10). The beds between the Hitch Wood Hardground and the base of the Top Rock span the *plana* Zone. The bulk of the Lewes Chalk falls in the *cortestudinarium* Zone, but its top is of early *coranguinum* Zone age (e.g. Normandy Farm [840 060], Winterborne Stickland). The flints within late *cortestudinarium* and early *coranguinum* Zone strata are characteristically very carious (Plate 10).

Blandford Chalk

The Blandford Chalk consists of 75 to 90 m of firm, white, flinty, fossiliferous chalk with thin marl seams. There are few exposures of it in the district, but many former exposures, particularly in the upper part, were recorded by Jukes-Browne and Hill (1904), White (1923), and the late Mr S C A Holmes during the 1930s.

The base of the Blandford Chalk is taken at the incoming of soft to firm, smooth-textured, white chalk. Flints in the basal part of the member are typically carious.

Fossils include sheet-like spreads of the inoceramid *Platyceramus* and common *Micraster* in the lower half, and the crinoids *Bourgueticrinus, Uintacrinus* and *Marsupites*, and the echinoid *Offaster*, in the higher strata; *Echinocorys* occurs throughout. The fauna indicates that the Blandford Chalk spans the *coranguinum* Zone to the upper part of the *pilula* Zone (Table 10). Strata of *coranguinum* Zone age occupy just over half of the succession, with strata of *socialis, testudinarius* and *pilula* zones age falling within the remainder.

Faunal evidence from three former exposures establish the age of the highest beds of the Blandford Chalk. Near Blandford, a cutting [9060 0677] on the old military sidings proved the upper part of the *pilula* Zone (White, 1923). The fauna quoted by White cannot be found, but his record of '*Echinocorys scutatus* var. *depressus*' from a pit [9060 0675] intersected by the cutting indicates the higher part of the *depressula* Subzone (Wood and Mortimore, 1988, fig.18). Material collected by Mr S C A Holmes, presumably from higher in the pit, includes *Echinocorys scutata truncata* and *Offaster pilula*, indicative of the lower belt of abundant *Offaster pilula* in the basal part of the *pilula* Subzone (Table 11). The junction of the two subzones must therefore have been exposed in the pit. The base of the succeeding Tarrant Chalk forms a good feature some 15 m higher than the pit; this suggests that the top of the Blandford Chalk lies in the higher part of the *pilula* Subzone. This conclusion is supported by the occurrence of a small *Offaster pilula* and a possible *Echinocorys elevata truncata* from a pit [9099 0435] near Manor Dairy Farm (see p.131).

Holmes (MS) recorded 2.4 m of hard splintery chalk with irregular layers of flint in a pit [9485 1133] on Launceston Down just east of the district, lying some 8 m below the top of the member. A marked bed of *Echinocorys* was present about 1.5 m from the top. Biostratigraphically important fossils collected by Holmes included eight small *Echinocorys* that compare well with the so-called 'small forms' of Gaster (1930), and included two *E.scutata cincta* sensu Gaster and five *E.scutata depressula* sensu Gaster. By themselves, these fossils would place the horizon unequivocably in the higher part of the '*Hagenowia* Horizon' of the *quadrata* Zone. However, a foraminiferal assemblage collected from close to the pit indicates the higher part of the *pilula* Zone. The absence of *Gavelinella usakensis* and the occurrence of *Reussella kelleri* indicates benthonic foram Zone B1 of Swiecicki (1980) in the *pilula* Zone (Table 11). A second sample, 8 m higher and on the feature break at the base of the Tarrant Chalk yielded an essentially similar microfauna, but included *Stensioeina pommerana*, a species that enters the record towards the the top of Subzone B1iii, at an horizon equivalent to the *Offaster pilula* belt at the top of the eponymous zone. Thus, it appears on foraminiferal evidence that this pit falls in the *cincta* belt, between the upper and lower *Offaster pilula* belts (Table 11), and that the top of the Blandford Chalk lies near the top of the *pilula* Zone.

Table 11 The development of the biostratigraphical subdivisions of the upper part of the Blandford Chalk, Tarrant Chalk and Spetisbury Chalk.

Brydone (1914, 1915)			Gaster (1930)				Bailey et al. (1983)				Swiecicki (1980)		Member (this account)
			Zone	Subzone	Horizon	*Echinocorys* belt	Zone	Subzone	Foraminifera biozone	Foraminifera assemblage biozone	Foraminifera Assemblage Zone		
Actinocamax quadratus Zone			*Actinocamax quadratus*	*Saccocoma cretacea*		'small forms' of var. *cincta* and *depressa*	*Gonioteuthis quadrata*	unnamed	*Bolivinoides culverensis* total range zone	*Pullenia quaternaria*	iii	B2	Spetisbury Chalk
													Tarrant Chalk
								Applinocrinus cretaceus		*Eouvigerina galeata*	ii		
								'*Hagenowia* Horizon'					
			Offaster pilula	*Echinocorys scutatus* var. *cincta*	*Hagenowia rostrata*	'large forms' of *E. scutatus*				*Gavelinella usakensis*	i		
Offaster pilula Zone	abundant *Offaster pilula* Subzone	upper belt of *O. pilula*			abundant *O. pilula*	var. *cincta*	*Offaster pilula*	abundant *Offaster pilula*		*Gavelinella cristata*	iii	B1	
													Blandford Chalk
		middle belt or *cinctus* belt											
		lower belt of *O. pilula*				var. *truncata*					ii		
	Echinocorys scutatus var. *depressus* Subzone			*Echinocorys scutatus* var. *depressa*				*Echinocorys depressula*					
							U. anglicus		*Bolivinoides strigillatus* partial range zone	*Gavelinella cristata*/ *Globigerinelloides rowei*	i		
Marsupites Zone			*Marsupites testudinarius*				*Marsupites testudinarius*						

Tarrant Chalk

The Tarrant Chalk crops out only around Blandford Camp in the south-east of the district. It consists of 25 to 30 m of soft to firm, white, flinty chalk that is lithologically indistinguishable from the chalk of the adjacent units. Its base is marked by a distinct feature break which is presumed to be formed by a hardground. No section has been seen across this boundary. Fossils include brachiopods, crinoids, the echinoid *Echinocorys* and the belemnite *Gonioteuthis quadrata;* they indicate that most of the Tarrant Chalk falls within the lower part of the *quadrata* Zone, with a small part at the base falling within the *pilula* Zone. The forms of *Echinocorys* present allow broad biostratigraphical units to be recognised (Table 11).

Spetisbury Chalk

The Spetisbury Chalk crops out only on the east side of the Tarrant Valley in the south-east of the district. There is no exposure, but brash and sections in adjacent districts indicate that, like the underlying chalk, the Spetisbury Chalk consists of firm, white, flinty chalk. The base is marked by a prominent feature which is probably formed by a hardground. Fossils include *Gonioteuthis quadrata,* crinoids, asteroids and a distinctive elongate, narrow-based *Echinocorys marginata* (Bristow, 1987); they are indicative of the *cretaceus* Subzone (Table 11).

DETAILS

Lower Chalk

West Melbury Chalk

Stoke Wake

Jukes-Browne and Hill (1903) found *'Mantelliceras mantelli'*, indicative of a horizon low in the West Melbury Chalk, in a pit [7719 0476] at Moots Copse, south of Bulbarrow Hill. As the locality lies some 10 m above the Upper Greensand/Lower Chalk junction, the determination of the ammonite is in doubt.

Okeford Fitzpaine–Iwerne Minster

In the railway cutting at Gain's Cross [c.8420 1025], alternating beds of hard and soft marly chalk with *Aucellina gryphaeoides* and *Discoidea cylindrica* were recorded (White, 1923).

The section at Stour Bank [c.8465 1070] was described by Mottram (1957) and, in more detail, by Kennedy (1970). Kennedy's section is as follows:

	Thickness m
West Melbury Chalk	
Chalk, marly, glauconitic, soft, gritty, grey-green, with harder concretionary masses, intensely burrowed. Small creamy phosphates occur throughout, and are common in the basal 15 to 20 cm. At the base, there are some glauconitised and phosphatised cobbles of glauconitic sandstone, and these are particularly common in depressions in the underlying irregular erosion surface	1.5
Upper Greensand (see p.109)	

Phosphatised *carcitanense* Subzone fossils, presumably derived from the Melbury Sandstone, abound in the West Melbury Chalk (see Kennedy, 1970). The field to the north-east [8467 1070] is covered with blocks of glauconitic shelly chalk and phosphatised fossils. Species additional to those listed in Kennedy (1970) include (determinations by Dr C W Wright): *Hyphoplites pseudofalcatus, Neostlingoceras carcitanense, Turrilites scheuchzerianus, Avellana cassis, Cucullea mailleana* and *Aucellina gryphaeoides.*

On the east side of Hambledon Hill, blocks of glauconitic shelly limestone [from 8473 1304] yielded *Boubeithyris diploplicata, Hyphoplites arausionensis, Mantelliceras saxbii* [also from 8466 1315], *Mariella* sp., *Schloenbachia varians* and *Discoides subuculus.*

A roadside exposure [8653 1468] at Iwerne Minster shows 2 m of alternating hard white chalk and marly chalk with *Inoceramus, Mantelliceras* and *Schloenbachia varians.*

Melbury Abbas to Shaftesbury

The uppermost beds of the West Melbury Chalk were formerly exposed in a quarry [8697 1909] on Melbury Hill (Jukes-Browne and Hill, 1903):

	Thickness m
Soil, containing a layer of hard chalk	0.46
Zig Zag Chalk	
Grey blocky chalk marl	0.76
Course of hard grey chalk	0.22
West Melbury Chalk	
Grey chalk marl, passing down into bluish sandy marl, with conchoidal fracture, coming out in ball-like lumps	3.05

South-east of Melbury Abbas, the West Melbury Chalk is about 5 m thick. A temporary exposure [8865 1961] revealed sandy chalk with *Inoceramus* and *Schloenbachia varians* resting on glauconitic, very chalky Melbury Sandstone (p.110). Nearby [8866 1958], at a slightly higher level, alternating beds of off-white marl and harder marly chalk with *Schloenbachia* occur.

The best section in the West Melbury Chalk was formerly exposed in the Melbury quarry [8753 2015] (Jukes-Browne and Hill, 1900) (see p.110).

A pit [9117 2207] to the east of Charlton exposes about 0.7 m of off-white chalk with *Inoceramus virgatus?* and *Entolium* sp., overlain by 0.4 m of mottled medium grey and buff marly chalk with *Schloenbachia* sp.

Brash of off-white and grey chalk within a dominantly buff marly sequence near Higher Berrycourt yielded *Birostrina* sp? and *Schloenbachia varians,* including the inflated hypernodose form *ventriosa* [9205 2125], a whorl fragment of an inflated form of *Mantelliceras* sp., such as characterises the upper part of the *carcitanense* Subzone (sensu Gale, 1989), from near the base of the member [9200 2133], *Mantelliceras* cf. *saxbii* [9263 2138] and *Schloenbachia varians* [9262 2128], and a large mould of *Inoceramus* ex gr. *virgatus,* indicative of the lower part of the *dixoni* Zone, from close to the junction with the Upper Greensand [9387 2155].

North-west of Ferne, off-white marly chalk with *Schloenbachia varians* and fragments of large nautiloids [e.g 919 225; 923 227] dips 5° NNW towards the Ferne Park Fault. North-east of Rowberry House [924 230], brash of buff and pale grey marly chalk with *Inoceramus* ex gr. *virgatus* [e.g. at 9249 2308] is common. Nearby [9265 2322], brash includes *'Inoceramus'* ex gr. *crippsi?, I.* ex gr. *virgatus, Panopea* sp. or *Pholadomya* sp, a distorted *Mantelliceras* and *Schloenbachia varians.* The fauna, which comes from about 1 to 2 m above the base of the Chalk, indicates the lower part of the *dixoni* Zone.

Zig Zag Chalk

Dogbury–Woolland

A pit [6590 0516] at Dogbury, exposes 7 m of soft, grey, laminated chalk. On Ridge Hill, a pit [6851 0576] shows 6 m of soft grey marly chalk.

In a pit [6910 0719] on Dungeon Hill, 0.3 m of grey marl of the Plenus Marls is overlain by 0.1 m of hard nodular chalk of the Melbourn Rock.

The Knole Quarry [7030 0519], Sharnhill Green, exposes between 25 and 30 m of soft marly chalk, overlain by 2 to 3 m of Plenus Marls. From the lower part of the face, *'Inoceramus' atlanticus,* indicating the *jukesbrownei* Zone, was collected. The base of the Chalk is described by Kennedy (1970) as buff or grey-green, glauconitic chalk resting on the Bookham Conglomerate; more recently, Dr H G Owen (written communication, 1991) excavated the lower part of the following section:

	Thickness m
Zig Zag Chalk	
Chalk, hard, blocky, white, with large ammonites and belemnites near the top	35.00
Chalk, blocky, buff, marly, with glauconite grains	2.00
Chalk, buff, marly, highly glauconitic, with brown phosphatic nodules; base more glauconitic; phosphatic nodules more common in the lowest 2 cm; rests on the irregular phosphatic top of	0.23 to 0.30
Upper Greensand (Bookham Conglomerate) (see p.109)	

The basal part of the Zigzag Chalk is exposed at Bookham Farm [7064 0415], just south of the district, where 4 m of soft white chalk, with a 1 m glauconitic basal bed, overlies the Bookham Conglomerate (p.109)(Figure 48). The basal bed contains the lightly glauconitised fauna: *Concinnithyris subundata*, *Grasirhynchia* cf. *grasiana*, *Avellana* sp., *Acanthoceras* sp., *Scaphites* cf. *equalis*, *Schloenbachia* sp., *Sciponoceras baculoides* and *Turrilites costatus* indicative of the *costatus* Subzone. A shelly horizon in off-white chalk, some 2.5 m higher, yielded '*Inoceramus*' cf. *atlanticus*, suggestive of the lower part of the *jukesbrownei* Zone.

On Bulbarrow Hill, exposures of soft grey chalk with *Euthymipecten beaveri* [7720 0568] and *Inoceramus pictus* [7737 0555] are characteristic of a level just below the Plenus Marls. Another locality [7729 0517] showed grey chalk with '*Inoceramus*' *atlanticus*, indicative of the *jukesbrownei* Zone.

Woolland to Belchalwell

The quarry [760 065] at Stoke Wake exposes a degraded section in 30 m of creamy white, rather soft chalk. White (1923) recorded '*Aequipecten beaveri* and *Inoceramus crippsi*'. Carter and Hart (1977) noted a boulder bed (?Bookham Conglomerate) at the base, and concluded that the Chalk Marl was absent.

Jukes-Browne and Hill (1903) recorded the following section in a pit near Woolland Church [7792 0684]:

	Thickness m
Soil and chalk rubble	0.31
Zigzag Chalk	
Chalk, white, blocky, much fracturing	0.91
Chalk, silty and crumbly with occasional hard lumps set in soft silty chalk	0.76
Chalk, grey, consisting of very hard lumps with soft silty chalk between, mottled with brown markings	1.83

Fossils in the lowest unit were '*Am. [Acanth.] rotomagensis?, Pecten orbicularis, Rhynchonella grasiana* and *Terebratula biplicata*'.

A quarry [7840 0640] on Chitcombe Down exposes 7 m of soft cream-coloured chalk with *Inoceramus* ex gr. *pictus*, probably *Inoceramus pictus bohemicus*, indicative of the Plenus Marls.

White (1923) recorded 'bluish grey marl of '*Varians*' aspect' in a pit [7897 0766] 60 m north of Ibberton church.

A pit [7932 0818] near the top of the member north-east of Ibberton exposes 4 to 5 m of soft, marly, fractured chalk with *Inoceramus* ex gr. *pictus* and either *I. pictus* or *I. ginterensis*. A pit [7955 0870] south of Belchalwell exposes about 20 m of soft, marly, burrowed chalk with *Turrilites costatus*, indicative of the *costatus* Subzone, from the floor of the pit (White, 1923). The junction with the underlying Greensand is only a short distance below the base of the quarry.

Okeford Fitzpaine–Shillingstone

A pit [8104 1027] on the north side of Okeford Hill exposes the junction of the Plenus Marls and Holywell Chalk (Jukes-Browne and Hill, 1903; White, 1923) (Plate 9). The section below is a combination of present-day exposure and that recorded by White (1923):

	Thickness m
Holywell Chalk	
Hard white chalk, much broken, shaly at base	1.83
Hard yellow conglomerate (0.15 m) consisting of chalk pebbles in a shelly calcitic matrix, passing down into hard rough nodular chalk	0.91
Several beds of firm white chalk, with hard yellowish nodules, separated by layers of grey shaly chalk	1.52
Very nodular hard chalk	0.38
Chalk, off-white with scattered pale green-coated nodules	0.30
Marl, greyish buff	0.04
Chalk, off-white with only a few scattered nodules	0.17
Chalk, off-white, very nodular, with orange or pale green-coated nodules; many cidarid radioles	0.50
Chalk, off-white with very few scattered nodules; fragment of *Inoceramus pictus*	0.15
Chalk, very nodular; nodules orange or greyish green-coated; one small *Inoceramus*	0.40
Plenus Marls	
Smooth-bedded chalk	0.23
Thin layer of grey marl	0.08
Blocky greyish white chalk	2.44

Only the 8 cm bed of marl was considered by White to be the Plenus Marls. The status of the 0.23 m bed of smooth-bedded chalk above the marl is uncertain. Lithologically it is part of the Zig Zag Chalk, and may be a unit younger than Bed 8 of the type Plenus Marls sequence. On the east side of the pit, there is a vertical, east–west-trending fault of unknown throw, downthrowing to the west.

An old pit [8133 0992] at the foot of Okeford Hill exposes (Figure 51):

	Thickness m
Chalk, white (buff to grey when wet), marly, firm; impersistent marl seam, up to 5 cm thick, 1.2 m from top, and another, less than 1 cm thick, 2 m down	2.20
Chalk, off-white, grading down into	1.20
Calcarenite, pebbly, with scattered green-coated nodules up to 2 cm across, and larger hard nodules up to 6 cm across; scattered fossils including *Terebratulina protostriatula, Gryphaeostrea* juv.? and *Pycnodonte?*	0.40
Marl, buff-grey with *Inoceramus*	0.15
Chalk, marly, hard, with *Inoceramus* ex gr. *pictus* and *Acanthoceras?;* grades down into	0.30
Chalk, marly, buff	0.90

The pebbly bed is probably Jukes-Browne Bed 7.

Just west of the above occurrence, there is an exposure [8127 0998] in the Plenus Marls (Figure 51).

A track [8196 1001 to 8163 1000] on the north side of Okeford Hill has exposures in buff-grey marl with thin beds of marly chalk with *Grasirhynchia grasiana* [8177 1008 to 8195 1002], off-white hard chalk with *G. martini* [8171 1008], off-

Plate 9 Junction (arrowed) of the nodular Melbourn Rock and smooth marly chalk and marls of the Plenus Marls, Okeford Fitzpaine.

white chalk with *Turrilites* sp. [8167 1007] and hard, off-white chalk with *Grasirhynchia martini* and *Acanthoceras* [8165 8004]. The chalk between the last two sections is interbedded with buff, grey and pale brown marl. These sections correspond to the upper part of the Chalk Marl and lie close to the base of the Zig Zag Chalk, which here rests on Upper Greensand.

The Plenus Marls are exposed in a gully [8234 0988] in Shillingstone pit, but because of talus, there is no clear section. Mortimore (1987, fig.5) showed a total thickness of 4.25 m of Plenus Marls.

Hambledon Hill

On the west side of Hambledon Hill, the Zig Zag Chalk is about 17 m thick. Farther south-east, it thins to about 15 m [around 8490 1135]. On the north and east sides of the hill, the thickness is about 30 m. At the foot of the north-west side, the basal beds consist of buff and grey marl with thin beds of marly chalk and harder beds of chalk [8438 1314 to 8447 1308]. Fossils from this tract include *Inoceramus tenuis, Plicatula inflata, Sciponoceras baculoides, Schloenbachia coupei* and *Turrilites costatus.* The common occurrence of *Inoceramus tenuis* in the brash suggests the Tenuis Limestone.

At the quarry entrance [8417 1312] on the north side of Hambledon Hill, a 0.3 m bed of brownish grey, silty, shelly marl is exposed near the base of the Zig Zag Chalk. The fauna includes *Antalis* sp., *'Cirsocerithium'* ex gr. *reussi, Entolium* sp., *Freiastarte* sp. and *Oxytoma seminudum.* The combination of lithology and fauna indicates that this bed corresponds to the 'Cast Bed'. On the east side of the pit, soft blue and grey sandy marls alternating with beds of hard grey chalk, dipping at 7°SE, were formerly exposed (Jukes-Browne and Hill, 1903). On the west side of the pit, and possibly separated by a fault, about 9 m of similar beds, overlain by 6 m of massive grey chalk, were seen. From blocks estimated to have come from several tens of metres above the base of the Chalk, Kennedy (1970) collected a fauna indicative of the *costatus* Subzone, with the brachiopods indicating the O. mantelliana Band at its top.

Iwerne Minster–Melbury Hill

South of Iwerne Minster, nodular chalk with *Inoceramus* found [8680 1384; 8696 1383] low down in the Zig Zag Chalk, may be from Jukes-Browne Bed 7.

Near Compton Down [8790 1926; 8790 1902], blocks of off-white, firm chalk with brachiopods, *Acanthoceras, Schloenbachia* and *Turrilites* occur. A quarry [8795 1884] formerly exposed 6m of chalk, grey and marly with '*Schloenbachia varians*' in the lower part, and whiter, with several beds of hard chalk, in the upper part (Jukes-Browne and Hill, 1903, p.105).

In an quarry [8697 1973] on the west side of Melbury Hill, Jukes-Browne and Hill (1903, p.105) and White (1923, p.59) recorded seven or eight courses of hard grey chalk in soft marly chalk. '*Schloenbachia varians*', associated with '*S. coupei*' and '*Cymatoceras deslongchampsianum*', was common, especially at one level [c.8698 1973]. In the upper tier, massive, firm, blocky chalk, alternating with layers of soft marly or shaly chalk occurred, which passed up into blocky, nearly white chalk. This was overlain by about 1.2 m of shaly Plenus Marls, and by the Holywell Chalk (see below).

On the south-east side of Melbury Hill, an exposure [8779 1946] shows a 20 cm bed of glauconitic nodular chalk — possibly Jukes-Browne Bed 7. Some 70 m west and about 10 m higher [8787 1947], the Plenus Marls comprise white marly chalk with common *Inoceramus pictus*, and are overlain by a 0.5 m bed of nodular chalk of the Holywell Chalk. Just west of the above section, steeply jointed, firm white chalk with a few scattered nodules overlies a silty marl with *Inoceramus pictus, Metoicoceras geslinianum* and *Sciponoceras* sp.

Zig-Zag Hill–Ferne

On Breeze Hill, the Zig Zag Chalk is 25 to 30 m thick. In the track up the hill, marly chalk of the Plenus Marls crops out [8903 1957] close to an exposure [8904 1955] of nodular chalk at the base of the Holywell Chalk.

A representative, but poorly exposed section in the basal beds of the Zig Zag Chalk at the foot of Zig Zag Hill [8911 2071] exposes the following section; Bed e was not seen resting directly on Bed d:

	Thickness m
ZIG ZAG CHALK	
g. Blocky, off-white, shelly chalk with common *Sciponoceras baculoides, Schloenbachia coupei, 'Inoceramus' reachensis? s.s.,* occurring as a talus	1.5
f. Silty, marly chalk with *Orbirhynchia mantelliana* fragment (?0. mantelliana Band)	c.0.3
Unexposed	c.0.9
e. Bluish grey silty chalk with scaphopods, *Entolium, Oxytoma seminudum, Acanthoceras* sp., *Plagiostoma globosum, Genicularia rustica,* etc. (?'Cast Bed')	0.4
Unexposed	c.0.5
d. Hard splintery chalk with *Inoceramus, Turrilites, Sciponoceras* (?Tenuis Limestone)	c.0.9
Unexposed	c.1.0
?WEST MELBURY CHALK	
c. Buff silty marl with *Turrilites costatus*	c.0.4
b. Buff off-white, hard chalk	c.1.4
a. Silty marl with *Orbirhynchia, Inoceramus, Plicatula* aff. *minuta, Grasirhynchia grasiana, Terebratulina* ex gr. *nodulosa*	0.3

Fossils from Bed d include *Concinnithyris subundata, Inoceramus* cf. *tenuis, Plicatula inflata, Acanthoceras* sp., *Calycoceras asiaticum?, Cunningtoniceras* sp., *Schloenbachia coupei, Sciponoceras baculoides* and *Turrilites costatus,* indicative of the *costatus* Subzone.

At the top of a pit [8870 2139] south-east of Cannfield Farm, there is off-white, blocky marly chalk with *Cymatoceras deslongchampsianum, Acanthoceras* sp., *Calycoceras* sp. and *Cunningtoniceras* sp., indicative of the *costatus* Subzone. Another pit [8875 2163] showed c.0.8 m of grey to off-white blocky chalk with '*Inoceramus*' cf. *atlanticus* in the lower part [8872 2165] and 2 m of blocky off-white chalk with *'Inoceramus'* ex gr. *atlanticus*, '*I*' cf. *longobardicus, Cymatoceras deslongchampsianum* and *Sciponoceras baculoides* in the upper part [8875 2163]. The ammonites indicate the *rhotomagense* Zone; the inoceramids suggest the *jukesbrownei* Zone, but since none is definitely *atlanticus*, possibly this section is lower than the *jukesbrownei* Zone.

On Zig Zag Hill [8939 2082], hard nodular chalk at the base of the Holywell Chalk (p.126), partially overlies and is partially in fault contact with pale grey marl, 7 to 40 cm thick, which in turn overlies 1 m of marly, off-white chalk with *Inoceramus pictus* and *Metoicoceras geslinianum*, which in turn overlies 6 cm of pale grey chalky marl, on 0.35 m of pale grey marly chalk (Plenus Marls). Farther east [9038 2075], firm off-white chalk with *Concinnithyris subundata* and *'Inoceramus' atlanticus,* both typical of the middle part of the Zig Zag Chalk, occurs.

On the partly fault-bounded spur extending north-westwards from Elliott's Shed [around 913 214], were found *Inoceramus pictus,* indicative of the *guerangeri* Zone, and, slightly lower [9158 2107], *Sternotaxis trecensis*? and a possible '*I*. *atlanticus*, indicative of the *jukesbrownei* Zone.

North-east of Ferne House, *Inoceramus tenuis, Schloenbachia* sp. and *Sternotaxis gregoryi* or *trecensis,* suggestive of the *costatus* Subzone, were obtained from the base of the Zig Zag Chalk [9368 2290]. A little to the east [9403 2288], off-white marly chalk, about 8 m above the base, yielded *Inoceramus tenuis* and *Acanthoceras* sp.

A pit [9350 2395] just beyond the district at the western end of White Sheet Hill must formerly have exposed almost all of the Zig Zag Chalk and the basal Holywell Chalk. Sections seen in 1989 are described by Bristow (1990b).

Ashcombe Bottom

Fossils from brash of pale grey chalk and marly chalk in Ashcombe Bottom include *Acanthoceras* sp., *Schloenbachia coupei* and *Sciponoceras baculoides* [9310 2050], a possible '*Inoceramus*' ex gr. *atlanticus, I.* cf. *tenuistriatus* sensu Keller, *Acanthoceras* sp? and *A.* sp. or *Calycoceras (Newboldiceras)* sp. [9310 2030], indicative of either the *acutus* Subzone or *jukesbrownei* Zone, probably the former, and *Acanthoceras* sp., *I. tenuis?* and '*I*'. *atlanticus*? [9315 2020]. This last fauna is contradictory, because it includes elements of both *costatus* Subzone and ?*jukesbrownei* Zone faunas.

Middle Chalk

HOLYWELL CHALK

Dungeon Hill–Buckland Newton–Sharnhill Green

In the quarry [6901 0706] on Dungeon Hill, 0.1 m of nodular chalk overlies 0.3 m of marl of the Plenus Marls. South-west of Buckland Newton, nodular chalk debris [6773 0468] with *Sciponoceras* occurs above Plenus Marls debris with *Metoicoceras geslinianum.*

Holywell Chalk is exposed at the top of the Knole Quarry [7026 0515], Sharnhill Green. It appears to consist of c.1 m of hard, nodular, ferruginous chalk, above 5 m of marly chalk with bands of marl (the Plenus Marls).

Stoke Wake–Delcombe Wood–Woolland–Ibberton–Belchalwell

In a track [7698 0584 to 7699 0584] on the south side of Bulbarrow Hill, there are exposures of hard chalk with *Mytiloides mytiloides*, indicative of the *labiatus* Zone, just above the Melbourn Rock. Slightly down slope, nodular chalk is exposed [7697 0584; 7708 0582].

Exposures in Delcombe Wood showed about 12 m of creamy shelly chalk with *Orbirhynchia* sp. and *Mytiloides* cf. *mytiloides* [from 7843 0542], indicative of the *labiatus* Zone, just above the Melbourn Rock. Melbourn Rock was seen by White (1923) just east [7938 0486] of Delcombe Farm (now Delcombe Manor).

A pit [7903 0748] near Ibberton exposes 1 m of cream-coloured, somewhat nodular chalk at the top of the Holywell Chalk, overlain by 4 m of white chalk with small spindle-shaped flints and scattered inoceramids [New Pit Chalk].

A pit [7949 0845] south of Belchalwell Street exposes fairly hard, cream-coloured, nodular chalk. In the upper part of another quarry [7958 0870], nodular chalk overlies softer marly chalk and marl of the Plenus Marls.

Okeford Fitzpaine–Shillingstone–Durweston

The section in the basal Holywell Chalk in the pit [8104 1027] at Okeford Fitzpaine is described above (p.123) (see Plate 9). Fossils from the upper part of the 'Melbourn Rock' include *'Inoceramus labiatus, Ostrea* and radioles of *Cidaris hirudo'* (White, 1923). The 'conglomeratic' band in the roadbank [8052 1061] (White, 1923, p.64) south of Okeford Fitzpaine, formerly thought to be the Melbourn Rock, is now regarded, from its topographical position, as a nodular bed in the Lewes Chalk.

The basal beds of the Holywell Chalk are well exposed in the track [8129 0993 to 8130 0989] up Okeford Hill (Figure 51)

At times, there are good exposures in the 14 m of Holywell Chalk in the Shillingstone pit [823 098]. The Melbourn Rock at the base is usually exposed at the top of a gully [8232 0988] and in a nearby bank [8231 0990]. According to Mortimore (Ms, 1987, fig.5), the Melbourn Rock consists of 3.75 m of alternating nodular chalk and thin marls. A bed with abundant *Sciponoceras* occurs 0.75 m from the top. The higher part of the Holywell Chalk consists of nodular chalk with common clay drapes and wisps, in beds up to 0.5 m thick, of firm white marly chalk and thin marls.

Nodular chalk with marly partings and *Mytiloides* is exposed along the track [8358 0912 to 8402 0906] on the north side of Enford Bottom. On the south side, an exposure [8415 0862] shows up to 6 m of nodular chalk with marly wisps. A pit [8519 0927] south of the A357 near Durweston, exposes a degraded section of alternating beds of marly, hard, nodular chalk and blocky, greyish chalk with common shell fragments.

Winterborne Houghton

Nodular chalk with *Mytiloides* occurs as brash on the south side of Coombe Bottom [8027 0709; 8031 0709] and around the Winterborne Houghton valley [8001 0472 to 8022 0467; 8125 0435 to 8135 0440]. East of the church, debris from a trench [8215 0438] included nodular chalk with rhychonellids, *Mytiloides* cf. *labiatus* and *M.* cf. *mytiloides*.

Hod Hill and Hambledon Hill

'Melbourn Rock' was recorded by the road [c.8615 1103] north-east of Hod Hill (White, 1923). A pit [8575 1224] on the east side of Hambledon Hill shows a 4.5 m section in alternating beds of hard nodular chalk (in beds 0.3 to 0.6 m thick) and firm non-nodular chalk (in beds up to 0.6 m thick); thin partings, up to 8 cm thick, of more marly chalk occur. The dip is 5°SW; the face is heavily jointed, with open joints up to 8 cm wide. In a pit [8412 1226] on the west side of Hambledon Hill, there are poor exposures of white, non-nodular chalk, with some grey- and orange-coated nodular chalk with *Mytiloides*.

Stourpaine to Melbury Hill

In an exposure [8757 1806] on Fontmell Down, a 0.28 m bed of hard nodular chalk, with small irregular pyrite nodules in the top 15 cm, forms the basal bed of the Holywell Chalk; it rests on 0.1 m of marly chalk, which in turn rests on a thin dark grey marl. The nodular chalk is succeeded by 1.5 m of hard, weakly nodular chalk with *Inoceramus pictus* cf. *bannewitzensis* and radioles of *Hirudocidaris* in the basal 0.37 m, and then by c.0.1 m of slabby, firm, only weakly nodular chalk with inoceramids, including *?Mytiloides*. The top of the section consists of 1 m of firm chalk with *Mytiloides*.

At the valley head between Melbury Hill and Compton Down, firm white chalk with *Sciponoceras*, overlying 0.5 m of nodular chalk, resting on Plenus Marls, is exposed by a track [8786 1947].

On the west side of Melbury Hill, a quarry [8700 1975] formerly exposed 6 m of Holywell Chalk (White, 1923, p.59).

Zig Zag Hill–Higher Berrycourt

An exposure [8903 1955] on Melbury Down and brash to the south-east [8905 1940 to 8945 1940] includes much nodular chalk and weak chalkstones; fossils near the top of the member [8941 1941] include common *Mytiloides mytiloides*, a transitional '*mytiloides–labiatus* form' and a possible *M. submytiloides*.

An exposure [8940 2083] on the south side of Zig Zag Hill shows:

	Thickness m
Chalk, firm, off-white, with scattered nodules; *Mytiloides* fragments	0.40
Marl, chalky, laminated	0.03–0.06
Chalk, firm, off-white	0.26
Chalk, firm, nodular, off-white	0.26
(several small faults downstepping westwards; succession below repeated?)	
Chalk, off-white, weakly nodular, with nodules up to 3 cm across	1.00
Marl, pale grey, with wispy lamination	0.1–0.15
Chalk, off-white, with only a few scattered nodules (up to 4 cm across)	0.25–0.30
Marl, wispy	0.01–0.02
Chalk, off-white	0.20
Chalk, nodular	0.22
Marl, pale grey	0.01
Chalk, nodular	0.08
Marl	0.01–0.02
Chalk, nodular	0.25
Marl, laminated, pale grey	0.05
Chalk, nodular	0.70

ZIG ZAG CHALK (Plenus Marls — see p.125)

Common fragments of nodular chalk and *Mytiloides mytiloides*, about 6 to 7 m higher stratigraphically, occur as brash [around 8945 2095] to the north-west.

A section [9314 2108] south-east of Higher Berrycourt Farm showed 0.5 m of hard, nodular, off-white to buff, orange-stained chalk with echinoids, overlying pale grey marly chalk; similar nodular chalk occurs at the top [9331 2109] of a pit to the east-south-east.

A pit [9366 1894] in Ashcombe Bottom exposes about 2 m of rubbly nodular chalk and weak chalkstones. Some of the nodules have pale grey and green coatings. A fossiliferous layer at the top yields *'Rhynchonella cuvieri'*, *Inoceramus* and *Mytiloides* sp.

North-east of Ferne, at the top of the Holywell Chalk, there about 15 m thick, rhynchonellids, associated with *Mytiloides* [9422 2304] are common.

New Pit Chalk

Buckland Newton to Ibberton

Exposures [7032 0510; 7028 0504] at the top of Knole Quarry consist of white, fairly hard chalk with inoceramids. On Bulbarrow Hill, soft white chalk with *Orbirhynchia* sp. was noted [7693 0585]. Exposures [7831 0545 to 7840 0541] along a track in Delcombe Wood show about 5 m of soft, creamy, white chalk with inoceramids. A pit [7904 0747] at Ibberton shows 4 m of soft shelly chalk, overlying 1 m of nodular Holywell Chalk.

Okeford Hill to Durweston

Exposures of up to 6 m of firm white chalk with *Mytiloides subhercynicus* in the upper part of the member, occur along the track [8132 0987 to 8134 0985] up Okeford Hill (Figure 51).

About 16 m of New Pit Chalk, consisting of firm white chalk in beds generally up to 1 m thick, interbedded with thin marl seams, occur in the Shillingstone pit [8230 0983]. The Malling Street Marl, taken by Mortimore (1986, fig.1) to mark the junction between his Holywell and New Pit beds, occurs some 13 m beneath the base of the Lewes Chalk (1987, figs.2 and 5). Other marl seams identified by Mortimore higher in the sequence, but questioned by Wray and Gale (1992), include New Pit Marl 1 and 2, some 8.3 and 6 m below the base of the Lewes Chalk respectively.

Winterborne Houghton to Winterborne Stickland

Firm white chalk with *Mytiloides* at the base of the New Pit Chalk was seen in tree roots [8011 0711] in Coombe Bottom. An exposure [8351 0528] at Winterborne Stickland shows about 1 m of firm white chalk in the upper part of the New Pit Chalk. The uppermost beds were presumably exposed in the old pit 500 m to the south, beneath Lewes Chalk (see p.128).

Stourpaine to Sutton Waldron to Compton Abbas

The railway cutting [8642 0865] south of Stourpaine previously exposed 6.1 m of blocky greyish white chalk beneath Lewes Chalk (see pp.128–129).

A pit [8739 1635] on the north-west side of Sutton Hill formerly exposed the following section (White, 1923):

	Thickness m
?Lewes Chalk	
Lumpy white chalk, with a 2.5 cm marl seam [Okeford Marl?] in the lower part, and small, round, yellow-skinnned grey flints. Fossils included '*Terebratulina lata, Ostrea vesicularis* and *Inoceramus* (bits)'	1.30
New Pit Chalk	
Massive, blocky white chalk with a few flints like those above; '*Ventriculites radiatus, Plocoscyphia* sp.'	3.66

White estimated that the top of the section lay 3 m below the 'Chalk Rock' [?Spurious Chalk Rock] seen in an adjacent quarry [8753 1638] (see pp.129–130). At present, there is poor exposure in about 2.5 m of firm, white, non-nodular chalk with marl seams and rare flints; fossils include *Mytiloides subhercynicus, Collignoniceras woollgari, Lewesiceras peramplum, Romaniceras* sp?. and *Sciponoceras* cf. *bohemicum*.

Melbury Hill–Charlton Down

Brash of firm white chalk on Melbury Down, high in the sequence, includes *Mytiloides* cf. *subhercynicus* [8937 1951], *M. subhercynicus, M.* cf. *hercynicus* and *Lewesiceras peramplum* [8930 1947], and, about 7 m lower stratigraphically [8928 1945], small *Mytiloides* sp.

A section [8958 1989] in the basal beds on the north-east side of Melbury Down exposes about 1 m of blocky weathering, firm, off-white chalk with several *Mytiloides mytiloides*. Firm white chalk with *Mytiloides* is exposed along the track [8975 1997 to 8985 1977] on the western side of Compton Down.

A section [8963 2078] on the track up Breeze Hill exposes 1 m of firm, off-white chalk with a line of pale grey burrow-form flints up to 5 cm across, on slabby chalk rich in oysters, *Mytiloides subhercynicus* (Plate 8h), *Collignoniceras woollgari* and *Lewesiceras peramplum*. To the east, a section [8965 2078] reveals 0.3 m of firm white chalk with small, pale grey flints, terebratulids, gastropods, *Inoceramus* spp. and *Lewesiceras peramplum*.

Berwick St John

Brash on a field [9435 2320] north of Berwick St John includes common *Mytiloides subhercynicus*. Close by, but a little lower, brash includes *Orbirhynchia* sp., *Mytiloides mytiloides* and *M.* cf. *labiatus* [9435 2320].

Upper Chalk

Lewes Chalk

Delcombe Wood to Ibberton

In Delcombe Wood, exposures of Spurious Chalk Rock show 0.3 m of very hard, splintery, white chalkstone [7888 0558] and very hard, splintery, patchily glauconitic chalk in the floor of a track [7837 0539].

South-east of Delcombe Wood, 0.8 m of carious-weathering, very hard chalkstone with small nodular flints crops out [7991 0483] along a sharply defined feature on the side of a combe. This bed is probably the Chalk Rock.

In a lane [7913 0752] south-east of Ibberton, 0.2 m of splintery chalkstone with an orange-stained top, probably the Spurious Chalk Rock, is exposed. Some 0.4 m of nodular to soft chalk, about 25 30 m higher, is exposed in a pit [7925 0739] and floor of a nearby track [7938 0755].

Okeford Hill–Shillingstone–Durweston–Bryanston

Some 16 m of Lewes Chalk are exposed in the track [8134 0984 to 8164 0977] up Okeford Hill (Figure 51). The Spurious Chalk Rock, in part fault repeated, is well exposed along 170 m of the track [8134 0984 to 8149 0976].

The Shillingstone pit exposes some 35 m of Lewes Chalk, from the Okeford Marl at the base up to a massive, 0.25 m thick flint (Shillingstone Flint), followed by a succession of nodular, in part carious flints and, at the top of the pit, a prominent 10 cm thick marl seam, possibly the Shoreham Marl (Figure 53). The Spurious Chalk Rock, usually well exposed in the top face, picks out a series of small faults with downthrows up to 2 m. Fossils from the upper part of the pit include *Micraster cortestudinarium*. In this upper part, beds of porcellanous chalk and flint are brecciated and calcareously cemented into a massive bed (see also below).

An exposure [8355 0964] on Shillingstone Hill, reveals a 0.15 m thick bed of chalkstone with small flints, on firm white chalk. Close by, White (1923) recorded 3.9 m of blocky chalk with close-set, regular courses of cavernous flints and *'Inoceramus, "Terebratula" carnea* and *Micraster coranguinum'*. A notable feature was a bed of massive rock, 0.6 to 1.35 m thick, of calcareously cemented chalk and flint, similar to the one in the Shillingstone pit. East of the track, but lower down, a pit [c.8366 0975] exposed 'lumpy chalk with spongeous and other flints', and in the lower part of the pit, a layer of flints cemented into a continuous band by hard tufa or travertine; *Micraster cortestudinarium* was found above the tufaceous layer (White, 1923). White thought that the *Micraster* came from some 3 to 3.6 m above the 'Upper Rock' [?Top Rock], which was exposed by the entrance to the pit. Lower down the track [?c.8367 0977], hard, nodular chalk with *'Holaster planus'* and *'Echinocorys'* was seen (White, 1923).

At the western end of Enford Bottom, a tabular flint [8333 0909; 8360 0890], 0.2 m thick, and high in the Lewes Chalk succession, probably correlates with the flint near the top of the Shillingstone pit and on Okeford Hill.

Lumpy chalk with a conspicuous marl seam, probably the Okeford Marl, was formerly exposed in a pit [8526 0922] alongside the A357 (White, 1923, p.63). Fossils included '*Inoceramus labiatus* var. *latus*, *Ostrea vesicularis* and *Terebratulina lata*', which suggested a correlation with the pit at Sutton Waldron (see pp.129–130). Green-coated, hard, nodular chalk brash on the pit floor was thought to be from the 'Chalk Rock' [Spurious Chalk Rock].

Some 350 m west-south-west of Durweston Church, hard chalk with green-coated nodules is overlain by about 7 m of hard, white chalk with sparse bioclastic debris [8550 0845]. A river cliff [8708 0765] at Bryanston School showed about 3 m of firm to locally hard, irregularly blocky or nodular chalk, with thin interbeds of non-nodular chalk containing traces and wisps of grey marl. Sparse, small, rounded, thin-rinded flints occur throughout.

Winterborne Valley

A pit [8149 0880] at Turnworth exposed 0.46 m of 'Upper Rock' with upright flints ('as at Iwerne Hill') with scattered '*"Terebratula" carnea* and *Inoceramus inconstans*', overlain by 4.5 m of nodular chalk with abundant flint nodules and seams of tabular flint (White, 1923). The fauna, *'Holaster placenta, Echinocorys scutatus, Inoceramus inconstans,* but no *Micraster'*, suggests the *cortestudinarium* Zone. The beds dip 7°SE. The top of the Lewes Chalk lies some 20 m higher.

Brash [8081 0653] south of South Down included *?Cremnoceramus* sp. and *Micraster decipiens*, indicative of the *cortestudinarium* Zone at about the level of the Hope Gap Hardground (Mortimore, 1986).

A pit [840 060] at Normandy Farm, Winterborne Stickland, has been recently re-excavated and shows a 5 m section in firm to hard, patchily nodular, flinty chalk (Bristow, 1992). The presence of *Micraster coranguinum* (1.5 m above the base of the section), the common occurrence of *Platyceramus* and rare *Volviceramus*, together with carious flints (Plate 10), indicates the basal part of the *coranguinum* Zone. The hardest chalks occur in the basal 1.2 m; they pass up into firm to hard chalk. Similarly, the weak to incipient nodularity decreases upwards. Firm white chalk occurs as brash in the field above the pit. Thus, the transitional boundary between the Lewes and Blandford Chalk occurs in the upper part of this pit.

A pit [8356 0477] in Winterborne Stickland, where the Lewes Chalk is about 40 m thick, formerly exposed (White, 1923, p.70):

	Thickness m
Rubble of nodular chalk, 'perhaps replacing the Upper Rock'	0.3–0.76
Marl	0.02
Nodular white chalk, with a regular course of flints near the top, and scattered flints below; '*"Terebratula" carnea, "T." semiglobosa, Inoceramus inconstans, Holaster planus* and *Micraster praecursor*'	3.05
Lower Rock [?Spurious Chalk Rock]: hard yellowish chalk with a layer of green and brown nodules at top, which is uneven and has a brown phosphatic crust, traversed by borings	0.30
Nodular greyish white chalk, with a few green nodules; passing down into lumpy, and then NEW PIT CHALK Blocky chalk	3.66

The Lewes Chalk near Winterborne Houghton is about 45 m thick. A section [8212 0425] near the church, just south of the district, revealed porcellanous glauconitised chalkstone, the Spurious Chalk Rock, about 2 m above firm white chalk of the New Pit Chalk. On the north side of the valley, a pit [8226 0494] showed spongy [?carious] and solid flints, bits of *Inoceramus* and occasional *Micraster*, indicative of the *coranguinum* Zone (White, 1923, p.72).

Hod and Hambledon hills

These hills are capped by up to 40 m of Lewes Chalk. Hard nodular chalk and chalkstone brash, some porcellanous, is common all over the summits. Glauconite-coated nodular chalk found [8553 1218; 8575 1162] near the base of the sequence on Hambledon Hill is probably the Spurious Chalk Rock.

Blandford to Iwerne Minster

A pit [8809 0750] on the north side of Blandford formerly exposed 6m of rough and lumpy white chalk with common flints, some cavernous, and thick (up to 12 mm) *Inoceramus* fragments (Jukes-Browne and Hill, 1904; White, 1923). Cavernous flints were also noted in the railway cutting [?8785 0830] near Nutford Farm; in the next cutting [?8660 0845], *Micraster cortestudinarium* was recorded (Jukes-Browne and Hill, 1904).

Hard, nodular chalk is widespread across France Down. The upper boundary of the Lewes Chalk was seen in a section [8725 0876 to 8735 0882] along a track, where hard, nodular chalk with tabular flints passes up into soft to firm white chalk with thick-rinded flints. An exposure [8766 0988] 450 m south-east of Ash, reveals hard, nodular and green-stained chalk, interbedded with firm or hard white chalk with large, pale brown flints.

Jukes-Browne (in White, 1923) described the following section in the Stourpaine railway cutting [8645 0865]:

Plate 10 Carious flints in the top of the Lewes Chalk, Normandy Farm, Winterborne Stickland.

	Thickness m
LEWES CHALK	
Loose nodular chalk with shaly layers, and a layer of shaly chalk at the base	3.05
Hard, rough, nodular chalk	0.91
Hard, compact, cream-coloured rock with a layer of green-coated nodules at top	0.61
NEW PIT CHALK	
Blocky, greyish white chalk	6.10

The bed with green nodules is almost certainly the Spurious Chalk Rock, although *Micraster praecursor* ('sutured') from loose nodular chalk, 1.5 m above the green nodular bed (White, 1923), suggests that it could be the Chalk Rock. Higher strata were exposed in the quarry [8640 0895] above the railway cutting, where Jukes-Browne and Hill (1904) recorded hard and rough, greyish white, ?phosphatic nodular chalk, with grey streaks and seams, and enclosing many scattered flints. It was thought that the chalk was of *cortestudinarium* Zone age, although the only fossil found was '*Echinocorys scutatus*'. White (1923, p.71), who recorded a slightly different sequence, found an ovate form of *Echinocorys scutatus* close to the top of the lumpy chalk, and *Micraster coranguinum* in the blocky chalk about 1.8 m higher, and thought that all the chalk exposed fell within the *coranguinum* Zone. However, Mr S C A Holmes collected *Micraster cortestudinarium*, 1.5 m below the top of the nodular chalk.

Glauconite-coated nodular chalkstone low down in the sequence [8675 1197] north-west of Everley Farm is probably the Spurious Chalk Rock.

East of Iwerne Minster, nodular glauconite-coated chalkstone was found in brash in two places [8754 1474; 8754 1461]. South-east of the latter, a pit [8765 1455], partially filled to a depth of 3 m, exposes 4.4 m of strata. The complete section was recorded by White (1923); Drs R G Bromley and A S Gale (personal communication) measured a partial section in 1978. Figure 54 is a compilation of all three sets of data. The 'Upper Rock', cropped out in the floor of the pit (White, 1923) and dipped 2.5° S. It is difficult to relate White's section to that currently exposed. White recorded an extensive fauna from the 'Upper Rock', includimg elements of the 'reussianum' fauna such as naultiloids and pleurotomariids. The list suggests the Hitch Wood Hardground.

Iwerne Minster–Sutton Waldron

Glauconite-coated nodular 'Lower Rock' [?Spurious Chalk Rock] was formerly exposed in the floor a pit [8753 1639] on Sutton Hill (White, 1923). The 'Upper [?Chalk] Rock' was well developed (0.45 to 0.6 m thick), but with few glauconite-coated nodules. Above it occurred about 3 m of nodular chalk with nodular and tabular flints, in the lowest 0.9 m of which were found '*Holaster placenta, Echinocorys scutatus, Micraster praecursor* and a form of *Micraster* between *M. leskei* and *M. praecursor*'. Above this bed, fossils were scarce, but the *Micraster* indicated to White that the junction between the *planus* and *cortestudinarium* zones lay between 1.8 and 2.7 m above the 'Upper Rock'.

Fontmell Down–Charlton Down

'Chalk Rock' was formerly exposed in a pit [885 191] on Spread Eagle Hill. On the south side of Compton Down, a pit [8843 1885], originally described by White (1923), has been recently cleaned. Additional details are provided by Drs R G Bromley and A S Gale (personal communication):

	Thickness m
Chalk rubble	0.30
Chalk, chalkstone and nodular chalk; knobbly flints in the upper part, thin tabular in the middle; '*Rhynchonella reedensis, Echinocorys scutatus, Holaster placenta* and *Micraster praecursor*'	2.00
Glauconitised pebbles and rolled *Micraster* in hard chalk	0.20
Hummocky, convoluted glauconitic hardground developed on chalkstone ('Upper Rock)	0.40
Nodular chalk, thin tabular flint in middle; fossils include '*Rhynchonella plicatilis, "Terebratula" carnea, Dimyodon nilssoni* and *Micraster praecursor*' (White, 1923)	0.40
Phosphatised hummocky surface with common phosphatic pebbles under and on it; 'Reussianum' fauna with '*Inoceramus costellatus*, trochids, *Micraster*, rhynchonellids, sponges' (R G Bromley, personal communication)	0.20
Glauconitic nodular chalkstone with phosphatic intraclasts	0.50
Pale green hummocky surface, with phosphatic glauconitised pebbles on surface, developed on hard, cream-coloured chalk; small, elongate carious flints in upper 0.3 m ('Lower Rock') '*"Terebratula" carnea, "T." semiglobosa, Holaster planus* and *Micraster leskei*'	0.76
Chalk, greyish white, nodular	0.30

The 'Upper' and 'Lower' rocks correspond respectively to the Top Rock and Chalk Rock.

Charlton Down–Win Green

Brash of chalkstone and nodular chalk, some glauconitised, is common on Breeze Hill [893 198]. Fossils include *Sporadoscinia alcyonoides?* [8942 1972], *Inoceramus* cf. *costellatus* [8920 1979] and *Sternotaxis plana* [8917 1972]. Buff and glauconitised nodular chalk with cf. *Bathrotomaria perspectiva* and *Spondylus spinosus*, from near the base of the Lewes Chalk, occurs on the north-west side of the hill [8950 2015]. Rubbly nodular chalk, with scattered glauconitised pebbles and '*Holaster planus*', was seen in a pit [8973 2073] beside the B3081 (White, 1923). A nearby section [8997 2065] showed c.0.4 m of white firm chalk with a 2 cm thick bed of grey tabular flint, overlain by c.0.5 m of firm white chalk with nodular flints, containing a thin-tested *Echinocorys* with adnate *Ancistrocrania parisiensis*. Farther east, *Micraster* cf. *cortestudinarium* was found in brash [9000 2065]. Some 170m south-east, brash included *Porosphaera sessilis, Plinthosella* sp. and *Micraster cortestudinarium* [9013 2055].

On the side of the road leading to Ludwell, there are three small quarries in Lewes Chalk. In the most southerly one [9186 2054], the Elliott's Shed locality of Bromley and Gale (1982), the following section was formerly exposed (White, 1923): at the base, the 'Lower Rock' [?the Hitch Wood Hardground] with a layer of large (up to 75 mm) bored glauconitic nodules encrusted with '*Serpulae* and *Ostreae*', overlain by 3 m of nodular chalk with '*Coscinopora quincuncialis, "Terebratulae", Inoceramus costellatus, I. inconstans, Echinocorys scutatus, Holaster planus* and *Micraster praecursor*'. A few green nodules were seen in the rubble at the top of the pit. The present day section extends slightly higher than that recorded by White:

	Thickness m
Chalk, firm, white, with flints	0.3
Convolute surface, strongly glauconitised [Top Rock] with glauconite-coated nodules on the surface	
Chalkstone, hard, with a few scattered glauconitised nodules	0.5
Unexposed	c.3.0
Chalkstone, very hard, with few fossils	0.2
Glauconitised pebble bed in floor of pit seen	

The chalkstone at the base of the section is thought to be the Hitch Wood Hardground (Bromley and Gale, 1982, p.302). In the second pit [9186 2058], *Hyphantoceras reussianum* was found in hard white chalk with small nodular flints. The third pit [9188 2060] exposed 0.2 m of hard, off-white chalk, capped by glauconitised nodules beneath a strongly glauconitised surface, and overlain by 0.5 m of firm white chalk with small flints.

Fragments of nodular chalk, some glauconite-coated, and chalkstone are common in brash around Ashgrove Farm [around 924 188]. *Cremnoceramus? rotundatus* in the nodular chalk [9246 1869] and *Micraster cortestudinarium* in the chalkstone [9404 1998] indicate the *cortestudinarium* Zone. *Sternotaxis plana* [9302 1859] and a large form of *Micraster leskei* [9363 1845] in brash of both nodular chalk and chalkstone are from the *plana* Zone.

BLANDFORD CHALK

Bulbarrow Hill to Okeford Hill to Winterborne Stickland

In the fault-bounded block near Okeford Fitzpaine, there are two pits [8080 1025; 8078 1009]. In the former, a section measured in 1990 showed:

	Thickness m
Talus and poorly exposed chalk	1.50
Chalk, firm, white, with carious flints	1.20
Chalk, firm, white	0.40
Marl, silty, laminated, pale grey; poorly preserved echinoid at base; planar top, irregular base	0.05–0.08
Chalk, hard, orange-stained, spongiform, with poorly preserved echinoids	0.20
Chalk, hard, with carious flints	0.20
Flint, tabular	0.05
Chalk, hard, with carious flints	0.20
Chalk, hard	0.50
Chalk, hard, with carious flints	1.20
Chalk, hard	0.30

The stratigraphical position of this pit is uncertain. White (1923) thought that it fell in the *coranguinum* Zone, and this is supported by the carious flints, which typify the lower part of the zone. The harder beds at the base could be the top part of the Lewes Chalk, with the marl seam being the Shoreham Marl (as in the Shillingstone pit) at the base of the *coranguinum* Zone, although the absence of *Platyceramus* and *Volviceramus* in the overlying 3.1 m of strata is surprising.

South of Okeford Hill, clay-with-flints covers much of the outcrop. On the west side of the valley, the outcrops dip about 4° eastwards into a fault that runs along the valley. Chalk thought to be of mid-*coranguinum* Zone age was formerly exposed in a pit [c.8302 0619] at Hedge End (White, 1923, fig.9). There, a north-easterly trending fault, with a high-angle hade cut the chalk. Pronounced shear cleavage and slickensiding, together with crushed flints, were evident on the hanging wall for several metres.

Blandford area

The railway cutting [?885 070] north-west of the old station exposed flinty, firm, white chalk with common *Micraster coranguinum* (Jukes-Browne and Hill, 1904). Sections [8975 0710; 8970 0715] beside the Blandford by-pass, showed soft to blocky white chalk with inoceramids, probably *Platyceramus*, and a small *Micraster*.

South of the Stour, a pit [8875 0486], about 600 m south-west of Lower Blandford St Mary Church, showed 0.2 m of soft to firm, patchily flinty chalk, beneath 0.6 m of medium brown, flinty, clayey silt. The foraminifera *Gavelinella cristata* and *Reussella kelleri*, and absence of *Gavelinella usakensis*, in a sample from this locality indicates benthonic foraminifera zone B1 of Swiecicki (1980) (Table 11), with *Osangularia cordieriana* suggesting the upper part of the zone (*pilula* macrofaunal zone).

In an exposure [8934 0456] in the railway cutting near Littleton, a single *Marsupites* calyx plate was recorded (Reid, 1899) in a coarse-grained, calcarenitic chalk that contains much bioclastic debris. The specimen is broken and smooth, and is possibly reworked. The foraminifera *Gavelinella cristata* and *Reussella kelleri* in a sample [8961 0419] of soft white crumbly chalk taken just south of the district indicates benthonic foraminifera zone B1 (*pilula* macrofaunal zone) (Table 11).

Cuttings (numbered 1 to 5, from west to east) along the siding to Blandford Camp provided good exposures in the upper part of the member (White, 1923, pp.72–74). Suggested modern fossil names are given in square brackets:

Cutting 1 [8916 0630]
About 6 m of indistinctly bedded, blocky, white chalk with numerous flints contains *Porosphaera globularis, Bourgueticrinus ellipticus, Cidaris perornata* [*Prionocidaris vendocinensis*], (ovate) *Echinocorys scutatus* [*E. scutata*] and *Micraster coranguinum*, indicative of the upper part of the *coranguinum* Zone (White, 1923), although the absence of *Conulus* suggests a slightly lower level.

Cutting 2 [8952 0658]
This cutting yielded *Bourgueticrinus* columnals and *Conulus conicus* [*C. albogalerus*], the latter suggesting the higher (Santonian) part of the *coranguinum* Zone. White stated that the cutting exposed the highest beds of the zone and noted that the aspect of the chalk and flint at the eastern end of the cutting suggested the *socialis* Zone.

Cutting 3 [9011 0682]
The sparse occurrence of *Uintacrinus* [*U. socialis*] in this cutting, together with *Terebratulina striata* [*T. striatula*] and *Bourgueticrinus ellipticus*, proved the *socialis* Zone.

Cutting 4 [9035 0680]
The eastern part of this cutting was assigned to the *testudinarius* Zone by White, who noted *Marsupites* (poorly represented) and a sparse fauna of medium-sized *Porosphaera globularis, Ventriculites radiatus* [*Rhizopoterion?*] and *Echinocorys scutatus* var. *elevatus* [*E. scutata elevata*]. The *Echinocorys* suggests the lower part of the *testudinarius* Zone, as does the scarcity of *Marsupites* (Wood and Mortimore, 1988, fig.18).

Cutting 5 [west end at 9050 0675]
White recorded *Caryophyllia cylindracea* and *Ostrea wegmanniana* [*Pseudoperna boucheroni* Woods *non* Coquand], suggestive of the topmost *testudinarius* Zone, the *anglicus* Zone or the *pilula* Zone, *Echinocorys depressula* Subzone, up to near the top of the range of *Echinocorys tectiformis* (Wood and Mortimore, 1988, p.62 and fig.18). The cutting intersected a small pit [9060 0675] which yielded *Echinocorys scutatus* var. *depressus* [*E. scutata depressula*], indicative of the higher part of the *depressula* Subzone (Wood and Mortimore, 1988). From this pit, Mr S C A Holmes collected *Porosphaera pileolus?, Echinocorys scutata truncata* and *Offaster pilula*, indicative of the lower belt of abundant *Offaster pilula* of the *pilula* Zone. Thus, the junction between the two subzones of the *pilula* Zone occurs in this pit, at about 15 m below the base of the Tarrant Chalk.

A pit [9099 0435] near Manor Dairy Farm lies within the upper part of this member, although the thickness of strata between the pit and the base of the Tarrant Chalk is uncertain. Fossils included *Plinthosella* sp., *Bourgueticrinus* sp., *Inoceramus balticus pteroides?* with adnate bryozoa, *Pseudoperna* sp.juv, poorly preserved echinoids, possibly *Echinocorys elevata truncata*, and one small *Offaster pilula*. The echinoids suggest the lower belt of abundant *Offaster pilula* at the base of the *pilula* Subzone; this is supported by the foraminifera *Gavelinella cristata, Reussella kelleri, Bolivinoides culverensis* and *Stensioeina exculpta gracilis*, and the absence of *Gavelinella usakensis*, in samples from this pit. The fauna is indicative of benthonic foraminifera Subzone B1iii sensu Swiecicki (1980) (Table 11).

Pimperne–Sutton Hill

A pit [9136 0954] north-east of the Stud House formerly exposed 3.6 m of soft chalk with nodular and oblique tabular flint (White, 1923). Fossils included *'Kingena lima, Ostreae, Spondylus spinosa, Micraster coranguinum* and *Bourgueticrinus ellipticus*'. The beds dipped 10°SE. A small fault, hading east, was picked out by a slickensided marl; a rude cleavage occurred in the beds beneath the fault plane.

Quarries east of Iwerne Hill [8915 1436] and north-east of Everly Hill Farm [8932 1294] exposed blocky white chalk with *Micraster coranguinum* (White, 1923).

A pit [8974 1209] in Handcock's Bottom exposed lumpy chalk with ragged flints (White, 1923). The flints, rich in the siliceous sponges *'Doryderma, Siphonia, Ventriculites* and *Plinthosella'*, also included *'Inoceramus* fragments and *'Lima (Limatula) decussata'*.

Ashmore–Tarrant Hinton

A pit [9409 1674] near Rookery Farm revealed 2.5 m of firm white chalk with common flints, varying from horned, rounded to semi-tabular. *Plinthosella* and *Micraster coranguinum*, indicative of the Santonian part of the *coranguinum* Zone, were found.

A pit [9172 1625] south of Well Bottom revealed, beneath clay-with-flints, 2 m of white chalk with irregular-shaped horned and finger flints. There was a prominent bed of flints 0.6 m above the base of the section. *Micraster senonensis*, indicative of the Santonian part of the *coranguinum* Zone, was found.

Firm white chalk with fragments of *Platyceramus mantelli*, indicative of the *coranguinum* Zone, were found in the roots of an overturned tree [9238 1157] north-west of Tarrant Hinton.

TARRANT CHALK

The most complete section in the Tarrant Chalk is in the pit [9427 0666] just east of the district, below The Cliff, Tarrant Rawston (Barton, 1992, fig.4). There, a 14 m section, which starts about 6 m above the base of the Tarrant Chalk, exposes firm white chalk with flint beds, scattered nodular flints and tabular flints. They are lenticular in part and up to 50 mm thick. About 2 m above the base of the section is an anastomosing marl seam, just above which is a fossiliferous bed with *Bourgueticrinus bacillus* and *Echinocorys* with adnate *Atreta boehmi*; a fragment of *Echinocorys*, was found 1.5 m above the marl seam. A loose *Inoceramus balticus* was also found. The *Echinocorys* are closest to the innominate, 'large forms' of Gaster, and compare with forms collected by Mortimore from the Brighton Railway section, above the Pepper Box Marls i.e. near the top of the range of the 'large forms' and within the Castle Hill Beds at the base of the Culver Chalk of Mortimore (1986). This suggests that the *Echinocorys* bed falls in the lower part of the *Hagenowia* Horizon at the base of the *quadrata* Zone, and that the marl seam at the base of the section is the Pepper Box Marl. This conclusion is supported by the foraminifera *Bolivinoides culverensis, Gavelinella usakensis, G. cf cristata* and *Reussella szajnochae praecursor*.

A single *Echinocorys scutata depressula* sensu Gaster was found 3.35 m above the marl seam. Thus, the change from the 'large' to the 'small' *Echinocorys* forms of Gaster occurs near the base of the section, just above the lower *Echinocorys* bed and gives added support to the section being in the higher part of the Castle Hill Beds (Young and Lake, 1988, fig.18). A higher *Echinocorys* bed, some 7 m above the marl seam, is characterised by small, round-based forms with a somewhat protruberant apical disc that resemble *Echinocorys scutata cincta* sensu Gaster.

A pit [9277 0530] in Ashley Wood, 1 km south-west of Tarrant Rushton, exposes the following section, approximately in the middle of the member:

	Thickness m
Chalk, white, rubbly	0.3
Flint band, tabular	0.1
Chalk, white, blocky	0.3
Marl seam	0.1
Chalk, white, blocky, with small, fractured and greatly distorted *Echinocorys* at base	0.8
Chalk, white, soft to blocky, with *Proliserpula ampullacea*	0.5
Flint band, impersistent	0.1
Chalk, soft	1.0

The basal bed yielded the foraminifera *Bolivinoides culverensis, Gavelinella usakensis* and *Pullenia quaternaria*, indicative of benthonic foraminifera Subzone B2iii of Swiecicki (1980) (upper part of the *quadrata* Zone) (Table 11).

SEVEN
Structure

A combination of surface mapping and geophysical interpretation of the concealed formations enables the structural history of the district to be divided into the following four broad periods, in each of which a single dominant tectonic regime can be identified:

1. Pre-Permian: Devonian and Carboniferous folded and thrusted sequences of contrasting facies (see Chapter 2) were emplaced during the ?late Carboniferous in a regime of strike-slip deformation (Holder and Leveridge, 1986), in which faulting occurs in two sets, one oriented east–west or ESE–WNW the other NW–SE or NNW–SSE.

2. Permian to early Jurassic: north to south extensional reactivation of east–west Variscan thrust faults resulted in fault-controlled thickness and facies variations in Permian-Triassic and Jurassic strata generally across east–west lines.

3. Cretaceous to Palaeogene: thermal relaxation subsidence reactivated normal movement in the steep NW–SE- trending strike-slip faults (Chadwick, 1986), resulting in thickness and facies variations in Cretaceous sequences across the NW–SE-trending faults.

4. Neogene: a compressional tectonic regime during the Neogene resulted in reversal of the movement on the normal faults trending approximately east-west and reactivation of north-north-west-trending strike-slip faults.

PRE-PERMIAN STRUCTURE

The structure of the lowest crustal rocks (p.5) is not known, but may resemble that of the Precambrian rocks of Brittany which consist of highly deformed, high-grade metamorphic rocks, the Pentevrian, overlain by less deformed, lower grade metamorphic rocks, the Brioverian. The middle crust, consisting mainly of Variscan fold belt rocks, is better known from its expression in Cornwall, Devon and Somerset, from boreholes and from features in the seismic profiles. In the southern part of the district, the rocks are probably strongly folded with attitudes ranging from upright to flat-lying, and isoclinal folding is probably common. Seismic profile evidence suggests the presence of large-scale folding overturned to the north (Chadwick et al., 1983). In the northern part of the district, the folding is less intense and analogous to that of the Mendips.

In the southern part of the Wincanton district and northern part of the Shaftesbury district, Chadwick et al. (1983) described thrusts dipping towards the south-south-east. Seismic reflection profiling east of the Quantock Hills, in the Somerton area [350 208] (Donato, 1988), suggested the presence of a concealed Variscan thrust dipping to the south-west. Bott et al.(1958) interpreted the Bouguer gravity gradients in the Exmoor area as indicating thrusting of the folded Devonian rocks of Exmoor over possible Carboniferous rocks of the foreland. Although it is unlikely that these thrusts and the thrust underlying the northern part of district are one and the same thrust, they are closely similar structures.

Some of the suggested thrusts underlie normal faults in the cover sequence (zone 1)(p.11). It is likely that normal movement on faults, such as the east–west-trending Mere and Cranborne faults (and the associated post-Variscan sedimentary basin in southern England) owe their existence to reactivation of the underlying thrusts, possibly flattening and coalescing with the thrusts at depth (Chadwick et al., 1983; Whittaker et al., 1986). There is also a strong possibility that strike-slip movement took place along the same east–west faults, thus explaining the juxtaposition of the three basement tracts (p.8) (Figure 6) in the district. The contrast between the Devonian and Carboniferous facies of the foreland and that of the Variscan orogenic belt in southern Britain (Freshney, 1980) and the comparison of the south-west England and German Rheno-hercynides (Holder and Leveridge, 1986) led these authors to propose large-scale strike-slip faulting. These east–west-trending faults in the Variscan, Exmoor and Foreland tracts are offset by north-west-trending strike-slip faults, such as the Poyntington Fault, which belongs to the same group of structures farther west cutting Variscan rocks at outcrop, such as the Sticklepath and Watchet faults, which show large displacements (Dearman, 1963; Whittaker, 1972). Many of these latter faults are believed to have had been initiated during the Variscan Orogeny or even earlier, and to have had a history of movement during the Mesozoic and Tertiary.

Although the Bouguer anomaly values, after compensation for the effects of the Mesozoic rocks, increase generally towards the south-south-west (Figure 55), there is a zone with a steeper gradient crossing the district, which appears to be part of a gravity gradient zone extending from Exmoor (Bott et al., 1958; Cornwell, 1986; Donato, 1988) east-south-eastwards towards the Isle of Wight. Just east-south-east of the district, the pronounced zone of gravity gradient coincides with the position of the Cranborne Fault Zone.

PERMIAN TO JURASSIC STRUCTURE

Many of the east–west-trending 'Variscan' faults were reactivated as normal faults in an extensional regime in Permian and Triassic times (Chadwick, 1986). In the

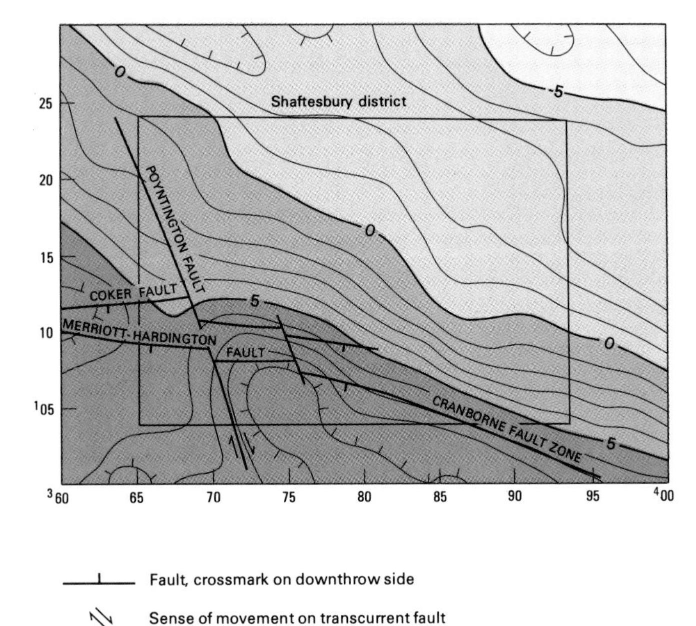

Figure 55 Bouguer gravity data compensated for the effect of the Mesozoic sedimentary rocks (i.e. stripped gravity) with contours at 1 mGal intervals, and selected faults.

Shaftesbury district and surrounding region, the most important faults of this type are the Mere and Bere Regis faults and the Cranborne Fault Zone (differentiated into the Coker and Merriott–Hardington faults in the western part of the Shaftesbury district). The area of thickest Permian and Triassic rocks lies south of the district in the Winterborne Kingston Trough. The Permian and earliest Triassic rocks (Aylesbeare Mudstone) are confined to the trough and, by comparison with similar deposits at outcrop in Devon (Edmonds et al., 1968; Whittaker,1975), their deposition was probably restricted by contemporaneous fault movement. The Sherwood Sandstone, higher in the sequence, is less constrained and spreads 2 to 3 km north of the Cranborne Fault Zone (Figure 11).

The distribution of the Mercia Mudstone Group is controlled to a lesser extent by the presence of the fault trough, but still thins markedly to the north (Figure 11) (pp.13–14). The presence of halite in the group is almost certainly governed by fault activity. Within the district, thick halite is developed south of the Cranborne Fault Zone within the Winterborne Kingston Trough, as in the Mappowder Borehole.

In the Jurassic, the influence of the east–west faults was less strong than in Permian and Triassic times. The Lias does not change markedly in thickness across the Cranborne Fault (Figure 11), but does thin northwards towards Fifehead Magdalen Borehole, particularly in the area to the north of Sturminster Newton. No fault has been seen at surface, but one has been mapped from the seismic data in this area. An isopachyte map for the Bridport Sands (Figure 16) shows some control by the Winterborne Kingston Trough because the formation thickens southward into it from around 60 to 70 m to 130 m. During deposition of the Inferior Oolite, the main depocentre moved north and subsidence may have been controlled by movements on the Mere and Poyntington faults (Figure 17). The Fuller's Earth shows a slight thickening over the Winterborne Kingston Trough (Figure 24), but during the deposition of the Forest Marble (Figure 26), little structural control is evident. The thickness of the Kellaways Formation (Figure 33) does not show much control by faulting within the district, but in the southern part of the Wincanton district, it is influenced by the Mere Fault. The Oxford Clay (Figure 35) shows the least thickness variation across faults in the district. By contrast, there are marked thickness variations in the Corallian Group (Figure 37). The Cranborne–Fordingbridge High underlies an area of thin (30 m) Corallian rocks bounded to the south by the Cranborne Fault, south of which the Corallian Group thickens to over 100 m in the Winterborne Kingston Trough. It also thickens into the Mere Basin to the north.

CRETACEOUS AND PALAEOGENE STRUCTURE

In mid Cretaceous (Aptian and Albian) times, extension ceased on the east–west-trending faults. Regional (thermal) subsidence ensued, with warping over the steep NW–SE-trending basement faults (Chadwick, 1986).

In the Hampshire Basin, north-north-west-trending structures influenced the location of depocentres and facies in the Palaeogene (Edwards and Freshney, 1987). These structures are gentle folds in the Palaeogene and Cretaceous rocks, and may overlie faults of the same trend. Although most of the control in thickness of the Cretaceous in north Dorset may have been by these north-north-west-trending faults acting in an extensional fashion, the east–west-trending Isle of Purbeck Fault Zone was active as a reverse fault in compression during the Palaeogene (Plint, 1982). This may indicate a change in the stress from extension in the Jurassic, through neutral regional subsidence between the early and late Cretaceous, to mild compression perhaps from as early as late Cretaceous.

The Mid-Dorset Swell

The Mid-Dorset Swell (Figure 47) was defined by Drummond (1970) as a positive structure with an axis trending NW–SE through central Dorset to the coast at Swanage, that controlled Albian and Cenomanian sedimentation. It was bounded to the north-east by the Wessex Trough, part of which falls within the district, and to the south-west by the Wessex Shelf which lies outside it.

New stratigraphical data allow a closer definition of the north-eastern boundary of the Mid-Dorset Swell (Figure 47). The north-eastern flank of the structure influenced the deposition and/or erosion of the Lower Greensand (?Aptian/Albian), part of the Upper Greensand (Albian), the West Melbury Chalk (Cenomanian) and the lower part of the Zig Zag Chalk (Cenomanian). During that time, the north-eastern flank migrated backwards and forwards, such that the area affected by it varied. It was during the late Albian that the Mid-Dorset Swell had its greatest areal effect, accounting for the absence of the Boyne Hollow Chert from Compton Abbas [870 180] southwards. The Cenomanian part (Melbury Sandstone) of the Upper Greensand has a more southerly extension of onlap, and occurs at Iwerne Minster [865 145] and farther south at Iwerne Courtney [860128], but is absent at Stour Bank, 2 km south-west. During the Mid-Cenomanian, the Mid-Dorset Swell ceased to be a positive structure. The hardgrounds and nodularity of the late Turonian to early Coniacian Lewes Chalk suggests that there was continuous subsidence over the area of the former Mid-Dorset Swell, but at such a rate that the sea remained shallow and current-scoured.

Borehole data, combined with the outcrop pattern of the Lower Chalk, provide the only evidence of the south-westerly trend of the Mid-Dorset Swell within the district. The West Melbury Chalk is present in the Quarleston [8594 0565], Spetisbury [8880 0268] and Shapwick [9428 0134] boreholes, but is absent in the Winterborne Kingston Borehole [SY 8470 9790].

Two explanations could account for the thinning and local disappearance of Albian and Cenomanian strata over the swell. First, the beds were deposited, uplifted by differential movement on deep faults and finally removed by erosion prior to deposition of the basal beds of the Chalk. The continuity of the Gault across the swell (Figure 47) suggests that such an explanation for the removal of Lower Greensand is unlikely, unless alternate periods of erosion and deposition are invoked. Second, syndepositional folding over buried faults that commenced in Aptian times resulted in convergent onlap of the Chert Beds and the West Melbury Chalk, and the overstep of the Zig Zag Chalk across West Melbury Chalk onto the Shaftesbury Sandstone toward the structural axis.

NEOGENE STRUCTURE

A major phase of compressional tectonics related to the Alpine Orogeny affected southern England in the Miocene. Normal faults, including the Isle of Purbeck and Mere faults, were reactivated as reverse faults. In the Isle of Purbeck, the fault did not break through the Cretaceous and Tertiary cover along much of its length, but the cover was deformed into a north-facing monocline above the line of the fault.

A faulted anticline and syncline affect Jurassic and Cretaceous rocks in the south-western part of the district; they were probably formed by the Miocene compressional movement along the Coker and Merriott–Hardington faults.

The north-north-west- and associated north-east-trending faults cut rocks as young as Upper Chalk, and were probably active in the Tertiary. The Poyntington Fault has a net dextral, post-middle Jurassic offset of approximately 3 km; the Tertiary component of this displacement is uncertain.

SURFACE STRUCTURE

Figure 56 shows the surface distribution of faults in the district; they can be divided into two main groups. The first, those with east–west or east-north-east trends, including the Templecombe, Coker and Merriott–Hardington faults, appears to have affected the thicknesses of the Permian, Triassic and some Jurassic strata. The second group trends mainly north-north-west, but includes associated north-north-east- and north-east-trending faults. The most important faults of this group are the Poyntington and Stalbridge Park faults, which show evidence of dextral strike-slip displacement.

East–west and related faults

TEMPLECOMBE FAULT [668 202 to 720 234]

The Templecombe Fault is a 7 km long SW–NE-trending fault complex, oriented parallel with and approximately 5 km south of the Mere Fault. For much of its length, the Templecombe Fault consists of two subparallel faults, less than 0.5 km apart, which define an asymmetrical graben with the larger displacement along the northern fault.

The Templecombe Fault marks the approximate northern limit of thick Fuller's Earth and Frome Clay successions. Within 2 km along strike north of the fault, both units rapidly attenuate and are less than half as thick to the north of Charlton Horethorne. The Forest Marble shows a marked reduction in the proportion of bioclastic limestone to mudstone north of the fault, though the Boueti Bed is unchanged.

FERNE PARK FAULT [915 222 to 943 228]

The roughly east–west, curved Ferne Park Fault and the sub-parallel Berwick St John Fault displace Cretaceous strata in the north-east of the district. The Ferne Park Fault can be seen from seismic evidence to throw down Jurassic strata to the south-east. At the the surface, the downthrow is c.5 to 10 m north-west. On the north side of the Ferne Park Fault, the dip in the Chalk is between 1° and 2°SE. South of the fault, the dip is about 2°N or NNE. The Berwick St John Fault does not appear to have any deep-seated expression; its maximum downthrow is about 10 m. South of the fault, the regional south-eastward dip is resumed.

COKER FAULT [650 115 to 770 090]

The Coker Fault trends approximately east–west, except where disturbed by later transcurrent faults. The fault has been mapped across the Yeovil district, where it shows southerly downthrows of c.110 m to the south of Yeovil. In the present district it has a southerly downthrow of c.100 m near Holnest [657 098], where it juxtaposes the higher part of the Oxford Clay against the Kellaways Formation. North-east of Holnest, the Coker fault is displaced dextrally to pass just south of Holwell [702 104], and has a southerly downthrow of approximately 70 m. Further dextral displacement north of Hazelbury Bryan shifts the fault farther south through

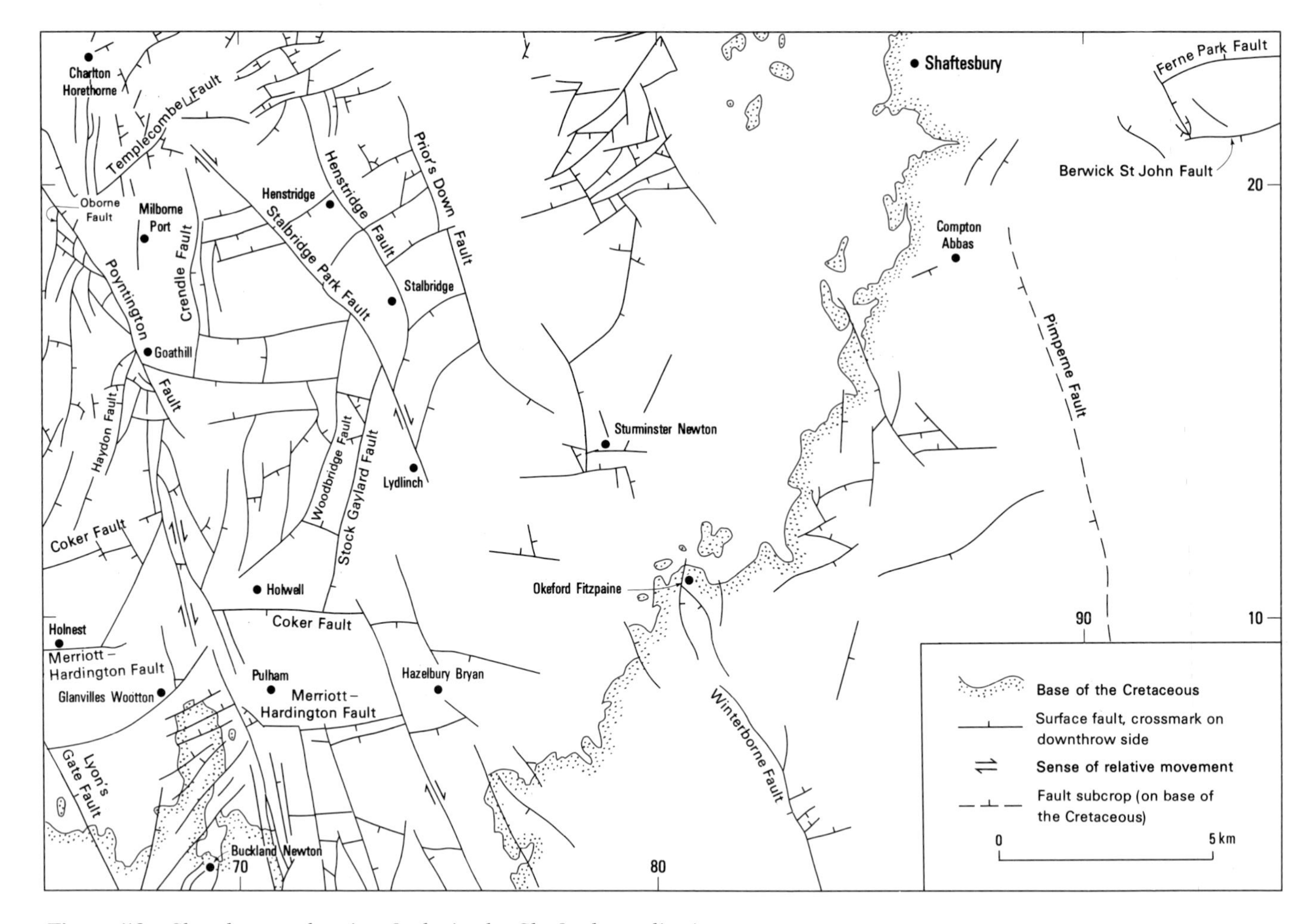

Figure 56 Sketch-map showing faults in the Shaftesbury district.

Kingston [750 095]. East-south-east of Kingston, the throw diminishes to about 10 m to the south. East of this, the fault cannot be traced in the Kimmeridge Clay.

The fault shows a similar diminution of southerly downthrow eastwards in the Inferior Oolite (Figure 18) from 100 to 80 m near Holnest, to 40 m near Holwell and to 10 m east of Hazelbury Bryan. The throw at the base of the Permian (Figure 10) appears to be over 200 m throughout the district.

Seismic data indicate that the fault dips at c.45°S in the Jurassic sequence and c.35°S in the Permo-Triassic sequence.

Merriott–Hardington Fault [650 091 to 770 070]

The Merriott–Hardington Fault runs subparallel to, but about 2.5 km south of the Coker Fault. It enters the district north of Holnest Park and extends eastwards into the Poyntington Fault complex. It re-emerges near West Pulham and continues eastwards, but is stepped southwards by a number of transcurrent faults to pass to the south of Hazelbury Bryan. Where it crosses the River Lydden, it splits into two parallel elements which continue almost as far as Hazelbury Bryan. Farther south, displacements by transcurrent faults take it to near Woolland where it disappears under the Gault, which is unaffected by it.

In the west of the district, the fault has a southerly downthrow of about 40 m; west of Hazelbury Bryan, the downthrow is over 100 m. At the top of the Penarth Group, the fault has a downthrow to the south of 50 to 100 m on the western margin of the district, which increases to c.150 m near Hazelbury Bryan.

It seems that both the Coker and Merriott–Hardington faults are the westerly extension of the Cranborne Fault complex, which defines the southern edge of the Cranborne–Fordingbridge High (Chadwick, 1986).

North-north-west-trending and related faults

Poyntington Fault [652 200 to 709 045]

The Poyntington Fault is an important structural element in the western part of the district (Figure 56). The trace of the fault has an average trend of 160°, and is broadly concave toward the west, with a 35° range of strike orientation. It is at least 20 km long, extending from Corton Denham in the north (Wilson et al., 1958) to the Cretaceous outcrop south of the present district, where the displacement is transferred to a NE–SW-trending fault. The Poyntington Fault is primarily a wrench fault with net dextral offset. Displacements associated with movement zones and splays along and adjacent to it (Figure 57) are described in terms of apparent vertical stratigraphical separation, because strike-slip displacements are difficult to quantify.

A seismic reflection profile shows the fault to dip west at between 65° and 90° within the Mesozoic cover succession; it dips at c.60° in the Variscan basement. An elongated north-north-west-trending gravity high lies to the east of the Poyntington Fault in the southern part of the district (Figure 55); it indicates the presence of high density basement rocks nearer the surface immediately to the east of the fault.

The structural morphology of the Poyntington Fault is analogous with strike-slip faults elsewhere (see Sylvester, 1988, and references therein). A braided geometry, in which arrays of faults oriented in narrow elongate zones, is evident in the Milborne Port district (Figure 57). Splays and lens-shaped faults of the type that occur in the western block of the Poyntington Fault, and regular changes in apparent vertical separation along the length of the fault, are typical of strike-slip faults. Strike-slip displacements are accommodated by differential vertical block movements within the braided fault zone. The vertical displacement in the area shown in Figure 57 varies between 25 and 150 m.

In the district, there is a greater density of faults with relatively large displacements on the western side of the Poyntington Fault. Several north-north-east-trending faults, with vertical throws greater than 30 m, bound horsts and grabens that terminate at the Poyntington Fault. For example, the Oborne Fault (Figure 57) defines the western edge of a 0.5 km wide horst capped by Inferior Oolite. Vertical throws along the Oborne Fault and along the eastern margin of the horst decrease about 10 m per 1 km toward the south. Relatively small throws (25 to 30 m) are recorded at the Poyntington Fault where it truncates the Oborne horst. The northern extension of the Frogden Fault [655 185], which has a vertical throw to the east (Kellaway, 1944), defines the western margin of a graben and lets down Lower Fuller's Earth around Oborne [655 185].

The Haydon Fault (Figure 57) separates a horst from the Haydon graben, in which Cornbrash and the lowermost Kellaways Formation are let down against older sequences. The largest vertical throw (c.150 m) on the Poyntington Fault, at its intersection with the Haydon graben, brings the Kellaways Formation and Upper Cornbrash into contact with the basal Fuller's Earth.

Faults with throws of 20 m or less and an average trend of 135°, within 15° of that of the Poyntington Fault, accommodate differential displacements within horst and graben blocks, and have dip-slip displacements toward the south-west in horsts, and the north-east in grabens (Figure 57).

West of the Poyntington Fault, the Glanvilles Wootton [678 082] to Buckland Newton [693 052] area is much affected by faults trending NE–SW. The most northerly faults of this set, around Glanvilles Wootton, show small southerly downthrows, while those to the south, in the area between Duntish [695 065] and Buckland Newton, mainly show northerly downthrows. The Corallian, Kimmeridge Clay and Cretaceous rocks in the area around Dungeon Hill [690 075] lie in a graben; this is truncated by the Poyntington Fault west of Pulham [700 083].

The ground east of the Poyntington Fault is less faulted (Figure 57). A series of north–south strike faults form

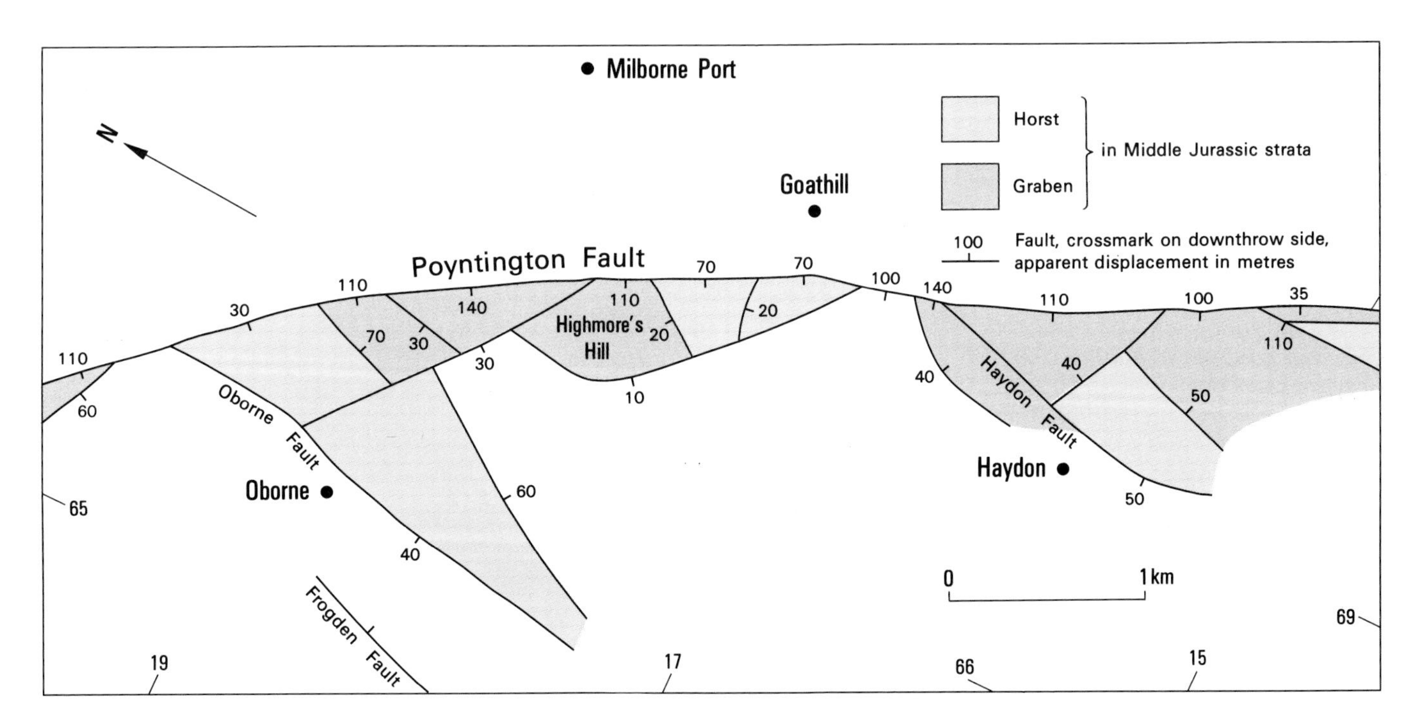

Figure 57 Mean apparent vertical displacement of Middle Jurassic sequences along part of the Pointington Fault.

small grabens in which Lower Fuller's Earth is downfaulted into the Inferior Oolite around Milborne Port. Farther east, the Fuller's Earth Rock is repeated by a series of faults, the most important of which, the Crendle Fault, has a markedly curved trace (Figure 56). The Crendle Fault is inclined east at a moderate angle and terminates to the north at a sub-vertical fault with an apparent downthrow toward the west. Most of the large strike faults to the east of the Poyntington Fault (e.g. Woodbridge Fault and Stock Gayland Fault, Figure 56), downthrow to the east or south-east.

The net dextral offset of identifiable stratigraphical markers, present on both sides of the Poyntington Fault, is c.3.5 km (range 2.5 to 4.5 km). The net dextral offset of the Coker Fault is c.2 km, which suggests that there has been at least this amount of strike-slip displacement. Any component of westerly dip-slip would tend to increase the apparent dextral strike-slip offset of easterly-inclined strata. Although the net displacement is dextral, the movement history of the Poyntington Fault may be complex, and may have included periods of sinistral strike-slip movement (see below).

The offset of the Upper Chalk outcrop suggests that the most recent movement along the Poyntington Fault was Tertiary in age. Tertiary deformation in southern Britain is widely believed to have involved a compressional tectonic regime, when Mesozoic extensional faults, such as the Mere Fault and, by analogy, the Templecombe Fault, were reactivated as north- or north-west-directed reverse faults (Chadwick et al., 1983). Similarly, there are numerous north-west or north-north-west-trending wrench faults of Tertiary age in the Devono-Carboniferous rocks of south-western England (Dearman, 1963).

Syndepositional displacements along the Poyntington Fault can be inferred from the isopachyte data for the Bridport Sands and Inferior Oolite (Figures 16 and 17), which show large thickness gradients parallel or sub-parallel to the fault. Evidence for Jurassic syndepositional movement along east–west extensional growth faults, like the Mere or Templecombe faults, suggests that any strike-slip movement on the transcurrent faults is dextral in sense. Analogy with the Sticklepath Fault (Holloway and Chadwick, 1986) in Devon, suggests Variscan strike-slip deformation along the Poyntington Fault.

Stalbridge Park Fault [723 176 to 735 158]

The Stalbridge Park Fault has an average trend of 145° and a mapped length of 10 km (Figure 56). Like the sub-parallel Poyntington Fault, 5 km to the west, the Stalbridge Park Fault is primarily a dextral wrench fault. The trace of the fault is broadly concave toward the west, and a seismic section shows it to be inclined in that direction. A braided fault pattern, of the type found along the Poyntington Fault, was not identified in the Stalbridge district.

Minimum dextral offset of the Stalbridge Park Fault can be estimated from the displacement of earlier north-north-east-trending faults. Where offset, the Woodbridge and Henstridge faults (Figure 56) are thought to be two strands of the same fault. The Stock Gayland and Prior's Down faults (Figure 56) are related to each other in the same way. The dextral offset of the two groups of structures along the Stalbridge Park Fault is 2.15 km.

The Stalbridge Fault cannot be traced in the higher part of the Oxford Clay any farther south than Lydlinch, and no large fault cuts the Corallian beds on the same line as this fault to the south-south-east. The nearest faults of this trend occur in the Hazelbury Bryan area; they possibly represent an extension of the Stalbridge Park Fault in which a transfer of its displacement takes place south-south-west of Lydlinch, perhaps along subsidiary north-east-trending faults. The faults in the Hazelbury Bryan area affect rocks as high as Chalk, but their apparent vertical displacements are small and it is difficult to trace them for any distance into the Chalk to the south of the district.

Lyon's Gate Fault [654 075 to 676 027]

The Lyon's Gate Fault (Figure 56) trends north-north-west across the south-west corner of the district; it displaces the base of the Upper Greensand by 10 m to the east. However, a lensoid outlier of Upper Greensand and Chalk is cut by this fault about 200 m north-north-west of the main outcrop, and the beds within it dip steeply towards the fault. The presence of this outlier and its steep dip suggest strike-slip faulting. North-north-west of Lyon's Gate, the fault truncates an anticline in the higher part of the Oxford Clay on its eastern side.

Winterborne Fault [820 080 to 840 040]

The straightness of the Winterborne valley, from its origin behind the Chalk scarp at the top of Okeford Hill [813 093], through Winterborne Stickland on the southern margin of the district, to just west of Winterborne Kingston in the adjacent Dorchester district, suggests that it is fault controlled. However, there is little good evidence for fault displacement. North of Winterborne Stickland, the Blandford Chalk dips at c.4° east-north-east on the west side of the valley; the east side of the valley is mostly in Lewes Chalk, cut by north-east-trending faults which let down small graben of Blandford Chalk. An old pit [8302 0619] in one such graben, lying close to the Winterborne Fault, formerly showed (White, 1923, p.73) a north-east-trending fault, with a high-angle, north-west hade. Pronounced cleavage and slickensiding, together with crushed flints, were evident on the hangingwall for several yards. These faults may be complementary splays off the Winterborne Fault.

The bounding faults of the graben of Blandford Chalk, just south of Okeford Fitzpaine [808 100], are probably closely associated with the Winterborne Fault. At Turnworth Down [816 090], at the head of the Winterborne valley, Lewes Chalk crops out on either side of the valley and there is no evidence of displacement; nor is there a significant feature break in the watershed between the Winterborne valley and the streams draining northwards.

Pimperne Fault [880 190 to 910 080]

A north-north-west-trending, westerly inclined fault is shown by reflection seismic surveys to be present below the Cretaceous of the Pimperne valley, as far north as Compton Abbas. The straightness of the valley suggests that it is fault controlled, although there is no evidence for surface displacement of the Chalk. The seismic data shows the fault to be downthrown to the west in the Jurassic rocks.

Folds

Folding is mainly restricted to the southern part of the district, south of the Coker Fault. The folds trend approximately east–west and are parallel to the faults. North of the Coker Fault, the regional dip of the Jurassic strata is c.2 to 3°SE. South of the fault, the dip reverses to about 1.5°N, and this dip continues as far south as an anticlinal axis close to the southern margin of the district. Dips obtained from depth-converted seismic maps for this area are as high as 3.5°N. The seismic data show more east–west faults than were proved in the surface mapping and demonstrate a tilted block between each pair of faults. The average dip across the blocks agrees with the estimated surface dip of 1.5°, derived from sporadic dip slopes. Cretaceous strata also appear to be affected by this structure, because the base of the Upper Greensand dips in a northerly direction between Buckland Newton and Glanvilles Wootton, and between Bulbarrow Hill and Ibberton. At the southern margin of the district, however, the dip is about 1 to 2°S, but this steepens to 4° just south of the district, possibly reflecting movements in pre-existing east–west faults in the Jurassic.

DETAILS

Templecombe Fault

A fault block adjacent to the Templecombe Fault and just north of the Laycock cutting shows atypical steep (18°) dips. At the southern end of an exposure [6732 2226 to 6734 2222] of Fuller's Earth Rock, 450 m north-north-west of Starve Acre, there are vertical calcite veins up to 50 mm wide, which trend N014°, N054° and N059°; another vein trending N035° dips 70° SE. Associated slickensides plunge 2 to 5°SW, and 1 to 2° to the north.

A section [6752 2184 to 6755 2180] in a quarry near Starve Acre exposed Fuller's Earth Rock with closely spaced and irregular joints. Slickensides in spar-filled major joints plunge 10° at N156° and, 9° at 275°, or are horizontal and trend N152°. A nearby quarry [6762 2174] exposes massive, irregularly jointed Fuller's Earth Rock that contains calcite veins with slickensides. A vein inclined 80° at N095° has slickensides plunging 3° at N005°, a vertical vein trending N068° has horizontal slickensides; another vertical vein trends N141°.

In an outcrop [6782 2120] beside Laycock Farm, the Fuller's Earth Rock has a dip of 8° at N326° related to the proximity of a fault. Surfaces with slickensides plunge 2 to 5° at N298°.

Poyntington Fault

Exposures close to the fault surface within the Fuller's Earth Rock at Goathill quarry [6712 1755] show that the hangingwall (western) block sequences are inclined 15°W within 80 m of the fault surface. Inferior Oolite within the footwall (eastern) block is gently folded within 100 to 150 m of the fault surface and inclined c.5°W.

A section [6535 2015 to 6536 2016] along the trackway east of Poyntington village shows Fuller's Earth Rock cut by 10 to 100 mm-wide vertical or subvertical calcite veins that trend N144°, N158° and N159°; there are also some small ramifying veins.

EIGHT

Quaternary

The Quaternary deposits of the district reflect the many oscillations of climate, ranging from cold periglacial to warm temperate. The deposits of clay-with-flints may have formed during one of the warm periods. During the more extreme cold periods, extensive ice sheets pushed southwards across England, but they did not reach the district. Nevertheless, it is probable that during these cold periods there were semi-permanent ice caps on the higher chalk hills. Marked changes in sea level were associated with these climatic oscillations. Meltwater rivers from the impermanent ice caps were graded to base levels that were at times considerably above or below that of today. These rivers carried large volumes of sand and gravel, which were deposited as a suite of fluvial deposits preserved as river terrace deposits. Periglacial conditions, which extended southwards beyond the ice sheet, were responsible for the formation of the older head and head deposits. Because of the lack of fossils or radiometrically dated material, it has not been possible to relate these older Quaternary deposits to the sequence of named stages recognised elsewhere in the British Isles.

During the Flandrian, the most recent of the Quaternary deposits, the alluvium, was laid down.

CLAY-WITH-FLINTS

The clay-with-flints of the district is of two types. The first consists dominantly of brown clay and angular flints and corresponds to the clay-with-flints as defined by Barrow (1919). The second consists of dense, clast-supported gravel with a sparse brown to reddish brown clay matrix. In addition to the dominant angular flint clasts, well-rounded flint, quartz/schorl, quartzite, subangular chert, sandstone, gritstone and, locally, chalkstone also occur. The well-rounded flints and other pebbles constitute up to 10 per cent of the clasts. In a few localities, such as one recorded by White (1923) in the district and others noted on the northern margin of the Dorchester district, quartz sands and pale coloured clays in solution pipes, are also mapped with the clay-with-flints. Because the junction between the differing types is gradational, they cannot be mapped as separate units (see below). In addition, mapping the edge of the gravelly clay-with-flints is difficult, because a wash of gravel trails downslope from it. In places, the boundary was delineated using a ground conductivity meter.

Clay-with-flints of the first type occurs mainly in the east of the district between between Ashmore [913 178] and Farnham Woods [940 160]. Smaller outcrops occur north [890 090] and east-south-east [930 055] of Blandford Forum. West of the River Stour, the deposit becomes more flinty, until west of a line drawn between Blandford Forest [830 090] and Winterborne Stickland [835 050], it passes into the dominantly gravelly type.

The clay-with-flints in the district is developed at altitudes of between 90 m and 270 m on dip slopes ranging from Holywell Chalk to Blandford Chalk, but mainly on the latter. The deposits vary in thickness from less than 1 m to over 4 m.

The origin of the clay-with-flints is uncertain. Hull and Whitaker (1861) noted that the deposits occur only on the Chalk surface, and later (Whitaker, 1864) concluded that the deposit is of many ages, that it may be forming at the present day, and that it was due in great part to the slow decomposition of the Chalk. Reid (1899), however, pointed out that the thickness of Chalk which would need to be dissolved to produce sufficient clay residue to form clay-with-flints was excessive, and that the bulk of the material was derived from Tertiary rocks. Wilson et al. (1958) suggested that the more gravelly deposits resulted from removal of the matrix by water.

The Tertiary deposits of the Hampshire Basin show an overall increase in sand and gravel, and a decrease in clay content towards the basin margin in the west. If the clay-with-flints is in part derived from Tertiary strata, this could explain why the clay-rich deposits occur in the east of the Shaftesbury district, and the more gravelly deposits in the west. Washing out of the more arenaceous matrix from the western outcrop could result in the very gravelly, clast-supported deposits.

Loveday (1960) proposed that clay from an overlying formation moved downwards by colloidal suspension to be deposited at the junction with the chalk because of the filtering effect of the chalk or a change in the chemical conditions at the junction. If the overlying formation is dominantly clay, then the downward movement can be purely mechanical, as suggested Hodgson et al. (1967). These authors also suggest that dissolution of the chalk and cryoturbation also took place. They proposed a cyclical development during the middle and late Pleistocene, with solution of the chalk and downward migration of the clay taking place in the warm interglacial periods, and mixing by cryoturbation taking place in the glacial intervals. The occurrence of clay-with-flints on chalk as low as Holywell Chalk suggests that southward tilting and extensive erosion took place prior to the deposition of Palaeogene deposits which were subsequently incorporated in the clay-with-flints.

DETAILS

Woolland to Turnworth area

A pit [7971 0605] to the east of Woolland Hill showed about 1.5 m of very angular flint gravel with a reddish brown clay matrix. Locally, as in another pit [9761 0602], there are rounded black flint pebbles. West of Winterborne Houghton, a pit [7923 0548] showed 1.5 to 1.9 m of angular flint gravel with a reddish brown clay matrix resting on soft white flinty chalk. North of Winterborne Houghton, the clay-with-flints is also flinty, and in places passes into a clayey flint gravel. Chalkstone derived from the Lewes Chalk occurs in the clay on the spur north of Houghton North Down [8002 0617 to 8053 0642] and near Turnworth [8142 0770; 8122 0828]. Locally, the deposits were worked for gravel. An old pit [8057 0558] on Houghton North Down formerly exposed 0.9 m of gravel in a dark buff matrix. Many other 'gravel' pits are marked by the Ordnance Survey on the topographical maps

White (1923) recorded a section at the Folly [c.842 082], west of Durweston, in white to pale grey silty clay with crimson stains, fine- to coarse-grained yellow sand and buff loamy sand, associated with angular flint.

Pimperne to Ashmore area

Clay-with-flints is extensive on the interfluves between Ashmore and Farnham. There are small outcrops farther south near Pimperne Wood [around 914 114]. The deposits consist of stiff brown clay with angular and unworn brown flints. The junction with the Chalk is commonly strongly cryoturbated. The base level of the deposits on any one interfluve generally has an easterly fall. Small pits [e.g. 9421 1610] have been opened in these deposits, some of which [e.g. 9423 1687] worked the clay for bricks (White, 1923, p.94).

A pit [9415 1650] at Rookery Farm, Farnham, exposed 0.9 to 1.5 m of brown flinty clay with an irregular cryoturbated contact on Chalk. A pit [9423 1687] north of Rookery Farm, noted by White (1923, pp.81–82), exposed up to 4.2 m of clay and 'sugary' sand with a seam of pebbles; the deposits 'occurred very irregularly'. The clay was taken to Tollard Green Kiln because it was better than the clay-with-flints there. In another pit [9173 1625], 1 to 1.5 m of brown flinty clay, with a strongly cryoturbated base, rested on Chalk.

OLDER HEAD

Patches of up to 2.5 m of unsorted, unstratified, clayey gravel with angular clasts occupy terrace-like features capping interfluves and small hills at several localities in the district. They were formerly mapped as 'Plateau Gravel'. In this account, they are regarded as having formed by solifluction and are thought to be the dissected remnants of a formerly more continuous spread of older Head.

The clasts appear to be of local origin. In the north of the district, around Sherborne Causeway [840 230] and northward into the Wincanton district, the clasts are angular and subangular chert and sandstone derived from the Upper Greensand. The upper surface of these deposits slopes gently northwards. In the central part of the district, at Knackers Hole [780 115], the clasts are unstratified whole and broken flints, subangular to subrounded chert and sandstone, well-rounded flint, quartz, quartzite and lydianite, in a matrix of greenish grey, orange-mottled, clayey, sandy loam with abundant glauconite grains and mica flakes (White, 1923). The well-rounded pebbles were probably originally derived from Palaeogene deposits, but were recycled via the clay-with-flints, in which they are locally a common component. The chert and sandstone are presumably derived from the Upper Greensand, although the nearest chert beds are 10 km to the north-east. Near Okeford Fitzpaine, terrace-like spreads [812 108] consist of unsorted chalky and flint clay.

HEAD

Head occurs in the district in the valley bottoms and to lesser extents on the valley slopes and as aprons descending from a gravel terrace. The valley slope deposits are locally, as for example west of Sutton Waldron [855 155], related to springlines. They result from the downhill movement (solifluction) and accumulation of weathered and unconsolidated rock debris under periglacial freeze-thaw conditions. Some of the deposits mapped as head, particularly those associated with springlines, are of colluvial origin and are probably still forming at the present day. Some of the principal spreads along the dry valleys on the Chalk probably include some waterlaid material. There is commonly a step in the surface of the head deposits where those along a tributary valley join the main valley.

The deposits are heterogeneous and vary in composition and thickness, but broadly reflect the upslope parent material. The maximum thickness proved is 4.9 m in a steep-sided valley cut into the Upper Greensand at Melbury Abbas [8853 2032]. Generally, the deposits near to the Cretaceous outcrops consist of angular and subangular flints set in a sandy clay matrix. In the north of the district, the clasts are commonly of chert derived from the Upper Greensand. Head in the dry valleys crossing the Chalk usually has a gravelly clay top, which commonly overlies a chalk gravel. A temporary section [9270 1487] in the side of a valley north-west of Bussey Stool Farm showed 1.5 m of stratified chalk rubble with scattered flints dipping parallel to the hillside. This is probably typical of much of the head on the lower valley sides on the Chalk outcrop; in the absence of exposure, such deposits cannot be distinguished from in-situ Chalk. In south-east England, such deposits of chalky head are referred to as Coombe Deposits; the term Coombe Rock is restricted to cemented chalky head (see Smart et al., 1966).

On the Jurassic outcrops, clasts of subrounded limestone and calcareous sandstone are more common than flint. Deposits formed on the slope below the escarpment of the Forest Marble may in part result from the disintegration of old landslips. Deposits up to 4 m thick occur where these solifluction sheets were trapped behind the ridge formed by the Fuller's Earth Rock. On the gentler slopes on the clay formations, particularly the Lower Fuller's Earth and Oxford Clay, there are extensive, but thin (up to 1.4 m) spreads of brown or orange-brown, mottled, gritty clay with pebbles, commonly with a basal gravel. Locally [e.g. 7845 1330], where the head deposits are argillaceous and calcareous, and associated with a spring issuing from the base of the Clavellata Beds, the terrestrial gastropods *Aegopinella nitidula,*

A? juvs., *Carychium minimum, Pomatias elegans, Vitrea crystallina* and vertiginid juvenile fragments have been found. *C. minimum, A. nitidula* and *V. crystallina* are normally associated with moist locations such as woods or marshy ground, and *P. elegans* is always found on highly calcareous soils. All occur in the district at the present time and there is nothing in the assemblage to indicate a different climate from that presently prevailing. Head deposits in these situations are probably of colluvial origin.

DETAILS

Charlton Horethorne area

Details of auger holes and sections proving up to 3.3 m of head deposits in the north-west of the district are given by Taylor (1990). A selection is given below:

Near Bradley Head [6610 2186]

	Thickness m
Topsoil, brown sandy	0.30
Clay, sandy, with small limestone fragments, medium brown	0.70
Clay, firm, buff	0.25
Clay, sandy, with much coarse limestone debris, on limestone	0.90

The Cleeve [6549 2324]

	Depth m
Topsoil	0.30
Sand, clayey, ochreous-brown with a few limestone fragments	0.70
Gravel comprising fairly well-rounded limestone pebbles, up to 70 mm across, in a clayey sand matrix the lower 1.3 m containing 50 per cent matrix	to 2.30

West of Cathill Lane [6698 2292]

	Depth m
Topsoil, sandy clay, brown	0.25
Clay, sandy, light brown, becoming smooth, plastic, light grey, buff to ochreous mottled	1.40
Clay, as above, becoming very oxidised, rusty brown and black mottled, firm	1.60
Clay, as before	1.80
Clay, gravelly, smooth, light grey, ochre-mottled, becoming firm	2.65
Clay of Upper Fuller's Earth	to 3.00

North of Hanglands Lane [6714 2287]

	Depth m
Topsoil, sandy, fine-grained, brown	0.20
Sand, fine-grained, slightly clayey, mid-brown	0.40
Clay, very sandy and silty, buff-brown, becoming clayey sand at 0.6 m; continuing brown mottled and quite plastic	2.00
Sand, fine-grained, silty, variably clayey, wet	2.30
Sand, becoming more clayey, buff to brown mottled	2.60
Sand, slightly clayey, with sandstone fragments (Forest Marble); water entering hole below 2.6 m	to 3.30

Kington Magna area

South-west of Kington Magna, a low-lying spread of gravelly sandy clay, 0.7 to 1.2 m thick, rests on Oxford Clay [around 755 225]. An auger hole [7673 2134] south-south-west of Kington Magna proved 1.4 m of sandy clay with small pebbles, on 0.35 m of sandy gravel, on Oxford Clay.

A 1 m deep ditch [7642 2155 to 7664 2151] near Coking Farm proved a highly irregular junction of head on Oxford Clay, with gravel-filled channels and gravel-free intervening highs. Lenses of gravel occur within the overlying clay. Cryoturbation 'flames' of Oxford Clay rise though the deposits, and there are overfolds in the head at the base. The clasts consist of flint, limestone and jasper.

Sturminster Newton to Belchalwell to Mappowder

North-east of Rixon, an auger hole [7954 1474] proved 2.15 m of dirty, brown, sandy clay, with scattered flints below 1.15 m, on a bed of flints. In the Coombs Valley south-east of Newton, pebbly oolitic clay with limestone fragments and Recent gastropods (see above) extends from a spring line at the base of the Clavellata Beds [around 7845 1330] to the stream bed.

A section [7991 0913] near Belchalwell Street showed 0.5 m of dark brown, clayey, flint gravel, with clasts up to 10 cm across, on 0.5 to 1.0 m of chalk and flint gravel, with clasts up to 20 cm across, in a chalky sand matrix, on Upper Greensand.

South-west of Mappowder, an auger hole [7280 0575] proved 2 m of orange-brown, extremely sandy clay to clayey sand with flints at the base, on more than 0.3 m of loose oolitic clay and oolitic debris. Near Pulham [around 712 088], a gently sloping spread of up to 1.5 m of clayey gravel and gravelly, extremely sandy clay occurs.

Sherborne Causeway to Manston

An auger hole [8333 2312] near Sherborne Causeway proved 0.4 m of sandy brown clay, on 2.1 m of soft to firm, mottled orange and grey, sandy clay with small sandstone and flint pebbles at 1.4 m.

South-west of Manston, the valley side sloping down to the River Stour is mantled by up to 1.4 m of brown clayey sand and sandy clay, with a gravelly base [around 805 150].

Shillingstone to Okeford Fitzpaine

Chalky and flinty sandy clay occupy the valleys draining from the chalk escarpment between Shillingstone [around 828 105] and Okeford Fitzpaine [around 818 105 and 811 106]. An auger hole [8240 1061] south of Shillingstone House proved 0.3 m of orange-brown flinty sand, on 1.5 m of sandy and gravelly clay, on 1.8 m of clayey sand with small scattered chalk and flint pebbles. A second auger hole [8236 1052] proved 0.4 m of dark brown pebbly sand, on 0.1 m of mottled orange and grey clayey sand, on 2.0 m of sandy mottled orange and grey clay with scattered flints and small chalk pebbles, becoming very glauconitic and sandy below 1.2 m, and very chalky at 1.8 m, on Upper Greensand.

Child Okeford–Durweston–Blandford Forum–Littleton

At Child Okeford, a south-west draining valley [around 833 123] is floored by chalky gravel. White (1923, p.86) referred to these deposits as Coombe Rock. He noted 1.5 m of compact unworn chalk rubble in a cutting north of the church, but in the lane bank 270 m west of the church, the upper part of the deposit was composed of well-rounded chalk pebbles with traces of bedding, unevenly covered with a few feet of brown loam.

A pit [?8603 0875] south-west of the mill at Durweston formerly exposed 0.46 m of dark grey 'mould', on 0.3 to 0.6 m of brown stony loam piped into 0.6 to 1 m of indistinctly stratified, but with some cross-bedded, compact, tufaceous chalk rubble with scattered flints, and containing lenses and seams of mixed chalk and quartz sand ['Coombe Rock'], on stratified river terrace deposits (White, 1923, fig.11). The beds dipped valleywards.

A section [8800 0820] just north of Nutford Farm, near Blandford Forum, showed 0.2 m of topsoil, over 0.5 m of brown silt and clay with small angular flints, on 0.5 m of very large angular flints. A section [8400 1106] west of Hanford showed 1 m of brown clayey sand, on 0.1 to 0.2 m of flinty and chalk gravel, on 0.1 to 0.2 m of chalky silt, on more than 0.1m of brown sandy clay.

At Littleton [890 053], White (1923, p.89) recorded a spread of fine, red-brown brickearth with small chips of flint mantling the slope between the railway and the alluvium.

Shaftesbury–Iwerne Courtney

At Shaftesbury, an auger hole [8621 2223] proved:

	Depth m
Sand, clayey, with sandstone clasts	1.2
Sand, fine- to medium-grained, weakly glauconitic	2.2
Sand, clayey, glauconitic, reddish brown, with sandstone clasts	3.0

At Melbury Abbas, a borehole [8853 2032] in the bottom of a steep-sided valley cut into Upper Greensand proved:

	Depth m
ALLUVIUM	
Clay, silty, soft, brown, mottled grey	0.70
HEAD	
Sand, fine-grained, clayey, silty, dark green, loose, becoming a very clayey slightly sandy silt downwards	4.90
UPPER GREENSAND	
Sandstone, fine-grained, glauconitic	1.08

East of Sutton Waldron, the head along Combe Bottom includes much chalk debris [around 867 158]; at one point [8720 1593], it consists of chalky silt. An auger hole [8656 1574] south of Manor Farm proved 0.6 m of clayey brown sand, on 1.1 m of chalky marl, on 0.4 m of very clayey glauconitic sand, on 0.1 m of chalky glauconitic sand. A second hole [8689 1590], some 370 m north-east, proved 1.1m of brown sand, on 0.4 m of brown clayey sand, on 0.2 m of brown chalky clay, on 0.4 m of glauconitic sand of the Upper Greensand.

South-west of Sutton Waldron, head deposits flank the west-facing slope that extends from the Upper Greensand across the Gault onto the Lower Greensand. The deposits are largely associated with springs issuing from the base of the Upper Greensand. Two auger holes [8559 1555; 8547 1553] in this tract proved similar sequences of 1.7 m of sandy, glauconitic, micaceous clay with scattered angular flints at the base, resting on Lower Greensand. A third hole [8541 1573] proved a thicker sequence of head comprising 0.6 m of flinty clayey sand, on 1.6 m of very sandy laminated clay, on 0.1 m of very sandy glauconitic clay with small angular flints and rounded chalk pebbles.

North-west of Iwerne Courtney, an embayment in the Gault is filled with more than 1.4 m of glauconitic chalky clay [around 848 137]. East of the village, at least 3 m of brown chalky clay occurs in the bottom of one of the dry valleys [c.8647 1296].

Ashmore–Stubhampton area

A section [9189 1500] on the east side of Ashmore Bottom showed 1.5 m of stratified chalk rubble with scattered flints. The dip of the deposit was roughly parallel to the hillside.

A section [9270 1487] north-west of Bussey Stool Farm showed 1m of flint gravel on 0.4 m of chalk gravel.

RIVER TERRACE DEPOSITS

River terrace deposits, laid down mainly by the rivers Stour, Lydden and Cale, and the Caundle, Bow and Manston brooks, with lesser spreads along some of their tributaries, occur at twelve principal levels within the district. In addition, some high-level spreads remain unclassified (Figures 58, 59 and 60). The numbered terraces range in height from 1 to 29 m above the alluvial plain. The First and Second terraces occur throughout the river valleys, but higher terraces are confined to the Caundle Brook, the River Lydden and the River Stour between its confluence [766 171] with the Lydden and Stourpaine [860 090] at the foot of the chalk escarpment. The restriction of terraces 3 to 12 to the north of the Chalk escarpment suggests that they were part of an eastward-flowing river, with a broad floodplain developed over the Oxford and Kimmeridge clay vales. The thalwegs of these terraces (Figures 58, 59 and 60) indicate that the chalk escarpment near Blandford had already been breached. The absence of terraces 3 to 12 where the Stour crosses the chalk outcrop may be due to the river being confined in a straight, narrow valley in which water velocities were too high to allow deposition. Downstream, terraces 3 to 12 reappear near Sturminster Marshall in the Bournemouth (329) district, in the area where the River Stour flows off the Chalk onto Tertiary deposits.

During the original geological survey, the deposits now regarded as river terrace deposits were grouped mostly as 'Valley Gravel', together with some occurrences of 'Plateau Gravel'. It is evident from White's (1923) descriptions that some 'Valley Gravel' is in part colluvial; during the recent survey, these deposits were mapped as head.

The deposits in the district consist mostly of subangular flint gravel, with small proportions of well-rounded flint pebbles derived from Tertiary deposits, chert from the Upper Greensand, Jurassic limestone, and chalk, set in a sandy or clayey sand matrix. In the north, the gravel is predominantly composed of limestone clasts from the Cornbrash, with some from the Forest Marble. Farther east, the gravels mainly comprise cementstone clasts derived from the Oxford Clay, commonly with many abraded gryphaeid shells and fragments. Chert appears to be restricted to the terraces of the River Cale; a source from the Upper Greensand north-east of the district is likely. Coarse chert gravels with gryphaeid debris form isolated remnants [741 233] up to 1.3 m thick, south-west of Pelsham Farm.

Most of the terrace deposits seem to be poorly sorted, with clasts generally less than 6 cm across, but some clasts exceed 10 cm. The lower terraces are commonly capped by a sandy clay or clayey sand. The thicknesses are unknown for most spreads. Deposits of the second terrace exceed 2.5 m to the east of West Orchard [840 160], and it is probable that some of the more recently worked sand and gravel deposits [764 169; 762 158; 817 129] are between 3 and 4 m thick.

No fauna or flint implement has been recovered from the terrace deposits within the district, the ages of which are unknown. Downstream, in the Bournemouth district, Second to Fourth River Terrace Deposits have yielded Aurignacian-type (uppermost Palaeolithic) flint implements (Calkin and Green, 1949), and are probably of Devensian age (Bristow et al., 1991).

DETAILS

High-level terraces of the Caundle Brook and rivers Lydden, Divelish and Stour

UNDIFFERENTIATED RIVER TERRACE DEPOSITS

High-level gravels occur at 137 m above OD (c.40 m above the floodplain) at Sharnhill Green [707 055], at 99 m above OD (47m above the floodplain) at the obelisk north of Lydlinch [7385 1555] and at Conygar Coppice [810 121], Okeford Fitzpaine, and at 77 m above OD (37 m above the floodplain) at Little Hanford [841 117].

NINTH RIVER TERRACE DEPOSITS

At Hargrove Farm [749 154], the deposits consist of up to 2 m of coarse flints.

EIGHTH RIVER TERRACE DEPOSITS

A section [7150 1314] in the spread east of Rowden Mill Farm along the Caundle Brook showed 0.3 m of flinty gravel, overlying 2.7 m of mottled orange and brown silty sand, on Oxford Clay.

The spread of sand and gravel partly in Tan-hill Copse and partly in the adjacent field [816 130] has been worked postwar for sand and gravel. There is no exposure, but the pits [8165 1300; 8170 1292] are 2 to 3 m deep.

SEVENTH RIVER TERRACE DEPOSITS

Deposits near Pulham are clayey and compact, but are probably less than 2 m thick. One section [7098 0856] showed over 1.5 m of very clayey orange gravel.

FIFTH RIVER TERRACE DEPOSITS

The spread to the south-east of Pulham [715 083] is probably less than 1 m thick. A small, partly filled pit [7169 0851] was less than a metre deep with Oxford Clay in the base.

Terraces along the higher reaches of the River Divelish lie at about 8 to 10 m above the floodplain and and are probably part of the Fifth Terrace suite. They are thin, usually less than 1 m thick, and composed of subangular flints with a clayey matrix. A small section [7786 0884] showed 0.8 m of coarse (clasts up to 0.15 m) flint gravel, resting on Kimmeridge Clay. Another section [7801 0823] exposed 0.8 m of clayey yellowish brown gravel, on Kimmeridge Clay.

Manston, Fontmell and Chivrick's brooks

SECOND RIVER TERRACE DEPOSITS

Extensive Second River Terrace Deposits occur along the Manston and Fontmell brooks, where the outcrops are over 1 km wide and extend along the brooks for over 2 km. The surface of the deposits is planar and tilted north-westwards, such that along the Fontmell Brook there is a prominent step up of about 1.5 m from the brook, with Kimmeridge Clay cropping out on the steeper slopes. From this step, the ground falls gradually north-westwards towards the Manston Brook, where there is only a slight feature break separating terrace deposits from alluvium on the south-east side of the stream. Deposits at the surface appear to consist dominantly of sandy clay with only a minor gravel content. A temporary exposure [8354 1635] at Winchell's Farm was still in sand and gravel at a depth of 2.5 m; another exposure [8414 1577] south-east of Winchell's Farm proved 2.4 m of sand and gravel above Kimmeridge Clay.

A section [8158 1524] in the new channel of the Manston Brook at Manston shows 1.8 m of terrace deposits arranged in two fining-upward cycles, each with a gravelly base and passing up into a sandy clay.

A section [7968 1440] along Chivrick's Brook close to its confluence with the River Stour formerly showed (White, 1923):

	Thickness m
Soil	'thin'
Loam, brown with seams of stone	0.3–0.6
Gravel, loamy, with clasts of chert and flint	0.3–0.6
Gravel, dark brown, with much rubbly oolite, pocketed into	0.3
CLAVELLATA BEDS	
Oolite, fine-grained flaggy	0.9

River Cale and Bow Brook

FIRST RIVER TERRACE DEPOSITS

The thickness of the terrace is variable, with 'islands' of clay emerging in the area of Henstridge airfield, and a thickness of 2.88 m being recorded [7403 2076] near Hackthorne Farm, Lower Marsh. At the western end of the terrace [7248 2353], at Vale Farm, 1.55 m of gravel is channelled into the underlying clay.

A ditch beside the A30 [7438 2126], west of Bow Bridge, exposed 1 m of fine-grained, slightly clayey sand, on 0.8 m of sandy clayey gravel, on Oxford Clay. The subrounded to subangular, medium gravel, contained clasts of Forest Marble limestone and sandstone, chert and fragments of large gryphaeids derived from the Oxford Clay. Traces of cryoturbation occur.

River Stour

FIRST RIVER TERRACE DEPOSITS

An auger hole [7696 2099] to the north-east of Factory Farm proved 0.3 m of sandy clay with pebbles, on 0.7 m of clayey sand with pebbles, which rested on 0.5 m of sandy flint gravel.

There are extensive spreads of First Terrace Deposits near Hammoon [806 144 and 820 140]. The surface deposits are particularly gravelly in the west [around 806 140] where they

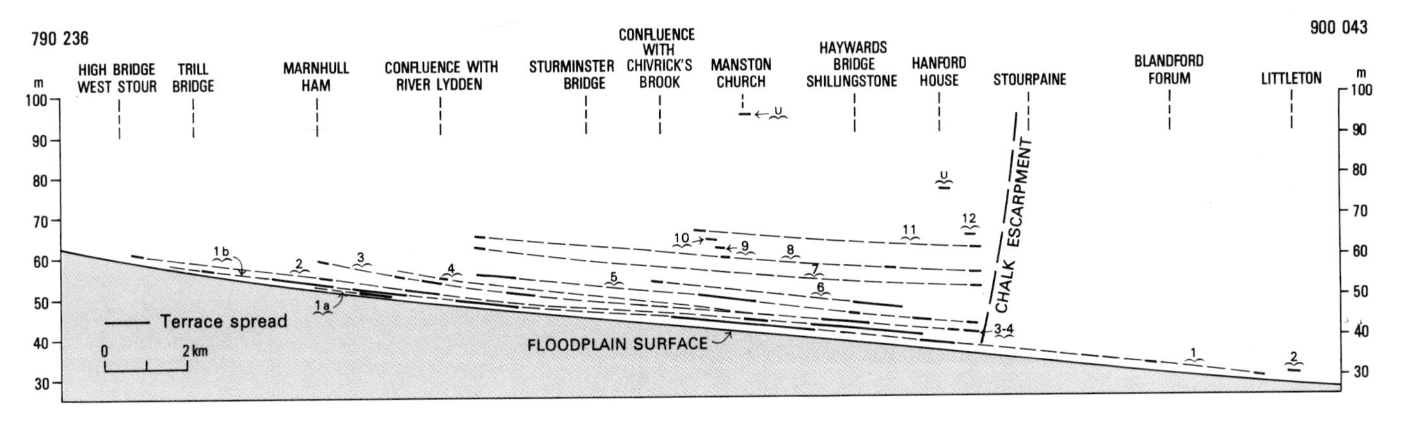

Figure 58 Profile of the River Stour and its terraces.

consist of angular flint clasts up to 10 cm across, and subangular and rounded limestone pebbles, generally less than 1.5 cm in diameter, set in a sandy clay matrix. A temporary exposure [8090 1367] north-east of Fiddleford proved 1.8 m of coarse gravel on 0.6 m of fine gravel.

River Lydden

FIRST RIVER TERRACE DEPOSITS

The deposits along the Lydden and tributaries generally consist of up to 1 m of reddish brown sandy clay to clayey sand, on up to 1.5 m of flint gravel which commonly has a clayey sand matrix.

A pit [7288 0974] near Lyddon House showed 0.4 m of clayey sand, on 1 m of clayey coarse gravel, on solid. Farther west, in the bank of the River Lydden, the following section [7160 0784] was seen:

	Thickness m
RIVER TERRACE DEPOSITS	
Sand, clayey with gravel at base	0.3
Sand, clayey, reddish brown	0.6–0.8
Clay, grey, sandy	0.1
Gravel	to over 0.2

Caundle Brook and River Cam

SECOND RIVER TERRACE DEPOSITS

Extensive areas of loam underlain by thin gravel deposits occur near Holnest. Similar deposits were mapped as Head in the Yeovil district. The deposits surround the small hills in the southern part of Holnest, where they appear to occur at at least two levels, with a separation of probably less than 2 m. An auger hole [6636 0934] proved 0.5 m of orange-brown, extremely sandy clay, on gravel. South of Holnest Park, an auger hole [6530 0841] penetrated 1.2 m of orange to buff, extremely sandy clay, on gravel with clayey sand. The gravel is generally less than 0.5 m thick.

FIRST RIVER TERRACE DEPOSITS

An auger hole [6805 1007] south-west of Sandhills showed 0.6 m of orange-brown and yellowish brown clayey sand on gravel. Another [6829 1012], revealed 0.5 m of brown loam, on 0.3 m of gravel, on Oxford Clay.

In the upper waters of the River Cam, auger holes [6542 0785; 6532 0760] west of Middlemarsh Common showed 1.5 and 0.6 m respectively, of orange-brown, extremely sandy clay on gravel.

ALLUVIUM

Alluvium is the floodplain deposit of the modern river system. It is most extensive along the River Stour, but it also occurs along all its major tributaries. It generally consists of an upper unit of mottled grey and brown, or orange-brown, commonly organic, silt, silty clay and clayey sand with scattered clasts, and a lower unit of sand and gravel. The thickness of the two units varies from 0.7 to 3.33 m and 0.1 to 0.6 m respectively. Locally [e.g. 7638 2024], interbedded sand and clay underlies the gravelly unit.

An extensive area, up to 0.8` km wide, of alluvial deposits of the River Cale and Bow Brook occurs on the low-lying Kellaways Formation and Oxford Clay outcrops in the north of the district. As a result of a feature in the Peterborough Member, the drainage has developed as two separate systems. That from the Forest Marble-Cornbrash dip slope forms wide alluvial flats extending south-east across the Kellaways Formation from Horsington Marsh [712 247] on the adjacent sheet to the north, to north of Combe Throop [725 238]. A second broad area of alluvial flats has developed on the Kellaways Formation east of Yenston [723 223 to 725 215]. The gravels consist of clasts of limestone, cementstones and flints. East of the Oxford Clay feature, chert occurs in the gravels, and peaty sands are present locally.

Locally, the alluvium of the River Stour contains a rich molluscan fauna. The fauna from the only sample collected (see below) is indicative of fluviatile deposition in hard water in a climate not significantly different from that of the present day.

DETAILS

River Cale and Bow Brook

North of Broadmead Lane [7265 2175], 0.3 m of dark brown clayey soil, overlay 1.7 m of sandy, brown clay, becoming grey ochre-mottled and very sandy below 1 m, on 0.1 m of water-

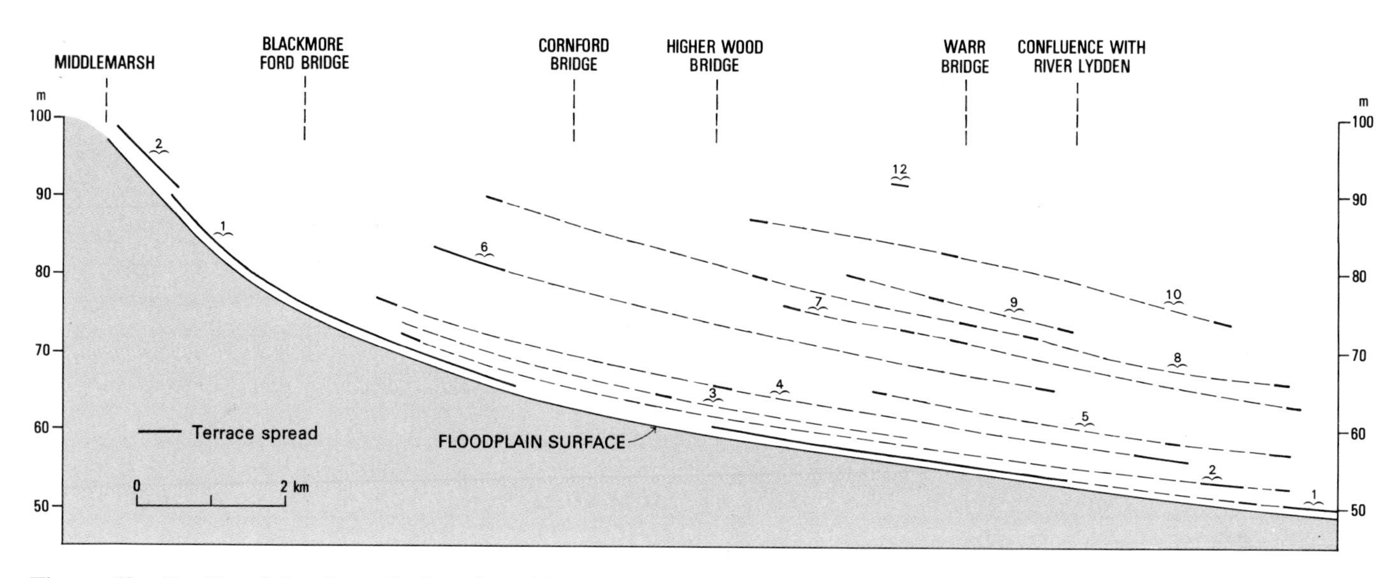

Figure 59 Profile of the Caundle Brook and its terraces.

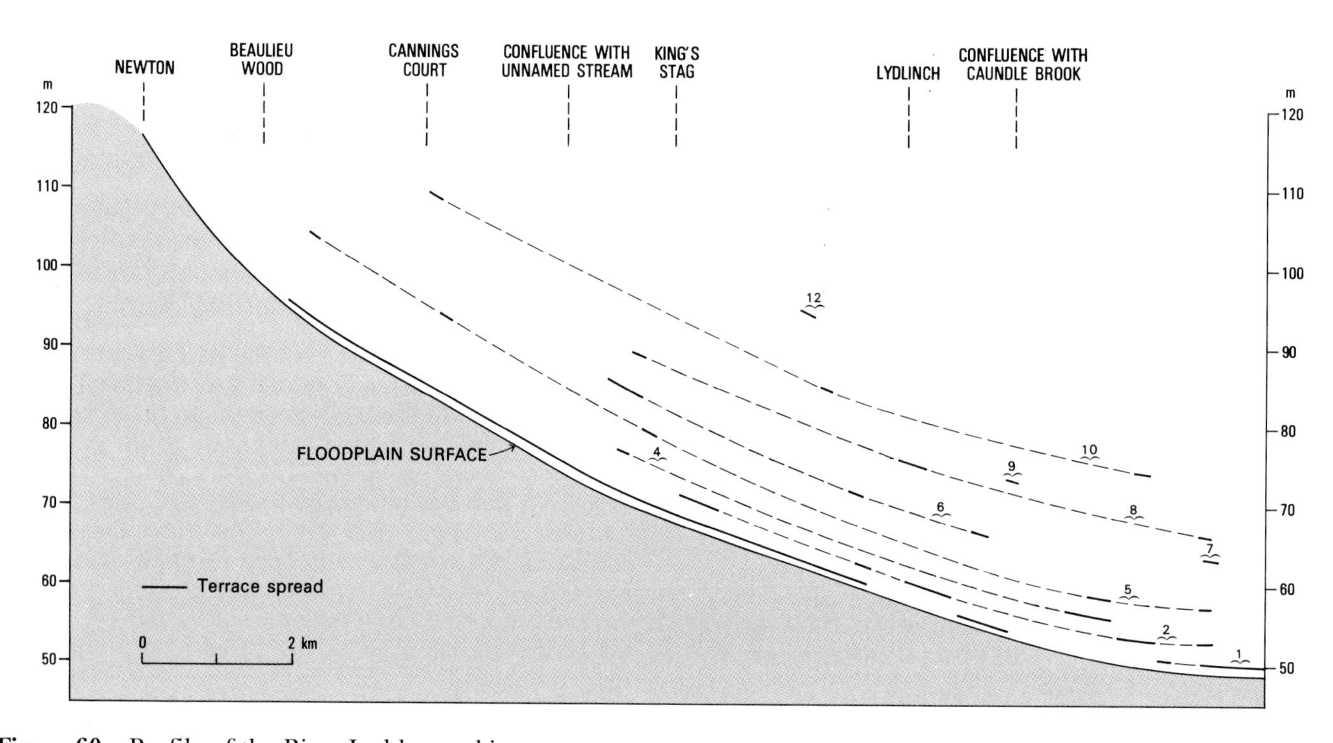

Figure 60 Profile of the River Lydden and its terraces.

logged sandy gravel, on Kellaways Formation. Farther north, beside the Bow Brook [7361 2140], 0.3 m of loamy, organic, brown soil, overlay 0.5 m of silty, fine-grained sandy, brownish buff clay, on 0.8 m of grey, ochreous-mottled clay, more sandy below 1.3m, on 0.1m of sandy, fine gravel, on Oxford Clay. A hole [7398 2275] on the bank of the River Cale [7398 2275] proved 0.3 m of silty, sandy, brown topsoil, on 1.1 m of grey, ochreous-mottled clay, on 0.6 m of sandy, clayey, ochreous gravel on Oxford Clay.

An exposure [7587 2172] in the bank of the River Cale revealed 0.8 m of brown sandy clay, on 0.7 m of clayey sand, on 0.05 m of fine-grained sandy clay. Farther upstream [7525 2207], 0.2 m of dark brown sandy clay, on 2.5 m of greyish brown sandy clay with brown mottling, limonite and manganese dioxide grains, becoming softer and sand-free downwards, on Oxford Clay, was exposed.

Boreholes [7577 2154; 7589 2158] for the A30 bridge crossing of the River Cale proved a maximum of 3.33 m of soft, organic, mottled greyish brown clay, resting on up to 0.17 m of limestone gravel:

	Thickness m	*Depth* m
Soil, dark brown, clayey	0.15	0.15
Clay, silty, brownish grey mottled, streaky limonitic	1.65	1.80
Clay, silty, with gritty manganese dioxide nodules, some selenite, race, bits of lignite, shells and limonitic pebbles; sandy below 1.95 m	0.20	2.00
Clay, grey, mottled, sticky; abundant angular to subangular flint clasts up to 2 cm across; limestone and shell fragments	0.30	2.30
OXFORD CLAY		
Clay, soft, selenitic, bluish grey	0.35	2.65

River Stour

An exposure [7903 2232] in the bank of the River Stour showed 0.7 m of brown sandy clay, on 0.15 to 0.4 m of flint and limestone gravel in a clay matrix, on orange-brown clayey sand of the Hazelbury Bryan Formation. Downstream, 1.5 m of pebbly alluvium, with clasts of flint, oolite and limestone were noted in the river bank [7796 2066] south of Fifehead Magdalen. Nearby [around 779 206], dredged alluvial material included much shelly clay.

An exposure [7638 2024] in the bank of the Stour close to its confluence with the Cale showed 0.7 m of clay with flint cobbles, on 0.2 m of gravel, on 0.5 m of interbedded sand and clay, on Oxford Clay.

In the large meander loop of Chivrick's Brook south-east of Marnhull, up to 1.2 m of soft brown clay was proved [7879 1761]. About 1.1 km downstream [7984 1725], 1.2 m of peaty clay overlies gravel.

Boreholes [7835 1353 to 7843 1355] near Town Bridge, Sturminster Newton, proved between 1 and 2.08 m of 'soft to firm brown clay with cobbles' above Corallian beds.

Near Fiddleford Mill [800 136], between 1.5 and 1.8 m of mottled orange and grey, structureless, silty and sandy clay are exposed. Nearby, a section [7998 1394] reveals 0.9 m of brown and mottled orange and brown sandy clay, above 0.6 m of gravel.

Downstream from Fiddleford, the alluvium consists of up to 1.8 m of brown shelly clay. A sample of the clay [8004 1391] yielded the following molluscs: *Ancylus fluviatilis, Anisus vortex, Armiger crista, Bathyomphalus contortus, Bithynia tentacula, Discus rotundatus, Hydrobia* sp., *Lymnaea peregra ovata, Segmentina nitida, Theodoxus fluviatilis, Trichia hispida, Trichia?, Vallonia pulchella, Valvata piscinalis, Anodonta* sp., *Pisidium amnicum, P. henslowanum, P. moitessierianum, P. subtruncatum, P.nitidum, P.* sp., *Sphaerium corneum* and *S.* sp. The fauna is indicative of fluvial deposition in hard water with an appreciable current.

Near Hammoon, 1.5 m of shelly brown clay was noted in the river bank [8168 1485]. South-east of Hammoon, 1 m of brown sandy clay, rests on 0.6 m of gravel, on Kimmeridge Clay [8239 1353].

At Hayward Bridge, north of Shillingstone, 2 m of brown sandy clay was noted above 0.4 m of gravel [8238 1204]. Some 500 m south-east of the bridge, 1.2 m of shelly brown clay are exposed [8265 1160]. Some 500 m downstream [8307 1132], 1.8 m of mottled orange and grey sandy clay with scattered small pebbles, rest on 0.3 m of smooth blue-grey clay, on gravel.

White (1923) noted shelly brown loam about 45 m below Haywards Bridge [c.824 120], from which Mr A S Kennard identified '*Arion* sp., *Agriolimax agrestis, Limnaea pereger, Planorbis crista, Bithynia tentaculata, Valvata piscinalis, Ancylastrum fluviatilis, Theodoxus fluviatilis* and *Pisidium* sp.'.

In the river bank [8413 1041] south-west of Hanford, 1.3 m of brown sandy clay with gravel stringers occur above 0.4 m of gravel.

Key, Twyford and Manston brooks

East of Margaret Marsh, up to 1.5 m of greyish brown silty clay rest on gravel [8316 1938]. North-west of Margaret Marsh, an auger hole [8100 1981] proved 1.5 m of mottled orange and grey silty clay, above 0.2 m of sandy clay with shell fragments and race, overlying Kimmeridge Clay. West of Margaret Marsh, up to 1.6 m of grey-brown clay were proved above gravel [8111 1891].

Near Manston, the Manston Brook has been diverted south-westwards through a 150 m wide tract of river terrace deposits into the River Stour. The course of the old river, which flowed 600 m southwards before joining the Stour, can be traced by the dry meander loops [around 820 150].

River Iwerne

A borehole [8622 1314] in the bed of the River Iwerne near Iwerne Courtney proved 0.4 m of stiff, brown, silty clay, above 4.1 m of firm and stiff, pale grey and brown, very sandy and gravelly clay to clayey sand and gravel, with clasts of flint and chalk.

River Divelish

A section [773 132] in the River Divelish showed 1.82 m of sandy, shelly clay, on 0.46 m of flint gravel, resting on Hazelbury Bryan Formation.

River Nadder

A borehole [9093 2287] at Ludwell proved 1.4 m of soft to firm, grey, brown and orange-mottled silty clay with a little gravel, on 0.4 m of soft greenish grey, mottled brown, clayey silt, which in turn rested on Upper Greensand.

LANDSLIP AND SLOPE DEVELOPMENT

The largest and most extensive landslips in the district are associated with the Gault in areas where the Upper Greensand escarpment is well developed. Less extensive slips are developed on the Oxford Clay beneath the escarpment of the Hazelbury Bryan Formation. Small slips also occur on the Frome Clay beneath the Forest Marble scarp, on the Forest Marble, Oxford Clay, Hazelbury Bryan Formation, Kimmeridge Clay and Lower Chalk.

The landslips associated with Gault and Upper Greensand are spectacular. They have a classic landslip morphology, with back scarp, undulating irregular topography, locally with enclosed ponds, and prominent toes (Frontispiece; Plate 11). Despite their large size and

Plate 11 Crescentic sand steps (5), cambers (4 and 7) and shallow rotational slips (arrowed) near Church Farm, Shaftesbury, viewed from the south-east. Nos. 4–6 are localities mentioned in the text (see also Figure 61).

proximity to a major urban area, the slips had not previously been recognised. During the recent survey, the slips were mapped and divided into formational types (Figure 61). The results of detailed investigations carried out as part of the present survey (Gostelow, 1991) are summarised here.

Figures 61, 62 and 63 show the surface geology and geomorphological features of the slopes. Three distinct elements or landforms can be recognised (Figure 63); they are i) the Shaftesbury Scarp, ii) the Shaftesbury Platform, and iii) the Shaftesbury Undercliff.

Shaftesbury Scarp

The scarp, 15 to 20 m high, with average slope angles of between 30 and 35°, can be traced continuously around Shaftesbury. The older part of the town lies on a flat surface above the escarpment at about 210 to 215 m OD. The scarp face, cut in the Shaftesbury Sandstone and capped by the Boyne Hollow Chert, is smooth, vegetated and divided into broad arcuate embayments up to 300 m across.

There is no evidence of current instability on the scarp. Several steep roads lined with buildings cross the feature where they connect the platform with Shaftesbury town centre (e.g. St John's Hill), but there is no visual sign of distress to structures or road pavements.

Shaftesbury Platform

The platform is 200 to 400 m wide and slopes from the base of the escarpment at between 3° and 7°. The inner edge, at the base of the Shaftesbury Sandstone, can be traced as a continuous feature at a constant elevation (195 m above OD) around Shaftesbury. The outer edge is irregular, but is usually marked by a distinct break of slope at about 145 m above OD, although its precise boundary is locally obscured by buildings.

The platform is developed on the uncemented to poorly cemented sands and sandstones of the Cann Sand. The platform topography varies from smooth, to gently undulating, to hummocky, with each type passing imperceptibly into another. For this reason, only the hummocky ground has been distinguished in Figure 61. Hummocky surfaces are more noticable near the outer edge where the slope angle steepens slightly. In that part of the platform, crescentic steps, about 1 to 2 m high, separated by 10 to 15 m of flat ground, occur locally. South-east of St James [859 223] (Figure 61, locality 8), they extend up to 100 m back from the edge of the platform. The platform is generally well drained, with little evidence of active springs at the inner edge, except near Castle Hill, Shaftesbury [8557 2287] (Figure 61, locality 3), where springs occur at 178 m above OD.

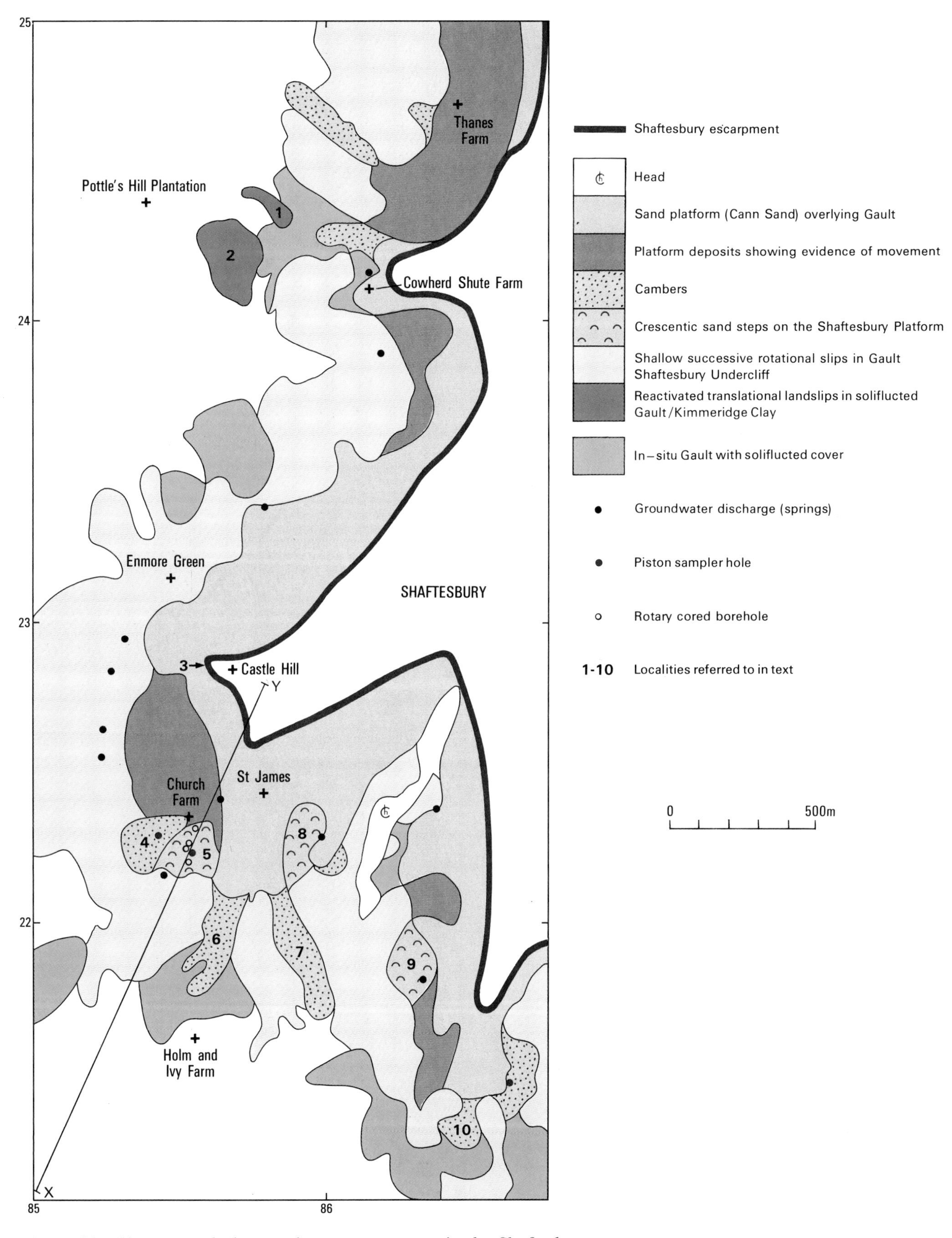

Figure 61 Slope morphology and mass movements in the Shaftesbury area.

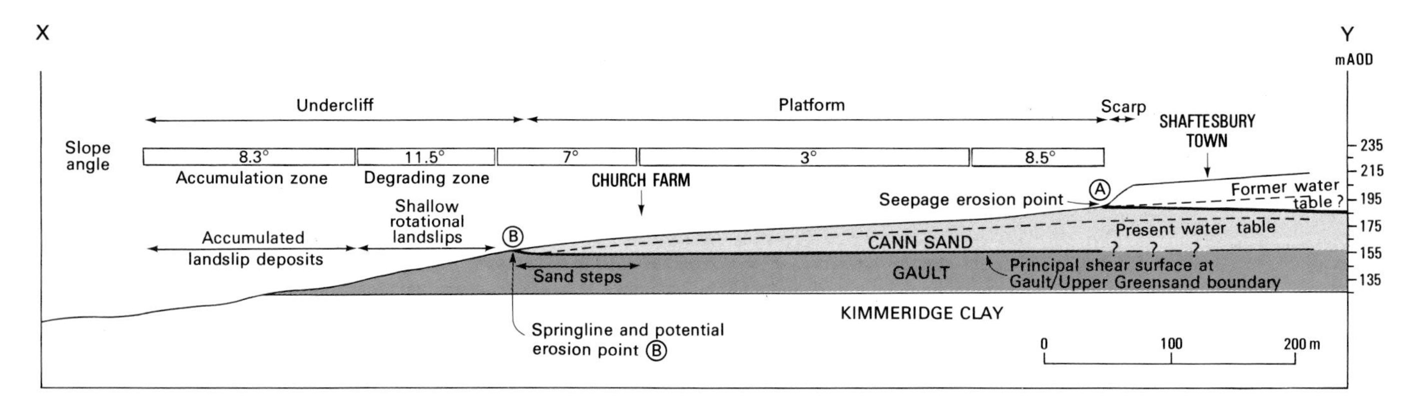

Figure 62 Section X–Y on Figure 61.

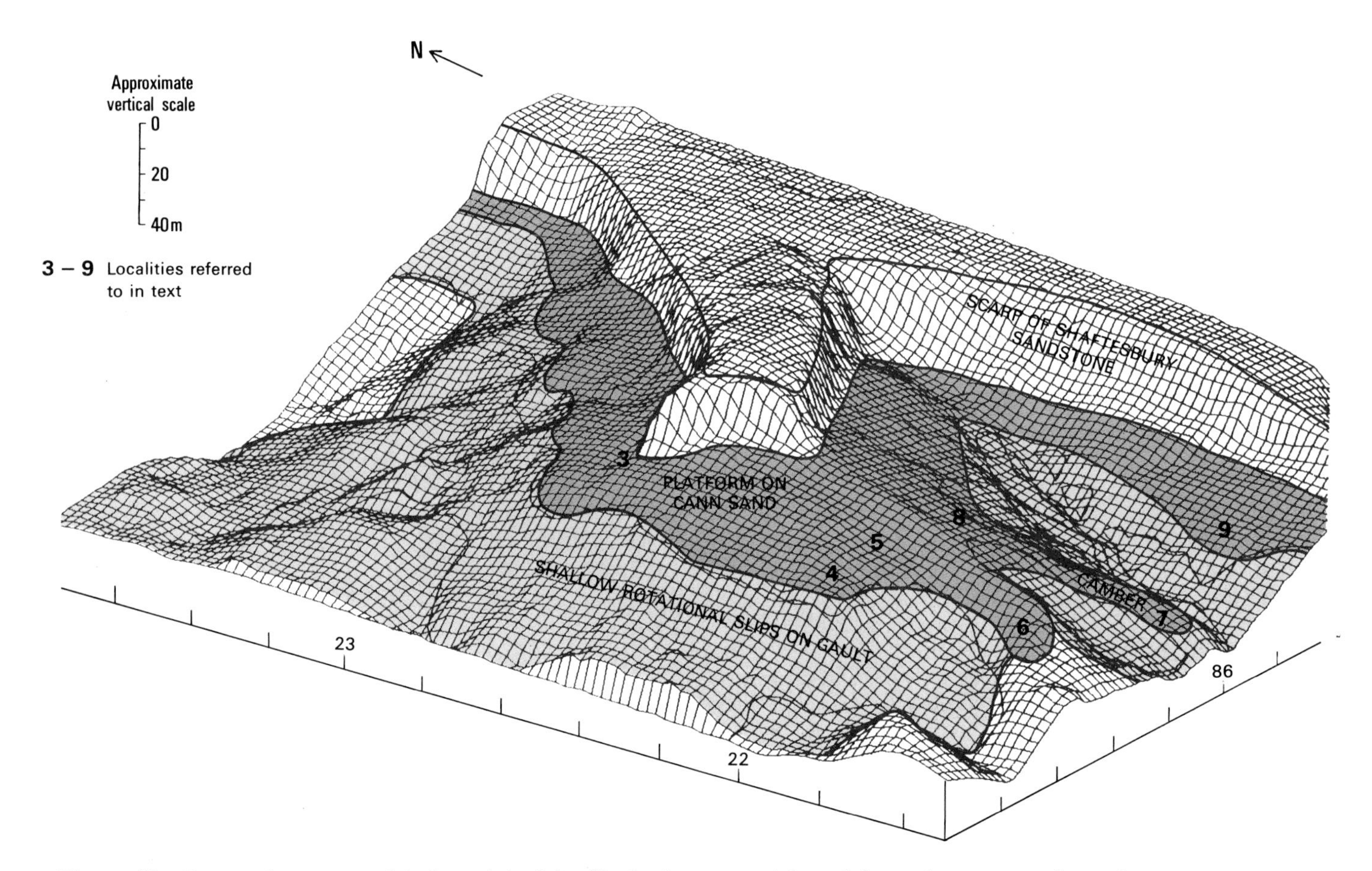

Figure 63 Isometric topographical model of the Shaftesbury area (viewed from the west-south-west).

Springs are, however, a common feature of the outer margin. In some places, between embayments, prominent spurs extend down to the vale where the undercliff is not fully developed. Similar spurs of smooth ground, where the platform is wider, but where the feature separating the platform from the undercliff is even less distinct, also occur (e.g. near St James [8545 2225] (Figure 61, locality 4; Plate 11); near St James's Common [856 219 (Figure 61, localities 6 and 7); 859 219]; and northwest of Cann [865 215] (Figure 61, locality 10)). In these situations, the Gault/Upper Greensand boundary falls towards the vale, suggesting that the Upper Greensand is cambered over the Gault. The gentler slopes carry the roads into Shaftesbury, e.g. that from Bedchester (Figure 61, locality 6; Plate 11).

Shaftesbury Undercliff

The outer edge of the platform coincides approximately with the Gault/Upper Greensand junction. The base of the undercliff varies in elevation from 110 to 120 m above OD, and in places includes part of the Kimmeridge Clay. The majority of the slopes are in landslipped Gault with a slope angle of 9° to 12°. There are few signs of present-day movement, but the presence of fresh back scarps, tilted trees and toe features indicate historically recent activity. At most localities, the undercliff can be divided into an upper, steeper (9° to 12°) degrading zone, and a lower accumulation zone which has lesser (7° to 10°) slope angles. The slides in the erosional zone are of shallow, successive rotational type; up to four separate slipped blocks, usually between 10 m and 20 m wide, occur. The depth of slipping is probably up to 6 m in places. Several small ponds have developed between such slips, particularly west of Shaftesbury [e.g. 8523 2263; 8526 2283; 8532 2294].

Some disturbed slopes on the slips below the undercliff are flatter and more convex, e.g. the slope below Cowherd Shute Farm [857 243] (Figure 61, locality 2) in the adjacent Wincanton district, where the overall angle between [856 243 and 869 243] is about 6.4°. This slope is interpreted as a reactivated translational slip. In addition, east of Pottles Hill Plantation [858 244] (Figure 61, locality 1), there are three translational slides on slopes of about 8.5°. Translational slipping is also present between Writh Farm [861 203] and the mill at Cann [869 209].

The extensively landslipped slopes of the undercliff are close to limiting equilibrium. Reactivation could easily be caused by rising ground-water levels in the overlying sand platform, or by cut and fill on the undercliff.

Three principal processes have acted individually, or in combination, to form the scarp, platform and undercliff. These are landslipping, seepage erosion and periglacial processes such as freeze-thaw (cryoturbation).

The steep scarp slopes at Shaftesbury are arcuate in plan and resemble deep-seated landslip backscarps. The initial development may have included some small-scale slipping of the Upper Greensand, but there is no evidence of large-scale slipped blocks or relict slip surfaces which pass across the Gault. The limited borehole evidence from the area suggests that the Gault/Upper Greensand contact beneath Shaftesbury and across the platform is continuous and not displaced by slip surfaces.

The long-term development of the Shaftesbury Scarp and Shaftesbury Platform has been controlled by the river system and the rates of lateral erosion, both of which were affected by Quaternary climatic changes. A combination of seepage erosion and periglacial processes was probably responsible for platform development. Hutchinson (1981) has reviewed the process of seepage erosion and has listed occurrences in the United Kingdom. The process is related to 'piping' failure in loose sands, in which individual grains are dislodged and removed by subsurface groundwater under a high hydraulic gradient. Closely spaced pipes in a sandy stratigraphical unit can cause the overlying strata to collapse or slide onto a shelf formed at the level of erosion. The sand and other debris moves across the shelf either by fluvial erosion or solifluction in a zone of high piezometric water pressures. The scarp face retreats by a combination of backsapping and removal of debris across the platform, the process being most effective where a loose, permeable sand overlies an impermeable clay and is overlain by stronger strata; such conditions are present at Shaftesbury. The Cann Sand is uniformly graded, loose and fine grained, and is of the type quoted by Hutchinson (1981) as particularly susceptible to this process.

The common occurrence of a layer of clayey gravel on the platform, and the lack of large sandstone blocks, suggests that cemented Upper Greensand rocks have been broken up by freeze-thaw processes. The movement of this head deposit across the gently sloping platform will have been promoted by solifluction. Head deposits presumably originally continued down to a slope base which has now largely been removed by more recent downcutting and landslipping on the undercliff.

The later phases of the development of the scarp, platform and undercliff are summarised in Figure 64, stages 1 to 4.

Stage 1 assumes an initial erosional slope of Gault and Upper Greensand which was formed by fluvial downcutting and erosion during the early Pleistocene. The water table is assumed to have been high and to have provided a hydraulic gradient sufficient to cause piping or back sapping in the lower portions of the Upper Greensand.

In Stage 2, the scarp collapsed and the debris was removed by fluvial and solifluction processes, possibly with the formation of secondary steps, such as those south-west [856 223] (Figure 61, locality 5) and south-east of St James [864 222] (Figure 61, locality 8) and north-north-east of Gears Mill [863 219] (Figure 61, locality 9). It is assumed that the point of erosion (A) at the base of the scarp moved back across the platform, because the debris on the shelf probably arrested the seepage-erosion process.

This migration of the erosion point led to Stage 3 in which the platform widened, and the escarpment decreased in height. Seepage point B may also have developed at this time, but the hydraulic gradient may have been too low for subsurface erosion to have taken place at the outer edge of the platform.

In Stage 4, the present situation, the water table has fallen to an elevation below point A and, for the most part, there is now no spring at this point; thus, further backsapping is prevented. Renewed toe erosion and oversteepening has caused landslipping on the Gault, and the main seepage point is now at B, leaving the Shaftesbury Scarp as an abandoned erosional slope. The present water table recorded in boreholes on the platform suggests that the hydraulic gradient (the difference in height of the piezometric head at two points divided by the distance between) at A is only about .043, which accounts for the current lack of subsurface erosion there.

Translational slipping of the Cann Sand on the Shaftesbury Platform

The sand steps on the Shaftesbury Platform are thought to have been caused by movement along low-angle shear surfaces that follow, or are close below, the Gault/Upper Greensand junction for most of their length. A similar mechanism is thought to explain comparable features in the Upper Greensand and Gault on actively eroding coastal cliffs on the Isle of Wight (Hutchinson, 1981).

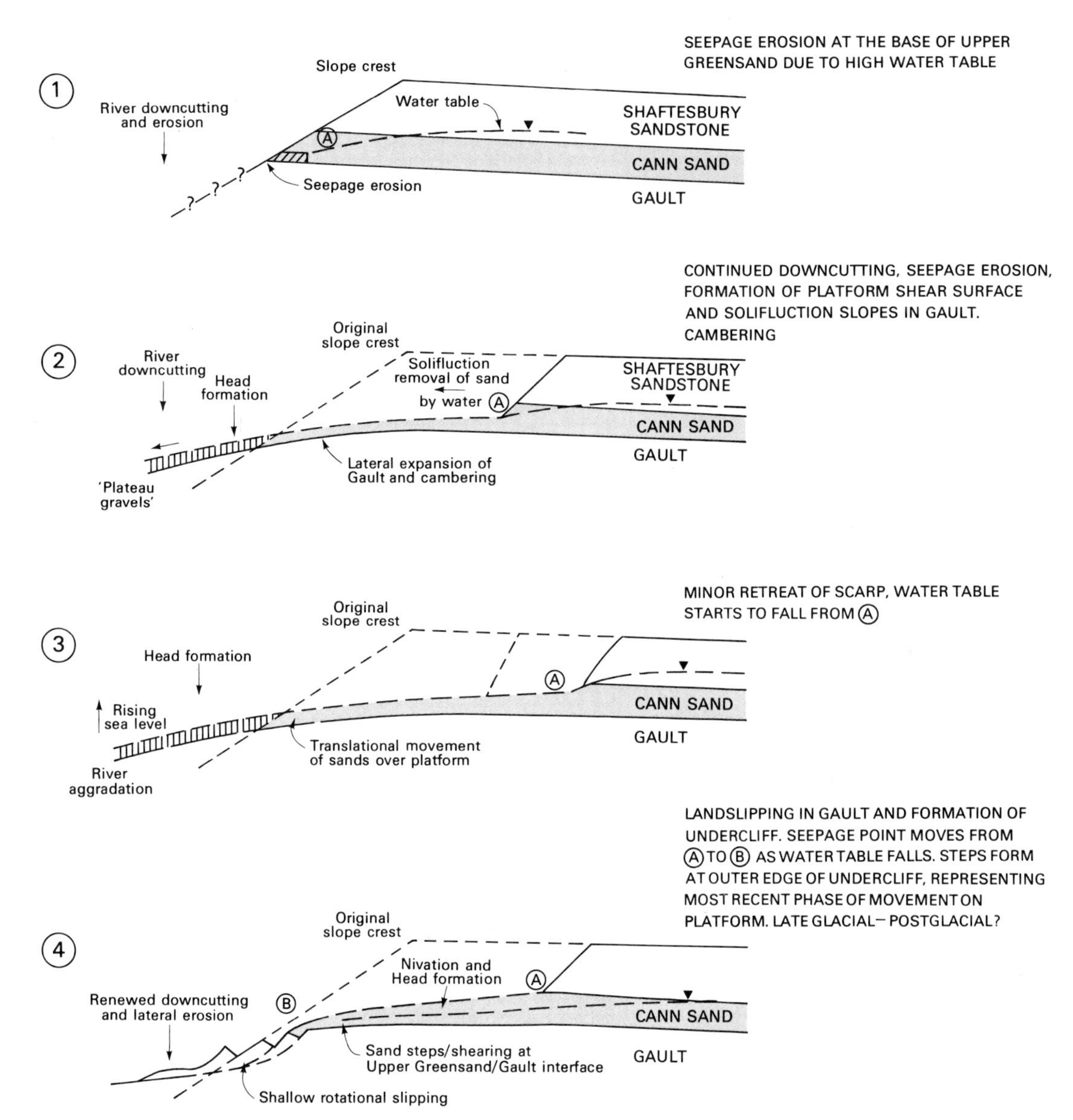

Figure 64 Shaftesbury undercliff — slope development.

Future stability may depend on variations in the water table below Shaftesbury, and also on changes in hydraulic properties at seepage faces, especially close to spring lines near the platform edge. Alterations, by cut and fill or building construction, may lead to a build up of water pressure in sand layers causing instability tens or hundreds of metres away. Piezometric instrumentation is recommended before any development takes place in this part of Shaftesbury.

DETAILS

The landslips are described here in the stratigraphical order of the failed parent rocks.

The steep slopes of Frome Clay below the Forest Marble are subject to landslip. Most slips are degraded and appear to be very old, probably predating the last phase of head formation. The slips were produced by sliding of parts of the limestone and sandstone beds of the Forest Marble on the underlying clays, probably associated with spring action and high groundwater levels. The presence of flowstone and solution cavities in the limestones in the quarry at the Rookery [6695 2420], just beyond the district, also provides evidence for a higher water table in the past.

The largest and possibly most recent slip occurs west of Wilkin Throop [676 232]. The main slip appears to have largely disintegrated into a shield-shaped area of disturbed ground, still with the trace of a back scarp, grading into a debris flow and a deposit of head at the toe.

About 0.5 km west of Templecombe and just south of the railway [7010 2215], an area of hummocky ground lies below limestone. Limestone cropping out on the valley side south of Chapel Lane, Yenston, runs eastwards into landslip, which forms an area of wet clayey ground with limestone debris. Willow trees growing on this slip have become tilted during growth. Areas of hummocky ground [700 187] are present on the west-facing slopes between the A30 [699 189] and Clayhanger [701 169]. Landslips occur in the upper part of the Frome Clay near Haydon [675 161], adjacent to the Poyntington Fault, and along the scarp of the Stalbridge Park Fault, such as near Sturt Farm [729 171].

Landslips are widespread in the Oxford Clay on the steep face below Corallian beds. North of Fifehead Magdalen, the slips [775 216 to 764 236] are associated with springs issuing from sands within the Hazelbury Bryan Formation. The slips appear not to be deep seated, but are probably a combination of shallow rotational and translational slips. They are best developed on the steeper slopes on the east bank of the River Stour between Hinton St Mary [777 160] and the northern margin of the district, and in the south-west below the scarp formed by the basal sand of the Hazelbury Bryan Formation [760 115 to 746 094]. The surface of the slips is characteristically hummocky, although the irregularities have locally been destroyed by ploughing [e.g. around 768 198]. Locally, a back scar is still visible [7685 1944], and in places [7697 1829 to 7708 1850] a prominent toe is present, although generally, there is little evidence of recent movement of the slips. Exceptions are those bordering the River Stour near Fifehead Magdalen [787 230; 787 220; 770 200] and north-west of Hinton St Mary [775 167].

Near Lyon's Gate, landslips are present on both sides of the small valley in which the village is situated. These slips are triggered by strong springs coming from the Lyon's Gate Bed and involve the Hazelbury Bryan Formation. Similar slips occur on the side of a small valley [653 063] about 1 km north-west of Lyon's Gate.

Small slips which involve the Kimmeridge Clay occur north and north-east of Cherry Orchard Farm [8430 2260; 850 230], near Guy's Marsh [around 8471 2015], where there is evidence of fresh movement [8400 2055; 8595 2010], and along the actively eroding unnamed stream that flows through Woodbridge [8455 2000 to 844 175], some of which display prominent back scars [8448 1917; 8360 1835], and near Okeford Common [around 785 115]. A slip in this last area, on the site of a recently cleared wood [7835 1105], is moving at the present day and has a well-developed back scar and tension cracks on its surface.

Slips related to springs issuing from the base of the Lower Greensand near Hartgrove have moved recently and have clearly defined toes [8412 1873; 8357 1847].

Extensive slips occur south-west and west of Lyon's Gate, one of which descends from the Lower Chalk [6515 0560], across the Upper Greensand and Gault, and onto the Hazelbury Bryan Formation below the Lyon's Gate Bed at Lyon's Gate. Eastwards along the Cretaceous scarp towards Dungeon Hill and Buckland Newton, there are numerous slips, some occupying combe-shaped areas [678 053; 681 057; 682 060]. One [760 060], west of Stoke Wake, which does not appear to be active, affects the Gault, part of the Upper Greensand and Lower Chalk.

Most of the slips in the Shaftesbury area are included in Figure 61 and have been discussed in the general account.

Duncliffe Hill is ringed by slips. These are most obvious on the northern and south-eastern side of the hill, beyond the wooded ground. These slips show signs of continuing movement and commonly have prominent toes [for example 8218 2270 to 8229 2293] and very hummocky poorly drained surfaces [e.g. 828 228; 829 229; 830 227]. The slips on the south-west side of the hill show no evidence of recent movement, possibly because the slopes have remained tree covered.

CAMBERING

The cambered base of the Upper Greensand on some smooth-surfaced spurs in the landslipped ground near Shaftesbury falls steadily away from the scarp. In places, the Upper Greensand cuts across the Gault to rest on Kimmeridge Clay. In the west of the district, the base of the Cornbrash limestone appears to be cambered downslope, where it overlies clay of the Forest Marble on some of the steeper east–west valley sides [7072 2247; 7165 2093].

Cambering probably developed under conditions of perennially frozen ground associated with a periglacial climate; it is often associated with overconsolidated clays with high lateral stresses. Outward flow of clay from beneath the cambers towards the valley bottom and its removal by stream erosion appears to be an essential part of the cambering process. Small-scale faulting is usually developed in the overlying strata.

VALLEY BULGING

Isolated folds with steep limbs in the floors of steep-sided, east-north-east-aligned valleys [e.g. 7072 2166] (Taylor, 1991) cut in the Forest Marble in the north-west of the district, an area of simple structure, are probably Pleistocene valley bulges.

SWALLOW HOLES

Scattered shallow, roughly circular depressions occur in the floor of some head-filled dry valleys on the Chalk, commonly in areas where clay-with-flints caps the interfluves. It is presumably the low pH values generated on the soils over these drift deposits which give rise to the acid conditions in the intermittent streams which flow along the valleys which dissolve the Chalk.

Swallow holes have been noted in the valleys between Stourpaine Down and Grammars Hill [8822 1282; 8741 1231; 8905 1154; 8748 1002], near Tarrant Gunville [9372 1279; 9371 1272; 9378 1278], near Tollard Royal [9404 1713] (White, 1923) and north-west and north-east of Ashmore [9321 2020; 9013 1953; 9072 1962; 9253 1952; 9404 1713].

NINE

Economic geology

BUILDING MATERIALS, WALLING STONE AND TILES

The massive, thick-bedded Sherborne Building Stone of the Inferior Oolite, between 5 and 8 m thick, was once extensively quarried. Its principal occurrence within the district is within the Oborne Horst. The stone forms part of the exterior walls and fabric of Sherborne Abbey (Fowler, 1938).

The flaggy limestones and sandstones of the Forest Marble have been worked for building and walling stone in many small quarries; the tile-like linsen beds have been used for roofing. The largest group of quarries, now largely filled, were those in the thick lens of coarsely bioclastic limestone capping Windmill Hill [7080 2107], west of Yenston. Other small quarries nearby [7070 2364; 7155 2085; 7075 2069; 7088 2070] worked limestone and sandstone. Flaggy limestones of the Forest Marble have been used locally for roofing tiles; larger flagstones have been used for bridging ditches and small streams on the Kellaways Formation and Oxford Clay lands to the east. The steps that lead from the nave to the choir in Sherborne Abbey are of Forest Marble from Highmore's Hill [668 180] near Milborne Port (Fowler, 1938).

The limestones of the Cornbrash have been quarried mainly between Horsington and Templecombe. The pits are now mainly filled or grassed over. It seems likely that the hard, bioclastic, sand-free limestone of the Lower Cornbrash was used mainly for limeburning, while the locally flaggy-weathering, sandy Upper Cornbrash, worked from shallow quarries [7050 2364; 7075 2330; 7115 2233] on the dip slope, was used for building stone.

Limestone from the Corallian beds has provided a source of building and walling stone for centuries. The older parts of Marnhull, Hinton St Mary and Sturminster Newton, and many of the other villages, are built of local stone. The principal limestones worked were the Todber Freestone, the Clavellata Beds and the Cucklington Oolite. White (1923, p.93) noted that blue shelly limestone (Clavellata Beds and Cucklington Oolite) was used for building, and stout flagstones (?Todber Freestone) for flooring and dry-stone walling.

The Cucklington Oolite, a flaggy bioclastic limestone, was exploited in three quarries [7715 2320; 7728 2295; 7715 2280] near Kington Magna in the north, and in several quarries [7652 1183; 7685 1080; 7686 1060; 7690 1035] near Fifehead Neville in the south; most are now overgrown.

Many quarries worked the Clavellata Beds and Todber Freestone; those at Stour Provost, Marnhull and Whiteway Hill are the best known, as many of them have been described and figured (Arkell, 1933; Blake and Hudleston, 1877; White, 1923; Mottram, 1957; Gutmann, 1970; Wright, 1981). Most are disused, but one [7948 1980] near Todber still operates on a small scale for building stone. A pit [7940 1785] at Whiteway Hill was opened in 1991; the cross-bedded Todber Freestone, together with the Clavellata Beds, are both exploited.

The Eccliffe Member at the top of the Clavellata Beds is similar to the Todber Freestone and was worked in quarries near East Stour [7955 2313] and in many quarries in the adjacent Wincanton district (Bristow, 1990a).

In the north-east of the district, one of the principal sources of building stone was the 'Ragstone' at the top of the Shaftesbury Sandstone and, on a more limited scale, the Melbury Sandstone. Near Shaftesbury, ragstone was worked in a large rambling quarry known as The Wilderness [866 223] (White, 1923) and in another quarry [8732 2224] on the south-east side of Shaftesbury (White, 1923). Several other quarries in the Shaftesbury Sandstone were also presumably opened for building stone [e.g. 8668 2442; 8700 2208; 8834 2270; 8831 2250; 9170 2337; 9187 2330; 9322 2245].

In a fairly extensive pit [8690 1935] near West Melbury, the Shaftesbury Sandstone was quarried as building stone, and the overlying Boyne Hollow Chert was worked for road metal. Ragstone, and possibly Melbury Sandstone, was dug from several pits [8582 1538; 8586 1535; 8600 1538; 8607 1542] south-west of Sutton Waldron. The Melbury Sandstone was also exploited for building stone [8752 2016; 8665 2008; 8357 2095; 8815 2109] (Jukes-Browne and Hill, 1900, p.160; H B Mottram, MS map, BGS; White, 1923, p.57).

SAND

Sand has been dug from the Forest Marble at Sandpit Cottage [6955 2345], south-west of Horsington and, more extensively, from shallow pits near Wilkin Throop [6853 2330; 6835 2340].

Pits [8083 1075; 803 102] in the Upper Greensand near Okeford Fitzpaine were probably worked for sand, although ragstone may also have been quarried.

ROADSTONE

Unstratified deposits of whole and broken flint, subangular to subrounded chert and sandstone pebbles, together with some well-rounded pebbles of flint, quartz, quartzite and lydianite of the older head were worked in several pits on Fifehead Common [780 119; 7785 1170; 7805 1155; 7815 1170; 7780 1128], but no section remains.

River terrace deposits have been worked on a small scale at three localities [7645 1695; 7625 1580; 817 219], but the pits are now grassed over.

Chert from the Boyne Hollow Chert has been used for road metal. Several of the larger quarries, such as The Wilderness [866 233], Shaftesbury, and several smaller quarries [9170 2337; 9187 2330; 9322 2245] exploited both ragstone and chert. There are a number of shallow excavations, mainly on the edge of the escarpment, which appear to have worked the chert beds for local use [8890 2375; 8912 2381; 8875 2312; 8790 2286; 9263 2461]. A pit [8690 1935] near Melbury Hill, and a smaller one [8637 1893] to the south-west, probably worked chert for roadstone.

Some of the chalkstones and nodular chalks of the Lewes Chalk were worked as roadstone. Pits near Elliott's Shed [9185 2055], on Compton Down [8850 1912; 8845 1885], near Sutton Waldron [8752 1649] and at Iwerne Minster [8765 1455] exploited such material. More recently, flinty chalk of the Blandford Chalk near Okeford Fitzpaine has been dug [808 102] for farm tracks.

Some of the plateau areas of the Chalk have very flinty soils [e.g. around 890 153], and some of the clay-with-flints is very gravelly, e.g. at Houghton North Down [803 058]; flint was probably taken from the fields for roadstone.

LIME

Hydraulic lime has been made from the Inferior Oolite; the remains of an old kiln [6632 1986] can be seen on Miller's Hill.

The Fuller's Earth Rock and the Lower Cornbrash have also been widely burnt for lime, and there are remains of kilns all along the outcrop.

Many pits have worked the Chalk for lime, particularly the Lower Chalk. Except for the Shillingstone pit [822 099], none appears to have been used for at least 70 years. The Shillingstone pit has been in existence for at least 100 years, but large-scale exploitation did not begin until about 1924. Originally, the Zig Zag Chalk was exploited, but now, the Holywell, New Pit and Lewes chalks are dug. Agricultural and hydrated limes are produced by The Shillingstone Lime and Stone Company; production in 1990 was 14 153 and 1008 metric tonnes respectively (Paul Simmons, personal communication). An old limekiln [9068 1262] west of Tarrant Gunville used the Blandford Chalk.

BRICK CLAY

Old pits at Bathpool Lane [7130 2440], at the base of the Mohuns Park Member, are said to have been worked for brick clay. Shallow pits at Middle Throop Farm [7195 2330] in thin terrace deposits and weathered Peterborough Member are said to have been dug to produce locally fired bricks for construction work on the Southern Railway. Poor-quality bricks containing race and gravel fragments were used in the nearby railway underpass arch and in barns at Middle Throop Farm.

The Weymouth Member was dug for brick clay in a small pit [7758 2030] on the north side of Marnhull (White, 1923) and, between 1842 and 1911, at Bagber [7555 1330] (Young, 1972, p.231).

Sandy clay of the Gault, particularly the lower strata, has been worked in several places for brick clay. The largest, and the last to close, was at Okeford Fitzpaine [815 109]. According to Young (1972), the pit commenced operations between 1839 and 1859, and ceased in 1939. One down-draught kiln of about 50 000 brick capacity was still standing in 1970. Another pit [848 150] in Gault, near Farrington, operative from 1838 to 1911 (Young, 1972), has been partially filled. Hand-moulded red bricks were seen by Young in an old shed in 1970. A third pit [8540 1635], now a garden, worked the Gault near Sutton Waldron from pre-1770 to 1901 (Young, 1972).

Clay-with-flints was worked in the Larmer Grounds [9422 1687] near Tollard Royal, and at Tollard Green [c.933 168] (White, 1923). The presence of many small pits on the outcrop of the clay-with-flints suggests that brickmaking, using clay from this deposit, was widespread.

IRON ORE

Ferruginous oolite in the Sandsfoot Grit at Broad Oak [7869 1320] has been compared to the iron-rich deposits at Abbotsbury in south Dorset and was considered as a potential source of iron ore. However, tests during the Second World War proved that its low iron content (Fe 27 per cent) (see p.85), thinness and limited lateral extent exclude it from economic exploitation.

HYDROCARBONS

The district lies within the Wessex Basin (Kent, 1949), the main oil-producing basin of onshore Britain. The presence of commercial oil accumulations at Kimmeridge, Wareham and Wytch Farm in south Dorset led to the drilling of exploration wells by Carless Exploration Ltd at Mappowder in 1984 and Fifehead Magdalen in 1985, and by the Gas Council at Spetisbury in 1984, just south of the district. All were dry.

Potential source rocks occur at several stratigraphical levels in the district. These are mainly the relatively thick, major mudstone divisions of the Jurassic (Lias, Oxford Clay and Kimmeridge Clay). Minor Jurassic mudstones are too thin and Cretaceous mudstones are too immature to be of consequence as source rocks. Similarly, potentially significant hydrocarbon source rocks are unlikely to be present in the Variscan basement rocks.

The Lias clays, Oxford Clay and Kimmeridge Clay have considerable source-rock potential in the Wessex Basin (Penn et al., 1987). The most richly organic parts of the Lower Lias, for example, contain up to 7 weight per cent (wt%) total organic carbon (TOC), those in the Peterborough and Stewartby members of the Oxford Clay have up to 12 wt% TOC, and those in the Kimmeridge Clay up to 20 wt% TOC. They all contain kerogens of Types II, III and II/III, each type being a potential source of hydrocarbons (Ebukanson and Kinghorn, 1985). The oil accumulation at Wytch Farm has been sourced mainly from the

Lias. The highest maturity levels occur in all three potential source-rock sequences in the axial regions of the main sub-basins of the Wessex Basin, in areas which experienced almost continual burial before the (Alpine) tectonic inversion event in Miocene times.

Most of the district overlies the area of the Cranborne–Fordingbridge High, which is flanked on the north by the Mere Basin (sub-basin) and on the south by the Winterborne–Kingston Trough (sub-basin) (Figure 4). The Kimmeridge Clay of the Winterborne Kingston and Cranborne boreholes is immature, with vitrinite reflectance values (Ro%) of 0.32 and 0.36 respectively; a Ro% of 0.42 was obtained from the Oxford Clay in the Cranborne Borehole. Lower Lias Ro% values of 0.46 and 0.56 from the Cranborne and Winterborne–Kingston boreholes, respectively (Ebukanson and Kinghorn, 1986a), indicate that it is a potential source rock beneath the district.

Structures filled by hydrocarbons generated prior to the Tertiary inversion, may have been disturbed by the inversion process itself. The best prospects in the district therefore are extensional horsts and tilted blocks overlain by flat-lying Cretaceous sediments (Penn et al., 1987).

The reservoir rocks in in south Dorset include the Sherwood Sandstone, Bridport Sands and Corallian Beds (e.g. Colter and Havard, 1981; Penn et al., 1987). Fractured Cornbrash limestones and Oxford Clay also produce hydrocarbons in the Kimmeridge Bay oilfield (Brunstrom, 1963; Selley and Stoneley, 1987).

The Sherwood Sandstone, although proved at Winterborne Kingston south of the district, is likely to have poor porosity and thus to be a poor potential reservoir beneath the district (Penn et al., 1987). The Bridport Sands and Corallian Beds have reservoir potential beneath the district; less likely reservoir rocks include fractured Cornbrash, Forest Marble, Pennard Sands and 'Rhaetian' sands.

WATER SUPPLY AND HYDROGEOLOGY

The district lies predominantly within hydrometric area 43 (draining southwards into the River Stour and its tributaries and a small area in the north-east draining into the Nadder), but includes small parts of areas 44 and 52 (Institute of Geological Sciences and Wessex Water Authority, 1979). Its water resources are administered by the Wessex Region of the National Rivers Authority. The average annual rainfall varies from less than 800 mm in the extreme north-west of the area, around Marnhull in the Stour Valley and around Margaret Marsh, to over 1100 mm on the higher ground west of Blandford Forum; the average annual evapotranspiration ranges between 500 and 525 mm.

The main aquifers within the district are the Chalk and Upper Greensand, which together supply 94 per cent of the groundwater licensed to be abstracted from the area. However, small supplies are used from most formations (Table 12). Out of the total of 15.8 million metres cubed per annum (Mm^3/a), public supply accounts for 6.3 Mm^3/a, cressbeds 5.8 Mm^3/a, fish farming 2.2 Mm^3/a, industry 0.9 Mm^3/a, agriculture including spray irrigation 0.6 Mm^3/a, and private water supplies less than 0.1 Mm^3/a. There is no abstraction from surface water sources for public supply. Rivers and one lake are licensed to supply 0.3 Mm^3/a, of which 81 per cent is used for spray irrigation.

Jurassic

All the Jurassic aquifers are of limited extent, are compartmentalised and locally juxtaposed by extensive faulting. Springs are therefore common along the boundaries of the aquifers with underlying less permeable formations and at faulted junctions. Water resources are limited, with both yields and water levels often falling significantly after only a few hours of pumping; in the fissured aquifers, a source can affect others not in the immediate vicinity.

The Bridport Sands form a good aquifer to the west of the district; however, their limited outcrop in this area means that they are only utilised here in combination with the overlying Inferior Oolite, with which they are in hydraulic continuity. Sources obtaining water from the sands need to be carefully designed to prevent the ingress of sand.

The oolitic and sandy limestones of the Inferior Oolite form an excellent local, fissured aquifer. The highest recorded yield is 22.7 l/s from a 24.4 m deep borehole at Bradley Head [6637 2117]. A second 254 mm diameter and 24.4 m deep borehole [6637 2119] provided 12.6 l/s for 14 days for only 2.9 m of drawdown. Both are used for public supply. However, these are exceptional and generally yields are of the order of 2 to 4 l/s. A 254 mm diameter borehole at the tannery in Milborne Port [6751 1882] yielded 3.8 l/s for 2 hours for 0.6 m drawdown. Two boreholes at Milborne Wick [6682 2013], 44.2 m and 45.7 m deep respectively, obtain water from both the Inferior Oolite and the underlying Bridport Sands for public supply. They provided 7.6 l/s and 16.8 l/s respectively on construction and had a

Table 12 Quantity of groundwater licensed to be abstracted by formation, in April 1991.

Formation	m^3/a (× 1000)
Superficial deposits	7
Chalk Upper Greensand	14 806
Gault	6
Lower Greensand	2
Kimmeridge Clay	1
Corallian Beds	86
Cornbrash Forest Marble	51
Fuller's Earth	7
Inferior Oolite Bridport Sands	836
TOTAL	15 802

Data supplied by Wessex Region, National Rivers Authority.

combined minimum yield of 0.93 Ml/d (equivalent to a continuous yield of 10.7 l/s).

Water from the Inferior Oolite is normally hard, but potable, although nitrate levels in excess of 10 mg/l (as N) have been recorded. Water from the combined Inferior Oolite and Bridport Sands sources is similar.

The Fuller's Earth and the overlying Frome Clay generally act as an aquiclude separating water in the Forest Marble from that in the Inferior Oolite. However, small local supplies are obtained from shallow large-diameter wells in the Fuller's Earth Rock, and an 168.6 m deep borehole at Stowell [6849 2180] which finished in the Bridport Sands obtained 1.3 l/s from the Fuller's Earth Rock.

Springs commonly issue at the base of the Forest Marble. One at Charlton Horethorne [6722 2376] had a yield of 2.9 l/s, but others are less than 1 l/s. A 112.8 m × 203 mm borehole at Henstridge [7244 1989] yielded 2.0 l/s from beneath Cornbrash. A shaft and borehole to a depth of 52.4 m at Templecombe [7100 2272] yielded 1.4 l/s for 1½ days for a drawdown of 10.1 m, from an inflow in the Forest Marble at 25.3 m. Other boreholes have yielded 0.1–1.5 l/s, but some were dry. The quality of water is very hard, with total dissolved solids contents of over 1000 mg/l having been recorded. A typical analysis is shown in Table 13. The high iron content (up to 5.6 mg/l) requires treatment.

Very small local supplies are obtained from the Cornbrash. A 1820 mm diameter and 9.1 m deep well at Bishop's Caundle [6998 1236] was fitted with a pump with a capacity of 0.6 l/s.

The Oxford Clay acts primarily as an aquiclude separating the Corallian aquifer from the Forest Marble. Small ferruginous supplies have been obtained from shallow wells into the weathered clay; the Kellaways Formation at the base has yielded small local supplies of poor quality water from small diameter boreholes.

Corallian strata form a multi-layered aquifer, with water in the permeable horizons separated by intermediate confining clay layers. Springs and overflowing boreholes are therefore common. The thicknesses of the individual limestone and sandstone aquifers are small, storage is hence limited and yields are relatively low, around 1 to 2 l/s from springs and shallow wells, and often less than 1 l/s from 150 mm diameter boreholes. The thickest aquifer is the Hazelbury Bryan Formation from which the best yields are obtained. The highest recorded yield was 5.7 l/s for 3 hours on construction in 1948 from an 1800 mm diameter 6.1 m deep well at Sturminster Newton [7888 1426]. A six-day test on a 152 mm diameter borehole, 52.4 m deep, which penetrated the Corallian sequence from the Clavellata Beds to the Hazelbury Bryan Formation, at Mappowder Court [7384 0585], yielded 0.5 l/s for 45.7 m drawdown. It struck water at 30.5 m and 45.7 m in the Hazelbury Bryan Formation, which then rose to 1.5 m from the surface. Another borehole (203 mm × 24.4 m) at Holnest Park [6528 0897] yielded 1.3 l/s for a day, with no noticeable drawdown, from the Hazelbury Bryan Formation in 1952. Water quality is normally good.

There is one licensed source for 4.5 m^3/d from a shallow well in Kimmeridge Clay at Twyford [853 187]. Another source (25.9 m deep) at Manston [816 150] yielded 'plenty' of water, but had a total dissolved solids content of 3465 mg/l. A 305 mm diameter and 39 m deep borehole at Stoke Wake [7574 0680] yielded 1.4 l/s for 3 hours in 1948. The iron content of water from the Kimmeridge Clay is generally high.

Cretaceous

The limited extent of both the outcrop and subcrop of the Lower Greensand means that it is of little use as an aquifer. However, shallow wells at Compton Abbas [859 185], Iwerne Minster [850 145] and Child Okeford [832 130], utilise its water from beneath the Gault.

The Gault confines water in the underlying Lower Greensand where it is present, but its primary hydrogeological significance is as an impermeable base to the Upper Greensand aquifer. Small local supplies, often from springs, are obtained from the Fontmell Magna Sand.

The Upper Greensand is the main aquifer of the district supporting seven public supply sources at Barton Hill [864 232], Donhead [8999 2298], Boyne Hollow [8730 2210 and 8749 2179], Berwick St John [9415 2182], Buckland Newton [6853 0597], Ibberton [7893 0759] and Cookwell Spring [8049 1061], as well as several minor sources. Springs are one of the main sources for public supply and watercress beds. They are not restricted to the junction with the Gault, but issue at various level within the formation (including the boundary with the Chalk) dependant on the local topography and hydrogeology. Those at Boyne Hollow yielded 14.2 l/s in 1961, the Cookwell Spring at Okeford Fitzpaine 25.2 l/s in 1948, and those at Buckland Newton in excess of 5 l/s. The two sources at Barton Hill yielded over 10 l/s and 6.9 l/s from a 49.4 m deep shaft and borehole, and a 38.4 m deep well with adits respectively. A 254 mm diameter borehole, 18.3 m deep (of which the top 15.2 m was in Upper Greensand) at Little Hanford, Child Okeford [838 119], yielded 0.6 l/s during a 2 hour test, for a drawdown of 3.2 m, giving a transmissivity value of 7 m^3/d. Yields of less than 1 l/s are not uncommon from small springs or boreholes at the Upper Greensand/ Chalk junction.

Sources obtaining water from the Upper Greensand require proper construction with a sand screen and gravel pack to limit the ingress of sand. The water is hard, but generally of good quality, with total hardness in the range 175 to 300 mg/l (as $CaCO_3$), of which about three quarters is temporary, chloride ion concentrations of less than 35 mg/l and iron levels of up to 1.5 mg/l. Nitrate concentrations are rising with time in response to changes in agricultural practice and have reached 6.1 mg/l (as N) at the Buckland Newton springs. Typical analyses are shown in Table 13.

The Chalk forms a soft, microporous limestone aquifer in which fissure flow predominates. Fissures are best developed beneath the valleys, and as yields depend on the number and size of fissures intersected, yields are also greatest in the valleys. The Chalk supports three abstractions for public supply at Stubhampton [914 141], Blandford [901 071] and Pimperne [905 084]. The

Upper and, to a lesser extent, the Middle Chalk form the best aquifers, however, they occupy the higher ground, and, as the water table follows the topography in subdued form, water is often not found until considerable depths (e.g. 83.8 m, only 2.2 m above the base of the Chalk at Ibberton [7925 0746]). The water table fluctuates seasonally in response to recharge. The Lower Chalk is more marly, and in consequence less well fissured; therefore yields are smaller. In some cases, the whole of the Chalk is dry or virtually dry, and boreholes commonly continue on down into the Upper Greensand. At Bulbarrow Hill [7792 0555], all the Chalk, and at Ibberton [7925 0746], all the saturated Chalk, were lined out.

The highest yield recorded in the district was 80 to 100 l/s for a drawdown of over 20 m from a 600 mm diameter borehole, 100 m deep into Upper Chalk near Blandford [902 072]; this is only one of several high-yielding public supply boreholes at this site. Yields are normally a few litres per second for 150 mm diameter holes, up to 12.5 l/s for 250 mm diameter and up to 30 l/s for 380 mm diameter boreholes. To provide their maximum yields, boreholes require acid treatment.

Chalk water is normally of good quality and of the calcium bicarbonate type, with total hardness in the range 170–275 mg/l (as $CaCO_3$). Chloride ion concentrations are less than 25 mg/l, and iron levels up to 0.4 mg/l. Nitrate levels up to 3.5 mg/l (as N) have been recorded at Stubhampton. A typical analysis is shown in Table 13.

Superficial deposits

Where clay-with-flints is present, run-off occurs, and so infiltration into the Chalk is reduced, but the streams disappear underground at the periphery of the deposit.

The terrace gravels are utilised for small supplies, particularly the ones at lower levels which are in hydraulic continuity with streams or underlying permeable formations. A large (910 mm) diameter, 5.5 m deep, well at Child Okeford [830 123] yielded 9.1 m^3/d in 1961, but there is no data on the pumping rate. A 1820 mm diameter, 9.1 m deep well into gravels overlying Chalk at Durweston [868 081] yielded 0.6 l/s. Water quality is similar to, but harder than, that of the surface sources with which the gravels are in hydraulic continuity. As they are shallow, the need to be protected from surface pollution.

Waste disposal

There is one operational waste disposal site in the district, taking domestic, commercial and industrial wastes; this is at Stourpaine [868 093] and is sited on the Chalk. There are several other landfill sites restricted to construction and inert wastes. Sites on the Forest Marble at Henstridge [717 185] took domestic, commercial and industrial wastes, and Henstridge Bowden [691 209] took domestic wastes in the past, but were closed in 1987 and pre-1974 respectively. The siting of potentially leachate-producing wastes on aquifers means that great care is required with site construction and monitoring to ensure that the leachate will not represent a hazard to surface or groundwater. New sites of this type will normally be lined with an impermeable layer of clay, an artificial membrane, or both. Domestic refuse can also produce methane, which requires collecting and burning off.

Table 13 Typical chemical analyses of groundwaters from the Shaftesbury district.

Location	Millborne Wick*	Bradley Head*	Templecombe†	Barton Hill*	Donhead†	Buckland Newton*	Blandford†
National Grid reference	ST 6682 2013	ST 6637 2119	ST 7100 2272	ST 8647 2320	ST 8999 2298	ST 6853 0597	ST 8856 0590
Type of source	Borehole	Borehole	Shaft and borehole	Wells and adits	Shallow well	Springs	Borehole
Aquifer	Inferior Oolite and Bridport Sands	Inferior Oolite	Forest Marble	Upper Greensand	Upper Greensand	Upper Greensand	Chalk
Date of analysis	13/3/1990	13/3/1990	26/1/1948	22/3/1990	30/4/1970	27/3/1990	7/11/1921
pH	7.3	7.4	7.9	7.5	7.3	7.4	
Electrical conductivity µmhos/cm	670	655	715	576	401	557	
Total dissolved solids mg/l			596				355
Total hardness ($CaCO_3$) mg/l	338	334	444	247	175	268	
Bicarbonate (HCO_3^-) mg/l	329	326	337	235	188	297	134
Sulphate (SO_4^{2-}) mg/l	35	33	181	40	27	15	33
Chloride (Cl^-) mg/l	20	19	19	32	13	20	24
Fluoride (F^-) mg/l	0.13	0.10		0.12	0.14	0.05	
Nitrate (NO_3-N) mg/l	8.2	8.4	9.2	4.8	3.7	6.1	9.1
Calcium (Ca^{2+}) mg/l	126	124	144	94	67	104	101
Magnesium (Mg^{2+}) mg/l	5.35	5.59	20.6	3.16	1.70	1.94	6.0
Sodium (Na^+) mg/l	9	10	36	13	9	8	19
Potassium (K^+) mg/l	0.7	0.6		5.1	1.5	0.8	0.0
Iron (total) mg/l	<0.020	<0.020	5.6	0.45	1.5	<0.02	trace
Manganese (total) mg/l			<0.01				
Silica (SiO_2) mg/l	3.6	3.7	5.6	6.0	7.6	4.8	

* Wessex Region National Rivers Authority
† National Well Record Collection

REFERENCES

Most of the references listed below are held in the Library of the British Geological Survey at Keyworth, Nottingham. Copies of the references can be purchased subject to the current copyright legislation.

ALLEN, D J, and HOLLOWAY, S. 1984. The Wessex Basin. *Investigation of the geothermal potential of the UK.* (London: British Geological Survey.)

ARKELL, W J. 1933. *The Jurassic System in Great Britain.* (Oxford University Press.)

— 1947. The geology of the country around Weymouth, Swanage, Corfe and Lulworth. *Memoir of the Geological Survey of Great Britain,* Sheets 341, 342, 343 and parts of 327, 328 and 329 (England and Wales).

— 1956. *Jurassic geology of the World.* (Edinburgh and London: Oliver & Boyd.)

— 1958. A monograph of English Bathonian ammonites. *Monograph of the Palaeontographical Society,* Pt. 8, 209–264.

— and DONOVAN, D T. 1952. The Fuller's Earth of the Cotswolds and its relation to the Great Oolite. *Quarterly Journal of the Geological Society of London,* Vol. 107, 227–253.

BAILEY, H W, GALE, A S, MORTIMORE, R N, SWIECICKI, A, and WOOD, C J. 1983. The Coniacian–Maastrichtian stages of the United Kingdom, with particular reference to southern England. *Newsletters on Stratigraphy,* Vol. 12, No. 1, 29–42.

BARNARD, T, CORDEY, W G, and SHIPP, D J. 1981. Foraminifera from the Oxford Clay (Callovian–Oxfordian) of England. *Revista Española de Micropalaeontologia,* Vol. 13, 383–462.

BARROW, G. 1919. Some future work for the Geologists' Association. *Proceedings of the Geologists' Association,* Vol. 30, 2–48.

BARTLETT, B P, and SCANES, J. 1917. Excursion to Mere and Maiden Bradley, Wiltshire. *Proceedings of the Geologists' Association,* Vol. 27, 117–134.

BARTON, C M. 1989. Geology of the Milborne Port district (Somerset and Dorset). 1:10 000 sheet ST61NE. *British Geological Survey Technical Report,* WA/89/60.

— 1990. Geology of the Stalbridge district (Dorset and Somerset). 1:10 000 sheet ST71NW. *British Geological Survey Technical Report,* WA/90/21.

— 1992. Geology of the Blandford Forum district (Dorset). 1:10 000 sheets ST 90 NW, ST 80 NE and ST 80 SE. *British Geological Survey Technical Report,* WA/91/81.

— and FRESHNEY, E C. 1991. Geology of the Lydlinch district (Dorset). 1:10 000 sheet ST71SW. *British Geological Survey Technical Report,* WA/90/22.

— IVIMEY-COOK, H C, LOTT, G K, and TAYLOR, R T. 1993. The Purse Caundle Borehole, Dorset: stratigraphy and sedimentology of the Inferior Oolite and Fuller's Earth in the Sherborne area of the Wessex Basin. *British Geological Survey Research Report,* SA 92/01.

BATE, R H. 1978. The Jurassic. Part II Aalenian to Bathonian. 213–258 *in* A stratigraphical index of British Ostracoda. BATE, R H, and ROBINSON, E (editors). *Geological Journal Special Issue,* Vol. 8.

BIRKELUND, T, and CALLOMON, J H. 1985. The Kimmeridgian ammonite fauna of Milne Land, central East Greenland. *Grønlands Geologiske Undersøgelse,* Bulletin 153.

— — CLAUSEN, C K, NØHR HANSEN, H, and SALINAS, I. 1983. The Lower Kimmeridge Clay at Westbury, Wiltshire, England. *Proceedings of the Geologists' Association,* Vol. 94, 289–309.

BLAKE, J F, and HUDLESTON, W H. 1877. On the Corallian rocks of England. *Quarterly Journal of the Geological Society of London,* Vol. 33, 260–405.

BLOODWORTH, A J. 1990. Mineralogy and petrography of borehole samples taken from landslipped Gault Clay, Shaftesbury, Dorset. *British Geological Survey Technical Report,* WG/90/36.

BOIS, C, CAZES, M, DAMOTTE, B, GALDEANO, A, HIRN, A, MASCLE, A, MATTE, P, RAOULT, J F, and TORREILLES, G. 1986. Deep seismic profiling of the crust in northern France. The ECORS project. 21–29 *in* Reflection seismology: a global perspective. BARAZNAGI, M, and BROWN, L (editors). *American Geophysical Union, Geodynamics Series,* Vol. 13, .

BOSWELL, P G H. 1924. The petrography of the Cretaceous and Tertiary outliers of the west of England. *Quarterly Journal of the Geological Society of London,* Vol. 79, 205–230.

BOTT, M H P, DAY, A A, and MASSON SMITH, D. 1958. The geological interpretation of gravity and magnetic surveys in Devon and Cornwall. *Philosophical Transactions of the Royal Society of London,* Series A. Vol. 251, 161–191.

BRISTOW, C R. 1987. *Geology of Sheets ST 90 SW and SE (Shapwick–Pamphill, Dorset).* (Exeter: British Geological Survey.)

— 1989a. Geology of the East Stour–Shaftesbury district (Dorset). *British Geological Survey Technical Report,* WA/89/58.

— 1989b. Geology of Sheets ST71NE and SE (Marnhull–Sturminster Newton, Dorset). *British Geological Survey Technical Report,* WA/89/59.

— 1990a. Geology of Sheets ST72NE and SE (West Stour–Cucklington district, Dorset). *British Geological Survey Technical Report,* WA/90/48.

— 1990b. Geology of Sheet ST92SW (Berwick St John district, Wiltshire). *British Geological Survey Technical Report,* WA/90/49.

— 1991. Geology of the Tollard Royal–Tarrant Hinton district. *British Geological Survey Technical Report,* WA/91/20.

— 1992a. Geology of Sheet ST81SW (Shillingstone), Dorset. *British Geological Survey Technical Report,* WA/91/03.

— 1992b. Geology of sheets ST 80 NW and SW (Turnworth–Milton Abbas, Dorset). *British Geological Survey Technical Report,* WA/92/20.

— and COX, B M. 1991. The Kimmeridge Clay of the Darknoll Brook, Okeford Fitzpaine, Dorset. *Proceedings of the Dorset Natural History and Archaeological Society,* Vol. 112, 99–103.

— — and WILKINSON, I P. 1989. A section through the Oxford Clay in North Dorset with notes on the Holnest and

King's Stag Brickpits and the Winterborne Kingston Borehole. *Proceedings of the Dorset Natural History and Archaeological Society*, Vol. 110, 137–140.

— Freshney, E C, and Penn, I E. 1991. Geology of the country around Bournemouth. *Memoir of the Geological Survey*, Sheet 329 (England and Wales).

— and Owen, H G. 1991. A temporary section in the Gault at Fontmell Magna in north Dorset. *Proceedings of the Dorset Natural History and Archaeological Society*, Vol. 112, 95–97.

Bromley, R G, and Gale, A S. 1982. The lithostratigraphy of the English Chalk Rock. *Cretaceous Research*, Vol. 3, 273–306.

Brunsden, D, and Jones, D K C. 1976. The evolution of landslide slopes in Dorset. *Philosophical Transactions of the Royal Society of London*, A, Vol. 283, 605–631.

Brunstrom, R G W. 1963. Recently discovered oilfields in Britain. *Proceedings of the 6th World Petroleum Congress, Frankfurt*, Vol. 2, paper 49.

Bryant, I D, Kantorowicz, J D, and Love, C F. 1988. The origin and recognition of laterally continuous carbonate-cemented horizons in the Upper Lias Sands of southern England. *Marine and Petroleum Geology*, Vol. 5, 108–133.

Brydone, R M. 1914. The zone of *Offaster pilula* in the South English Chalk. Part 1. *Geological Magazine*, Vol. 51, 359–369.

— 1915. The *Marsupites* Chalk of Brighton. *Geological Magazine*, Vol. 52, 12–15.

Buckman, J. 1875. On the Cephalopoda Bed and Oolitic sands of Dorset and part of Somerset. *Proceedings of the Somerset Archaeological and Natural History Society*, Vol. 20, 140–164.

— 1879. On the so-called Midford Sands. *Quarterly Journal of the Geological Society of London*, Vol. 35, 736–743.

— 1889. On the Cotteswold, Midford and Yeovil Sands, and the Division between Lias and Oolite. *Quarterly Journal of the Geological Society of London*, Vol. 45, 440–474.

Buckman, S S. 1893. The Bajocian of the Sherborne District: its relation to the subjacent and superjacent strata. *Quarterly Journal of the Geological Society of London*, Vol. 49, 479–522.

— 1910. Certain Jurassic (Lias-Oolite) strata of south Dorset and their correlation. *Quarterly Journal of the Geological Society of London*, Vol. 76, 52–89.

— 1918. The Brachiopoda of the Namyau Beds, Northern Shan States, Burma. *Palaeontologia Indica*, New Series, Vol. 3, 1–300, pls. 1–21.

— 1927. Jurassic chronology: III — Some faunal horizons in Cornbrash. *Quarterly Journal of the Geological Society of London*, Vol. 83, 1–37.

Calkin, J B, and Green, J F N. 1949. Palaeoliths and terraces near Bournemouth. *Proceedings of the Prehistorical Society*, Vol. 15, 21–37.

Callomon, J H. 1955. The ammonite succession in the Lower Oxford Clay and Kellaways Beds at Kidlington, Oxfordshire, and the zones of the Callovian Stage. *Philosophical Transactions of the Royal Society of London*, Vol. B239, 215–264.

— 1964. Notes on the Callovian and Oxfordian stages. *Comptes rendus et memoires Colloque Jurassique Luxembourg 1962*, 269–291.

— 1968. The Kellaways Beds and Oxford Clay. Chapter 14, 264–290 in *The geology of the East Midlands*. Sylvester-Bradley, P C, and Ford, T D (editors). (Leicester University Press.)

— 1991. Temporary sections — Frome by-pass 1989–90. *Geologists' Association Circular*, No. 884, 29.

— Dietl, G, and Page, K N. 1989. On the ammonite faunal horizons and standard zonations of the lower Callovian Stage in Europe. *2nd International Symposium on Jurassic Stratigraphy (Lisboa 1987)*, Vol. 1, 359–376.

Carter, D J, and Hart, M B. 1977. Aspects of mid-Cretaceous stratigraphical micropalaeontology. *Bulletin of the British Museum (Natural History) (Geology)*, Vol. 29, 1–135.

Casey, R. 1956. Notes on the base of the Gault in Wiltshire. *Proceedings of the Geologists' Association*, Vol. 66, Pt. 3, 231–234.

Cave, R. 1977. Geology of the Malmesbury district. *Memoir of the Geological Survey of Great Britain*, Sheet 251 (England and Wales).

— and Cox, B M. 1975. The Kellaways Beds of the area between Chippenham and Malmesbury, Wiltshire. *Bulletin of the Geological Survey of Great Britain*, Vol. 54, 41–66.

Chadwick, R A. 1985. Permian, Mesozoic and Cenozoic structural evolution of England and Wales in relation to the principles of extension and inversion tectonics. 9–25 in *Atlas of onshore sedimentary basins in England and Wales: Post-Carboniferous tectonics and stratigraphy*. Whittaker, A (editor). (Glasgow: Blackie.)

— 1986. Extension tectonics in the Wessex Basin, southern England. *Journal of the Geological Society of London*, Vol. 143, 465–488.

— Kenolty, N, and Whittaker, A. 1983. Crustal structure beneath southern England from deep seismic reflection profiles. *Journal of the Geological Society of London*, Vol. 140, 893–911.

— Pharaoh, T C, and Smith, N J P. 1989. Lower crustal heterogeneity beneath Britain from deep seismic reflection data. *Journal of the Geological Society of London* Vol. 146, 617–630.

Chidlaw, N, and Campbell, M J A. 1988. Sedimentation patterns in the Cornbrash Limestone Formation of the Cotswold Hills. *Proceedings of the Geologists' Association*, Vol. 99, 27–42.

Cifelli, R. 1959. Bathonian Foraminifera of England. *Bulletin of the Museum of Comparative Zoology Harvard College*, Vol. 121, No. 7, 265–368.

Cifelli, R. 1960. Notes on the distribution of English Bathonian Foraminifera. *Geological Magazine*, Vol. 91, 33–42.

Coleman, B. 1981. The Bajocian to Callovian. 106–124 in *Stratigraphical atlas of fossil foraminifera*. Jenkins, D G, and Murray, J W (editors).

Colter, V S, and Havard, D J. 1981. The Wytch Farm Oil Field, Dorset. 493–503 in *Petroleum geology of the continental shelf of North-West Europe*. Illing, V C, and Hobson, G D (editors). (London: Institute of Petroleum.)

Cope, J C W. 1972. Dorset natural history reports, 1971 (geology): Gas pipeline trench Yeovil—Mappowder. *Proceedings of the Dorset Natural History and Archaeological Society*, Vol. 93 for 1971, 39–40.

— 1978. The ammonite faunas and stratigraphy of the upper part of the Upper Kimmeridge Clay of Dorset. *Palaeontology*, Vol. 21, Pt. 3, 469–533.

— Getty, T A, Howarth, M K, Morton, N, and Torrens, H S. 1980a. A correlation of Jurassic rocks in the British Isles. Part One: Introduction and Lower Jurassic. *Special Report of the Geological Society of London*, No. 14, 73pp.

— DUFF, K L, PARSONS, C F, TORRENS, H S, WIMBLEDON, W A, and WRIGHT, J K. 1980b. A correlation of Jurassic rocks in the British Isles. Part Two: Middle and Upper Jurassic. *Special Report of the Geological Society of London*, No. 15. 109pp.

— HALLAM, A, and TORRENS, H S. 1969. Guide for Dorset and south Somerset: International Field Symposium on the British Jurassic, Excursion No. 1., Geology Department, Keele University.

CORNWELL, J D. 1986. The geological significance of some geophysical anomalies in western Somerset. *Proceedings of the Ussher Society*, Vol. 6, 383–388.

COX, B M. 1990. Wincanton Sheet (297): Kimmeridge Clay of the Gillingham Brickpit and nearby localities, Dorset. *British Geological Survey Unpublished Report*, WH/90/100R.

— and GALLOIS, R W. 1979. Description of the standard stratigraphical sequences of the Upper Kimmeridge Clay, Ampthill Clay and West Walton Beds. *Report of the Institute of Geological Sciences of London*, No. 78/19, 68–72.

— — 1981. The stratigraphy of the Kimmeridge Clay of the Dorset type area and its correlation with some other Kimmeridgian sequences. *Report of the Institute of Geological Sciences, London*, No. 80/4.

DAVIES, D K. 1967. Origin of friable sandstone–calcareous sandstone rhythms in the Upper Lias of England. *Journal of Sedimentary Petrology*, Vol. 37, 1179–1188.

— 1969. Shelf sedimentation: an example from the Jurassic of Britain. *Journal of Sedimentary Petrology*, Vol. 39, 1344–1370.

— ETHERIDGE, F G, and BERG, R R. 1971. Recognition of barrier environments. *Bulletin of the American Association of Petroleum Geologists*, Vol. 55, 550–565.

DEARMAN, W R. 1963. Wrench-faulting in Cornwall and South Devon. *Proceedings of the Geologists' Association*, Vol. 74, 265–87.

DELAIR, J B. 1966. New records of dinosaurs and other fossil reptiles from Dorset. *Proceedings of the Dorset Natural History and Archaeological Society*, Vol. 87 for 1965, 57–66.

DONATO, J A. 1988. Possible Variscan thrusting beneath the Somerton Anticline, Somerset. *Journal of the Geological Society of London*, Vol. 145, 431–438.

DOUGLAS, J A. 1951. A new structure in the Forest Marble of Oxford. *Geological Magazine*, Vol. 88, 169–174.

— and ARKELL, W J. 1928. The stratigraphical distribution of the Cornbrash. I. The south-western area. *Quarterly Journal of the Geological Society of London*, Vol. 84, 117–178.

DRUMMOND, P V O. 1967. The Cenomanian Palaeogeography of Dorset and Adjacent Counties. Unpublished PhD thesis, University of London.

— 1970. The Mid-Dorset Swell. Evidence of Albian–Cenomanian movements in Wessex. *Proceedings of the Geologists' Association*, Vol. 81, 679–714.

DUFF, K L. 1980. Callovian correlation chart. 45–60 *in* A correlation of Jurassic rocks in the British Isles. Part Two: Middle and Upper Jurassic. COPE, C J W (editor). *Special Report of the Geological Society of London*, No. 15.

EBUKANSON, E J, and KINGHORN, R R F. 1985. Kerogen facies in the major Jurassic mudrock formations of southern England and their implication on the depositional environments of their precursors. *Journal of Petroleum Geology*, Vol. 8, 435–62.

— — 1986a. Maturity of organic matter in the Jurassic of southern England and its relation to the burial history of the sediments. *Journal of Petroleum Geology*, Vol. 9, 259–280.

— — 1986b. Oil and gas accumulations and their possible source rocks in southern England. *Journal of Petroleum Geology*, Vol. 9, 413–428.

EDMONDS, E A, WRIGHT, J E, BEER, K E, HAWKES, J R, WILLIAMS, M, FRESHNEY, E C, and FENNING, P J. 1968. Geology of the country around Okehampton. *Memoir of the Geological Survey of Great Britain*, Sheet 324 (England and Wales).

EDMUNDS, F H. 1938. A contribution on the physiography of the Mere district, Wiltshire, with report of field meeting. *Proceedings of the Geologists' Association*, Vol. 49, 174.

EDWARDS, R A, and FRESHNEY, E C. 1987. The geology of the country around Southampton. *Memoir of the Geological Survey of Great Britain*, Sheet 315 (England and Wales).

EDWARDS, W, and PRINGLE, J. 1926. On a borehole in the Lower Oolitic Rocks at Wincanton, Somerset. *Summary of Progress of the Geological Survey for 1925*, 183–188.

EKDALE, A A, BROMLEY, R G, and PEMBERTON, S G. 1984. *Ichnology: The use of trace fossils in sedimentology and stratigraphy.* (Tulsa, Oklahoma: Society of Economic Paleontologists and Mineralogists.)

ENSOM, P. 1985. *Pygurus blumenbachii* Koch and Dunker (Echinoidea) and *Stylina* sp. (Anthozoa) from the Corallian of North Dorset. *Proceedings of the Dorset Natural History and Archaeological Society*, Vol. 107, 179.

FOWLER, J. 1938. *The stones of Sherborne Abbey.* 32pp. [Sherborne.]

— 1944. Highmore's Hill outlier, Sherborne. *Proceedings of the Dorset Natural History and Archaeological Society*, Vol. 65, 157–162.

— 1957. The geology of the Thornford pipe-trench. *Proceedings of the Dorset Natural History and Archaeological Society*, Vol. 78, 51–57.

FRESHNEY, E C. 1990. Geology of Sheets ST70NW and NE (Mappowder–Woolland, Dorset). *British Geological Survey Technical Report*, WA/90/51.

— 1992. Geology of sheets ST 60 NE and 61 SE (Glanvilles Wootton–Bishop's Caundle, Dorset. *British Geological Survey Technical Report*, WA/92/82.

— 1993. Geology of sheets ST60 SE, 70 SW and SE (Cerne Abbas–Milton Abbas, Dorset). *British Geological Survey Technical Report*, WA/93/29.

— and TAYLOR, R T. 1980. The Variscan foldbelt and its foreland. Section 9, 49–57 in *United Kingdom: Introduction to general geology and guides to excursions 002, 005, 093 and 151.* OWEN, T R (editor). International Geological Congress, Paris 1980. [Institute of Geological Sciences, London.]

FULLER, N G. 1983. On Upper Jurassic (Callovian and Oxfordian) ostracods from England and northern France. PhD thesis, University College, London.

FÜRSICH, F T. 1976. The use of macroinvertebrate associations in interpreting Corallian environment. *Palaeogeography, Palaeoclimatology, Palaeoecology*, Vol. 20, 235–256.

GALE, A S. 1989. Field meeting at Folkestone Warren, 29th November, 1987. *Proceedings of the Geologists' Association*, Vol. 100, 73–82.

— 1990. A Milankovitch scale for Cenomanian time. *Terra Nova*, Vol. 1, 420–425.

— WOOD, C J, and BROMLEY, R G. 1987. The lithostratigraphy and marker bed correlation of the White Chalk (late Cenomanian–Campanian) in southern England. *Mesozoic Research*, Vol. 1, 107–118.

Gallois, R W. 1979. Oil shale resources in Great Britain. *Unpublished IGS Report for the Department of Energy.* 2 vols. (London: Institute of Geological Sciences.)

— and Cox, B M. 1976. The stratigraphy of the Lower Kimmeridge Clay of Eastern England. *Proceedings of the Yorkshire Geological Society,* Vol. 41, 13–26.

— and Medd, A W. 1979. Coccolith-rich marker bands in the English Kimmeridge Clay. *Geological Magazine,* Vol. 116, 247–260.

— and Morter, A A. 1982. The stratigraphy of the Gault of East Anglia. *Proceedings of the Geologists' Association,* Vol. 93, 351–368.

Gaster, C T A. 1930. Chalk Zones in the neighbourhood of Shoreham, Brighton and Newhaven, Sussex. *Proceedings of the Geologists' Association,* Vol. 39 (for 1929), 328–340.

Gatrall, M, Jenkyns, H C, and Parsons, C F. 1972. Limonitic concretions from the European Jurassic with particular reference to the 'snuff-boxes' of southern England. *Sedimentology,* Vol. 18, 79–103.

Gostelow, T P. 1991. Geological processes and their effect on the engineering behaviour of the Gault–UGS: An example from Shaftesbury, Dorset. *Report of the British Geological Survey,* WN/91/9.

Green, G W, and Donovan, D T. 1969. The Great Oolite of the Bath area. *Bulletin of the British Geological Survey,* No. 30, 1–63.

Gutmann, K. 1970. The Corallian Beds at Todber and Whiteway Hill in north Dorset. *Proceedings of the Dorset Natural History and Archaeological Society,* Vol. 91, 123–133.

Hahn, W. 1968. Die Oppeliidae Bonarelli und Haploceratidae Zittel (Ammonoidea) des Bathoniums (Brauner Jura) im südwestdeutchen Jura. *Jahreshefte des Geologischen Landesamts Baden-Württemberg,* Vol. 10, 7–72.

Hallam, A. 1970. *Gyrochorte* and other trace fossils in the Forest Marble (Bathonian) of Dorset, England. 189–200 in *Trace fossils.* Crimes, T P, and Harper, J C (editors). (Liverpool: Seel House Press.)

— 1975. *Jurassic environments.* (Cambridge: Cambridge University Press.)

Hodgson, J M, Catt, J A, and Weir, A H. 1967. The origin and development of Clay-with flints and associated soil horizons on the South Downs. *Journal of Soil Science,* Vol. 18, 85–102.

Holder, M T, and Leveridge, B E. 1986. Correlation of the Rhenohercynian Variscides. *Journal of the Geological Society of London,* Vol. 143, 141–147.

Holloway, S. 1982. Bruton No.1. Geological well completion report. *Deep Geology Unit Report,* No. 82/8. (London: Institute of Geological Sciences.)

— 1983. The shell-detrital calcirudites of the Forest Marble Formation (Bathonian) of southwest England. *Proceedings of the Geologists' Association,* Vol. 94, 259–266.

— 1985. Lower Jurassic: the Lias. 37–40 in *Atlas of onshore sedimentary basins in England and Wales: post-Carboniferous tectonics and stratigraphy.* Whittaker, A (editor). (Glasgow & London: Blackie.)

— Chadwick, R A. 1984. The IGS Bruton Borehole (Somerset, England) and its regional structural significance. *Proceedings of the Geologists' Association,* Vol. 95, 165–174.

— and Chadwick, R A. 1986. The Sticklepath–Lustleigh fault zone: Tertiary sinistral reactivation of a Variscan dextral strike-slip fault. *Journal of the Geological Society of London,* Vol. 143, 447–452.

— Milodowski, A E, Strong, G E, and Warrington, G. 1989. The Sherwood Sandstone Group (Triassic) of the Wessex Basin, southern England. *Proceedings of the Geologists' Association,* Vol. 100, 383–394.

House, M R. 1961. The structure of the Weymouth anticline. *Proceedings of the Geologists' Association,* Vol. 72, 221–238.

Hounslow, M W. 1987. Magnetic fabric characteristics of bioturbated wave-produced grain orientation in the Bridport–Yeovil Sands (Lower Jurassic) of Southern England. *Sedimentology,* Vol. 34, 117–128.

Hudleston, W H. 1886. Excursion to Sherborne and Bridport. *Proceedings of the Geologists' Association,* Vol. 9, 187–199.

Hudson, J D, and Palmer, T J. 1976. A euryhaline oyster from the Middle Jurassic and the origin of the true oysters. *Palaeontology,* Vol. 19, 79–93.

Hull, E, and Whitaker, W. 1861. The Geology of parts of Oxfordshire and Berkshire (Sheet 13). *Memoir of the Geological Survey of Great Britain.*

Hutchinson, J N. 1969. A reconsideration of the coastal landslips at Folkestone Warren, Kent. *Geotechnique,* Vol. 19, 6–38.

— 1981. Damage to slopes produced by seepage erosion on sands. Abstract of paper presented at the International Seminar on water-related exogenous geological processes and prevention of their negative impact on the environment, Alma ata, USSR.

Institute of Geological Sciences and Wessex Water Authority. 1979. *Hydrogeological map of the Chalk and associated minor aquifers of Wessex.* (London: Institute of Geological Sciences.)

Ivimey-Cook, H C. 1982. Biostratigraphy of the Lower Jurassic and Upper Triassic (Rhaetian) rocks of the Winterborne Kingston borehole, Dorset. 97–106 *in* The Winterborne Kingston borehole, Dorset, England. Rhys, G H, Lott, G K, and Calver, M A (editors). *Report of the Institute of Geological Sciences,* No. 81/3.

— 1989. Bajocian, Aalenian and Toarcian strata in the Stowell Borehole, Templecombe, Somerset. *British Geological Survey Technical Report,* WH/89/175R.

— 1991. The Middle Jurassic (Cornbrash and Forest Marble) of the Combe Throop Borehole; 1" 313, Shaftesbury. *Report of the British Geological Survey,* WH/91/301.

Jeans, C V, Merriman, R J, Mitchell, J G, and Bland, D J. 1982. Volcanic clays in the Cretaceous of Southern England and Northern Ireland. *Clay Minerals,* Vol. 17, 105–156.

Jefferies, R P S. 1963. The Stratigraphy of the *Actinocamax plenus* Subzone (Turonian) in the Anglo-Paris Basin. *Proceedings of the Geologists' Association,* Vol. 74, 1–33.

Jenkyns, H C, and Senior, J R. 1991. Geological evidence for intra-Jurassic faulting in the Wessex Basin and its margins. *Journal of the Geological Society of London,* Vol. 148, 245–260.

Jones, L E, and Sellwood, B W. 1989. Palaeogeographic significance of clay mineral distributions in the Inferior Oolite (Mid Jurassic) of Southern England. *Clay Minerals,* Vol. 24, 91–105.

Jukes-Browne, A J. 1891. Notes on an undescribed area of Lower Greensand or Vectian in Dorset. *Geological Magazine,* (3), Vol. 8, 456–458.

— and Hill, W. 1900. The Cretaceous rocks of Britain. 1, The Gault and Upper Greensand. *Memoir of the Geological Survey of Great Britain.*

— — 1903. The Cretaceous rocks of Britain. 2, The Lower and Middle Chalk of England. *Memoir of the Geological Survey of Great Britain.*

— — 1904. The Cretaceous rocks of Britain. 4. Upper Chalk of England. *Memoir of the Geological Survey of Great Britain.*

KANTOROWICZ, J D, BRYANT, I D, and DAWANS, J M. 1987. Controls on the geometry and distribution of carbonate cements in Jurassic sandstones: Bridport Sands, southern England and Viking Group, Troll Field, Norway. 103–118 *in* Diagenesis of sedimentary sequences. MARSHALL, J D (editor). *Special Publicationof the Geological Society of London,* No. 36.

KELLAWAY, G A. 1938. Report on recent excavations in the Sherborne District. *Unpublished Report of the Geological Survey of Great Britain.* 8pp.

— 1944. Report on the underground structure and hydrology of the Oborne-Milborne Wick area. *Unpublished Report of the Geological Survey of Great Britain.* 6pp.

— and WELCH, F B A. 1948. *British regional geology: Bristol and Gloucester District* (2nd edition). (London: HMSO.)

— and WILSON, V. 1941. An outline of the geology of Yeovil, Sherborne and Sparkford Vale. *Proceedings of the Geologists' Association,* Vol. 52, 131–174.

KENNEDY, W J. 1970. A correlation of the Uppermost Albian and the Cenomanian of South-west England. *Proceedings of the Geologists' Association,* Vol. 81, 613–677.

KENOLTY, N, CHADWICK, R A, BLUNDELL, D J, and BACON, M. 1981. Deep seismic reflection survey over the Variscan Front of southern England. *Nature, London,* Vol. 293, 451–453.

KENT, P E. 1949. A structure contour map of the surface of the buried pre-Permian rocks in England and Wales. *Proceedings of the Geologists' Association,* Vol. 60, 87–104.

KNOX, R W O'B. 1982a. Clay mineral trends in cored Lower and Middle Jurassic sediments of the Winterborne Kingston borehole, Dorset. 91–96 *in* The Winterborne Kingston borehole, Dorset, England. RHYS, G H, LOTT, G K, and CALVER, M A (editors). *Report of the Institute of Geological Sciences,* No. 81/3.

— 1982b. The petrology of the Penarth Group (Rhaetian) of the Winterborne Kingston borehole, Dorset. 127–134 *in* The Winterborne Kingston borehole, Dorset, England. RHYS, G H, LOTT, G K, and CALVER, M A (editors). *Report of the Institute of Geological Sciences,* No. 81/3.

— MORTON, A C, and LOTT, G K. 1982. Petrology of the Bridport Sands in the Winterborne Kingston borehole, Dorset. 107–121 *in* The Winterborne Kingston borehole, Dorset, England. RHYS, G H, LOTT, G K, and CALVER, M A (editors). *Report of the Institute of Geological Sciences,* No. 81/3.

LAKE, R D, YOUNG, B, WOOD, C J, and MORTIMORE, R N. 1987. Geology of the country around Lewes. *Memoir of the Geological Survey of Great Britain,* Sheet 319 (England and Wales).

LAMPLUGH, G W, WEDD, C B, and PRINGLE, J. 1920. Special reports on the mineral resources of Great Britain. Vol. 12 — Iron ores (contd.). — Bedded ores of the Lias, Oolites and later formations in England. *Memoir of the Geological Survey of Great Britain.*

LLOYD, A J. 1962. Polymorphinid, miliolid and rotaliform foraminifera from the type Kimmeridgian. *Micropaleontology,* Vol. 8, 369–383.

LOTT, G K, and STRONG, G E. 1982. The petrology and petrography of the Sherwood Sandstone (?Middle Triassic) of the Winterborne Kingstone borehole, Dorset. 135–142 *in* The Winterborne Kingston borehole, Dorset, England. RHYS, G H, LOTT, G K, and CALVER, M A (editors). *Report of the Institute of Geological Sciences,* No. 81/3.

LOVEDAY, J. 1962. Plateau Deposits on the southern Chiltern Hills. *Proceedings of the Geologists' Association,* Vol. 73, 83–102.

MARTIN, A J. 1967. Bathonian sedimentation in Southern England. *Proceedings of the Geologists' Association,* Vol. 78, 473–488.

MCKERROW, W S. 1953. Variation in the Terebratulacea of the Fuller's Earth Rock. *Quarterly Journal of the Geological Society of London,* Vol. 59, 97–124.

MELVILLE, R V, and FRESHNEY, E C. 1982. *British regional geology: Hampshire Basin and adjoining areas.* (London: HMSO for Institute of Geological Sciences.)

MORTER, A A. 1982. The macrofauna of the Lower Cretaceous rocks of the Winterborne Kingston borehole, Dorset. 35–8 in The Winterborne Kingston borehole, Dorset, England. RHYS, G H, LOTT, G K, and CALVER, M A (editors). *Report of the Institute of Geological Sciences,* No. 81/3.

— and WOOD, C J. 1983. The biostratigraphy of Upper Albian–Lower Cenomanian *Aucellina* in Europe. *Zitteliana,* Vol. 10, 515–529.

MORTIMORE, R N. 1983. The stratigraphy and sedimentation of the Turonian Campanian in the southern province of England. *Zitteliana,* Vol. 10, 22–41.

— 1986. Stratigraphy of the Upper Cretaceous White Chalk of Sussex. *Proceedings of the Geologists' Association,* Vol. 97, 97–139.

— 1987. Upper Cretaceous Chalk in the North and South Downs, England: a correlation. *Proceedings of the Geologists' Association,* Vol. 98, 77–86.

— and POMEROL, B. 1987. Correlation of the Upper Cretaceous White Chalk (Turonian to Campanian) in the Anglo-Paris Basin. *Proceedings of the Geologists' Association,* Vol. 98, 97–143.

— and WOOD, C J. 1986. The distribution of flint in the English Chalk, with particular reference to the 'Brandon Flint Series' and the high Turonian flint maximum. In *The scientific study of flint and chert; papers from the Fourth International Flint Symposium,* Vol. 1. SIEVEKING, G de G, and HART, M B (editors). (Cambridge: Cambridge University Press.)

MORTON, A C. 1982. Heavy minerals from the sandstones of the Winterborne Kingstone borehole, Dorset. 143–148 *in* The Winterborne Kingston borehole, Dorset, England. RHYS, G H, LOTT, G K, and CALVER, M A (editors). *Report of the Institute of Geological Sciences,* No. 81/3.

MOTTRAM, B H. 1957. Whitsun Field Meeting at Shaftesbury. *Proceedings of the Geologists' Association,* Vol. 67 (for 1956), 160–167.

MUIR-WOOD, H M. 1936. A monograph on the Brachiopoda of the British Great Oolite Series. Part 1, The Brachiopoda of the Fuller's Earth. *Monograph of the Palaeontolographical Society.*

NEWTON, R B. 1897. An account of the Albian fossils lately discovered at Okeford Fitzpaine, Dorset. *Proceedings of the Dorset Natural History & Antiquarian Field Club,* Vol. 18, 66–99.

OWEN, H G. 1971. Middle Albian stratigraphy in the Anglo-Paris Basin. *Bulletin of the British Museum (Natural History) (Geology),* Supplement 8, 1–164.

— 1976. The stratigraphy of the Gault and Upper Greensand of the Weald. *Proceedings of the Geologists' Association,* Vol. 86 (for 1975), 475–498.

PAGE, K N. 1988. The stratigraphy and ammonites of the British Lower Callovian. Unpublished PhD Thesis, University College London.

— 1989. A stratigraphical revision for the English Lower Callovian. *Proceedings of the Geologists' Association*, Vol. 100, 363–382.

PARSONS, C F. 1976. A stratigraphical revision of the *humphriesianum–subfurcatum* Zone rocks (Bajocian Stage, Middle Jurassic) of southern England. *Newsletters in Stratigraphy*, Vol. 5, 114–142.

PENN, I E. 1982. Middle Jurassic stratigraphy and correlation of the Winterborne Kingston borehole, Dorset. 53–76 *in* The Winterborne Kingston borehole, Dorset, England. RHYS, G H, LOTT, G K, and CALVER, M A (editors). *Report of the Institute of Geological Sciences*, No. 81/3.

— COX, B M, and GALLOIS, R W. 1986. Towards precision in stratigraphy: geophysical log correlation of Upper Jurassic (including Callovian) strata of the Eastern England Shelf. *Journal of the Geological Society of London*, Vol. 143, 381–410.

— CHADWICK, R A, HOLLOWAY, S, ROBERTS, G, PHAROAH, T C, ALLSOP, J M, HULBERT, A G, and BURNS, I M. 1987. Principal features of the hydrocarbon prospectivity of the Wessex-Channel Basin, UK. 109–118 in *Petroleum geology of Northwest Europe*. BROOKS, J, and GLENNIE, K (editors). *(London:* Graham Trotman.)

— and WYATT, R J. 1979. The stratigraphy and correlation of the Bathonian strata in the Bath–Frome area. 23–88 *in* The Bathonian strata of the Bath–Frome area. *Report of the Institute of Geological Sciences*, No. 78/22.

PERRIN, R M S. 1971. *The clay mineralogy of British sediments.* (London: Mineralogical Society.)

PLINT, A G. 1982. Eocene sedimentation and tectonics in the Hampshire Basin. *Journal of the Geological Society of London*, Vol. 139, 249–254.

PRICE, R J. 1977. The stratigraphical zonation of the Albian sediments of north-west Europe, as based on foraminifera. *Proceedings of the Geologists' Association*, Vol. 88, 65–91.

PRINGLE, J. 1909. On a boring in the Fullonian and Inferior Oolite at Stowell, Somerset. *Geological Survey of Great Britain, Summary of Progress for 1908*, 83–86.

— 1910. On a boring at Stowell, Somerset. *Geological Survey of Great Britain, Summary of Progress for 1909*, 68–70.

PULTENEY, R. 1813. *Catalogues of the birds, shells, and some of the more rare plants of Dorsetshire.*

DE RAFF, J F M, BOERSMA, J R, and VAN GELDER, A. 1977. Wave generated structures and sequences from a shallow marine succession. Lower Carboniferous, County Cork, Ireland. *Sedimentology*, Vol. 24, 451–485.

REID, C. 1899. The geology of the country around Dorchester. *Memoir of the Geological Survey of Great Britain*, Sheet 328 (England and Wales).

RHYS, G H, LOTT, G K, and CALVER, M A (editors). 1982. The Winterborne Kingston borehole, Dorset, England. *Report of the Institute of Geological Sciences*, No. 81/3.

RICHARDSON, L. 1916. The Inferior Oolite and contiguous deposits of the Doulting–Milborne Port district (Somerset). *Quarterly Journal of the Geological Society of London*, Vol. 58, 719–752.

— 1919. The Inferior Oolite and contiguous deposits of the Crewkerne district (Somerset). *Quarterly Journal of the Geological Society of London*, Vol. 74, 145–173 (for 1918).

— 1928a. Wells and springs of Somerset. *Memoir of the Geological Survey, England.*

— 1928b–1930. The Inferior Oolite and contiguous deposits of the Burton Bradstock–Broadwindsor district, Dorset. *Proceedings of the Cotteswold Naturalists' Field Club*, Vol. 23, 35–68, 1928; 149–185, 1929; 253–264, 1930.

— 1929. The country around Moreton in Marsh. *Memoir of the Geological Survey of Great Britain*, Sheet 217 (England and Wales).

— 1932. The Inferior Oolite and contiguous deposits of the Sherborne district, Dorset. *Proceedings of the Cotteswold Naturalists' Field Club*, Vol. 24, 35–85.

RIDING, J B, and THOMAS, J E. 1988. Dinoflagellate cyst stratigraphy of the Kimmeridge Clay (Upper Jurassic) from the Dorset coast, southern England. *Palynology*, Vol. 12, 65–88.

ROBINSON, N D. 1968. Lithostratigraphy of the Chalk Group of the North Downs, southeast England. *Proceedings of the Geologists' Association*, Vol. 97, 141–170.

— 1987. Upper Cretaceous Chalk in the North and South Downs, England: a reply. *Proceedings of the Geologists' Association*, Vol. 98, 87–93.

ROSS, M S. 1987. Observations along Wessex Water Authority Reinforcing Main: Marnhull to Kington Magna, April–October 1985. *Proceedings of the Dorset Natural History and Archaeological Society*, Vol. 108. 89–102.

ROWE, A W. 1900. The zones of the White Chalk of the English Coast. 1 — Kent and Sussex. *Proceedings of the Geologists' Association*, Vol. 16, 289–368.

SELLEY, R C, and STONELEY, R. 1987. Petroleum habitat in south Dorset. 139–148 in *Petroleum geology of Northwest Europe*. BROOKS, J, and GLENNIE, K (editors). (London: Graham and Trotman.)

SHANNON, P M. 1991. The development of Irish offshore sedimentary basins. *Journal of the Geological Society of London*, Vol. 148, 181–189.

SHEPPARD, L M. 1981. Bathonian ostracod correlation north and south of the English Channel with description of two new Bathonian ostracods. 73–89 in *Microfossils from recent and fossil shelf seas.* NEALE, J W, and BRASIER, M D (editors). (Chichester: Ellis Horwood.)

SMART, J G O. 1955. Notes on the geology of the Alton Pancras district, Dorset. *Bulletin of the Geological Survey of Great Britain*, No. 9, 42–49.

— BISSON, G, and WORSSAM, B C. 1966. Geology of the country around Canterbury and Folkestone. *Memoir of the Geological Survey of Great Britain*, sheets 289, 305 and 306 (England and Wales).

SMITH, D B, BRUNSTROM, R G W, MANNING, P I, SIMPSON, S, and SHOTTON, F W. 1974. A correlation of Permian rocks in the British Isles. *Journal of the Geological Society of London*, Vol. 130, 1–45.

SPATH, L F. 1923–1943. A monograph of the Ammonoidea of the Gault. Volume 1, parts 1–7, 1923–1930; Volume 2, parts 8–16, 1931–1943. *Monograph of the Palaeontographical Society.*

SYLVESTER, A G. 1988. Strike-slip faults. *Bulletin of the Geological Society of America*, Vol. 100, 1666–1703.

SYLVESTER-BRADLEY, P C. 1957. The Forest Marble of Dorset. *Proceedings of the Geological Society of London.* No. 1556, 26–28.

— 1968. The Inferior Oolite Series. 211–226 in *The geology of the East Midlands.* SYLVESTER-BRADLEY, P C, and FORD, T D (editors). (Leicester University Press.)

— and Hodson, F. 1957. The Fuller's Earth of Whatley, Somerset. *Geological Magazine,* Vol. 94, 312–325.

Sun, S Q. 1989. A new interpretation of the Corallian (Upper Jurassic) cycles of the Dorset coast, Southern England. *Geological Journal,* Vol. 24, 139–158.

— 1990. Facies-related diagenesis in a cyclic shallow marine sequence: the Corallian Group (Upper Jurassic) of the Dorset coast, southern England. *Journal of Sedimentary Petrology,* Vol. 60, 42–52.

Swiecicki, A. 1980. A foraminiferal biostratigraphy of the Campanian and Maastrichtian chalks of the United Kingdom. Unpublished PhD thesis, Plymouth Polytechnic.

Sykes, R M, and Callomon, J H. 1979. The *Amoeboceras* zonation of the Boreal Upper Oxfordian. *Palaeontology,* Vol. 22, 839–903.

Talbot, M R. 1973. Major sedimentary cycles in the Corallian beds (Oxfordian) of southern England. *Palaeogeography, Palaeoclimatology, Palaeoecology,* Vol. 14, 293–317.

Taylor, R T. 1990. Geology of the Charlton Horethorne district (Somerset and Dorset). 1:10 000 sheet ST62SE. *British Geological Survey Technical Report,* WA/89/69.

— 1991. Geology of the Templecombe district (Somerset). 1:10 000 sheet ST72SW. *British Geological Survey Technical Report,* WA/91/31.

Torrens, H S. 1967. Field meeting in the Sherborne–Yeovil district. *Proceedings of the Geologists' Association,* Vol. 80, 301–324.

— 1968. In M R House 'Geology'. *Proceedings of the Dorset Natural History and Archaeological Society,* Vol. 89, 41–45.

— 1969. International Field Symposium on the British Jurassic. Excursion No.1. Guide for Dorset and south Somerset, A1–A71.

Townson, W G. 1975. Lithostratigraphy and deposition of the type Portlandian. *Journal of the Geological Society of London,* Vol. 131, 619–638.

Tresise, G R. 1960. Aspects of the lithology of the Wessex Upper Greensand. *Proceedings of the Geologists' Association,* Vol. 71, 316–339.

- 1961. The nature and origin of chert in the Upper Greensand of Wessex. *Proceedings of the Geologists' Association,* Vol. 72, 333–356.

Warrington, G, and Scrivener, R C. 1990. The Permian of Devon, England. *Review of Palaeobotany and Palynology,* Vol. 66, 263–272.

Welch, F B A, and Crookall, R. 1935. *British regional geology: Bristol and Gloucester district.* (London: HMSO for British Geological Survey.)

Whitaker, W. 1864. Geology of parts of Middlesex, Hertfordshire, etc. (Sheet 7). *Memoir of the Geological Survey of Great Britain.*

— and Edmunds, F H. 1925. The water supply of Wiltshire from underground sources. *Memoir of the Geological Survey of Great Britain.*

— and Edwards, W. 1926. Wells and springs of Dorset. *Memoir of the Geological Survey of Great Britain.*

White, H J O. 1923. Geology of the country south and west of Shaftesbury. *Memoir of the Geological Survey of Great Britain,* Sheet 313 (England and Wales).

Whittaker, A. 1972. The Watchet Fault — a post-Liassic transcurrent reverse fault. *Bulletin of the Geological Survey of Great Britain,* No. 41, 75–80.

— 1975. A postulated post-Hercynian rift valley system in southern Britain. *Geological Magazine,* Vol. 112, 137–149.

— (editor). 1985. *Atlas of onshore sedimentary basins in England and Wales,* 1–68. (Glasgow: Blackie.)

— and Chadwick, R A. 1984. The large scale structure of the Earth's crust beneath southern Britain. *Geological Magazine,* Vol. 121, 621–624.

— — and Penn, I E. 1986. Deep crustal traverse across southern Britain from seismic reflection profiles. *Bulletin of the Geological Society of France, B,* 55–68.

— Holliday, D W, and Penn, I E. 1985. Geophysical logs in British stratigraphy. *Special Reportof the Geological Society of London,* No. 18.

Wilson, R C L. 1968a. Upper Oxfordian palaeogeography of southern England. *Palaeogeography, Palaeoclimatology, Palaeoecology,* Vol. 4, 5–28.

— 1968b. Carbonate facies variation within the Osmington Oolite Series in southern England. *Palaeogeography, Palaeoclimatology, Palaeoecology,* Vol. 4, 89–123.

Wilson, V, Welch, F B A, Robbie, J A, and Green, G. 1958. Geology of the country around Bridport and Yeovil. *Memoir of the Geological Survey of Great Britain,* Sheets 312 and 327 (England and Wales).

Wood, C J. 1991. The stratigraphy of the Upper Chalk in the Blandford Forum–Sturminster Marshall area, Dorset, with particular reference to the geology of the Tarrant Valley. 1:50 000 sheets Shaftesbury (313), Ringwood (314) Dorchester (328) and Bournemouth (329). *British Geological Survey Unpublished Report,* WH/91/123R.

— Bigg, P J, and Medd, A W. 1982. The biostratigraphy of the Upper Cretaceous (Chalk) of the Winterborne Kingston borehole, Dorset. 19–27 *in* The Winterborne Kingston borehole, Dorset, England. Rhys, G H, Lott, G K, and Calver, M A (editors). *Report of the Institute of Geological Sciences,* No. 81/3.

— and Mortimore, R N. 1988. Biostratigraphy of the Newhaven and Culver Members. 58–65 *in* Geology of the country around Brighton and Worthing. Young, B, and Lake, R D. *Memoir of the British Geological Survey,* Sheets 318 and 333 (England and Wales).

Woodward, H B. 1888. Notes on the Midford Sands. *Geological Magazine,* (3), Vol. 5, 470.

— 1893. The Jurassic rocks of Britain, Vol.3. The Lias of England and Wales (Yorkshire excepted). *Memoir of the Geological Survey of the United Kingdom.*

— 1894. The Jurassic rocks of Britain. Vol. 4. The Lower Oolitic rocks of England (Yorkshire excepted). *Memoir of the Geological Survey of the United Kingdom.*

— 1895. Jurassic rocks of Britain, Vol. 5. The Middle and Upper Oolitic Rocks of England (Yorkshire excepted). *Memoir of the Geological Survey of Great Britain.*

Wray, D S, and Gale, A S. 1992. Geochemical correlation of marl bands in Turonian chalks of the Anglo-Paris Basin. 211–226 *in* High Resolution Stratigraphy. Hailwood, E A, and Kidd, R B (editors). *Geological Society Special Publication,* No. 70.

Wright, J K. 1980. Oxfordian correlation chart. 61–76 *in* A correlation of Jurassic rocks in the British Isles. Part two: Middle and Upper Jurassic. Cope, C J W (editor). *Special Report of the Geological Society of London,* No. 15.

— 1981. The Corallian rocks of north Dorset. *Proceedings of the Geologists' Association,* Vol. 92, 17–32.

— 1982. Reply by the author [to Talbot, G R. 1982. The Corallian rocks of north Dorset]. *Proceedings of the Geologists' Association*, Vol. 93, 312–313.— 1985. A new exposure of the Corallian Beds in north Dorset. *Proceedings of the Dorset Natural History and Archaeological Society*, Vol. 106 (for 1984), 168.

— 1986. A new look at the stratigraphy, sedimentology and ammonite fauna of the Corallian Group (Oxfordian) of south Dorset. *Proceedings of the Geologists' Association*, Vol. 97, 1–21.

Wright, T. 1856. On the palaeontological and stratigraphical relations of the so-called 'Sands of the Inferior Oolite.' *Quarterly Journal of the Geological Society*, Vol. 12, 292–325.

— 1860. On the subdivisions of the Inferior Oolite in the south of England, compared with the equivalent beds of that formation in Yorkshire. *Quarterly Journal of the Geological Society*, Vol. 16, 1–48.

Young, B, and Lake, R D. 1988. Geology of the country around Brighton and Worthing. *Memoir of the British Geological Survey*, Sheets 318 and 333 (England and Wales).

Young, D. 1972. Brickmaking in Dorset. *Proceedings of the Dorset Natural History and Archaeological Society*, Vol. 93, 213–242.

APPENDIX

Geological Survey photographs

Copies of the photographs are deposited for reference in the British Geological Survey library at Keyworth, Nottingham NG12 5GG. Prints and slides may be purchased. The more recent photographs listed below (numbers commencing 15149) were taken by Mr T Cullen and are available in colour and black and white. They belong to Series A.

2093	Floodplain of the River Stour, Bryanston Park, Blandford.
2094–7	Drift-filled channel in the Chalk. Cutting on W D Railway-siding, Blandford.
2098–0	Sections in Valley Gravel, Durweston
2101–2	Panoramic view of River Stour in flood, near Durweston
2103	Section in Upper Chalk, old quarry, Shillingstone Hill
2104	Chalk escarpment at Hambledon Hill
2105	Chalk escarpment at Hambledon Hill with Duncliffe Hill in distance
2106–7	Panoramic view of Chalk escarpment at Hambledon and Hod Hills
2108	Floodplain of River Stour and flood-arches, with Hambledon Hill in distance, showing earthworks of Camp
2109	View of north-west end of Hambledon Hill, showing earthworks
2110	Pebbly Gravel [Clay-with-flints], Okeford Hill (summit)
2111	Quarried spur of Chalk Downs, near Okeford Fitzpaine
2112	Section in Melbourn Rock. Quarry on Chalk Downs, at end of narrow spur, near Okeford Fitzpaine
2113–4	Plateau Gravel [Older Head), 1.5 mile south-south-west of Sturminster Newton
2115	Section in Corallian Beds, Sturminster Newton
2116	Section in Corallian Pisolites, Sturminster Newton
2117	False-bedded Corallian Limestone [Todber Freestone]. Railway-cutting north-west of Sturminster Newton Station
2118	False-beeded Oolitic Limestone [Todber Freestone], overlain by 'Ragstone' [Clavellata Beds]. Railway-cutting north-west of Sturminster Newton Station.
2119–20	Cornbrash. Rubbly arenaceous limestone with bands of gritty marl. Quarry near Stalbridge Station.
2121–3	Bowden Limestone in Forest Marble. Quarry, Henstridge Bowden
2124–5	Lynchets, Ibberton Hill, 1 mile east of Woolland
2126	View of broad combe in Chalk escarpment, south of Woolland
2127	View of country about Stoke Wake, looking west
2128	Earthworks of Rawlsbury Camp from Bulbarrow Hill
2129	Chalk escarpment and low-lying country near Mappowder
2130–1	Angular Flint Gravel, Ridge Hill, Buckland Newton
2132	Dogbury Gate. A gap in the joint escarpment of the Upper Greensand and the Chalk, 2 miles west of Buckland Newton.
2133–4	Panoramic view of Chalk escarpment from Shillingstone Hill
2135–6	Section in Corallian Limestone, 400 yards east of Marnhull Church
2137	Section in Corallian Limestone, Todber
2138	Small section of Lower Greensand near Bedchester
2139	Spur of Upper Greensand, near Compton Abbas
2140–1	Lynchets, near Compton Abbas
2142	Spur of Upper Greensand, near Compton Abbas
2143–5	Sections in Upper Greensand, showing the Chert Beds. Quarry, near Compton Abbas
2146	View of Chalk escarpment at Melbury Hill
2147	Section in Upper Greensand showing Chert Beds, Boyne Hollow, Shaftesbury
2150–3	Sections in Fuller's Earth Rock. Railway cutting (L & SW Railway.), Laycock
2154–5	Panoramic view of the characteristic landscape of the Fuller's Earth country near Milborne Port, with Hanover Hill in background
15149	Bridport Sands: bioclastic limestone bed, near Poyntington.
15150	Inferior Oolite: Rubbly Beds, Crackment Hill.
15151	Inferior Oolite: junction of the Rubbly Beds and Crackment Limestones, Oborne.
15152	Inferior Oolite: Sherborne Building Stone, Oborne.
15153	Inferior Oolite: Rubbly Beds, Milborne Wick.
15154	Fuller's Earth Rock, Laycock railway cutting.
15155	Lower Fuller's Earth, Laycock railway cutting.
15156	Fuller's Earth Rock scarp tilted and offset by faulting, between Milborne Wick and Stowell.
15157	Fuller's Earth Rock: junction of the Ornithella Beds and Milborne Beds, Goathill Quarry.
15158–59	Forest Marble: calcareous sandstone flags, Toomer Hill.
15160	Forest Marble: cross-bedded sandstones, Toomer Hill.
15161	Forest Marble limestone, Quarry Farm, Toomer Hill.
15162	Lower Cornbrash, Templecombe station.
15163	Cornbrash dip slope offset by faulting, Slades Hill, Templecombe.
15164	Lower Oxford Clay feature, Mohuns Park Farm, Henstridge.
15165–69	Junction of Clavellata Beds and Todber Freestone, Todber.
15170	Shaftesbury Sandstone and Boyne Hollow Chert, Ferne Park.
15171	Ragstone at the top of the Shaftesbury Sandstone of the Upper Greensand, Melbury Abbas.
15172	Junction of the Melbourn Rock and Plenus Marls, Zig Zag Hill.
15173	Plenus Marls, Zig Zag Hill.
15174–75	Glauconitised hardground (?Top Rock) within the Lewes Chalk, Elliott's Shed.
15176	Chalk scarp near Belchalwell.
15177	Hambledon Hill.
15178	Plenus Marls overlain by Melbourn Rock, Okeford Hill.

15179–80 Spurious Chalk Rock, Okeford Hill.
15181 Tabular flint in the Lewes Chalk, Okeford Hill.
5182 Junction of the Melbourn Rock and Plenus Marls, Okeford Fitzpaine.
15183 Lower part of the Blandford Chalk, Okeford Fitzpaine.
15184 Carious flints in the uppermost Lewes Chalk, Winterborne Stickland.
15185 Pit in uppermost Lewes Chalk, Winterborne Stickland.
15186 Scarp in the Tarrant Chalk, Tarrant valley.
15187 Hambledon Hill.
15188 Chalk scenery from Hambledon Hill.
15189 Chalk escarpment, White Sheet Hill.
15190 Cryoturbated junction of Clay-with-flints and Upper Chalk, Rookery Farm, Farnham.
15191 Landslips affecting the Upper Greensand and Gault, Thane's Farm, Shaftesbury.
15192 Landslipped Upper Greensand, Holyrood Farm, Shaftesbury.
15193 Poyntington Fault on Vartenham Hill.
15194 Poyntington Fault above Poyntington.
15195 Fuller's Earth Rock scarp east of Milborne Wick.
15196 The village pond, Ashmore.
15197 Sturminster Mill.
15198 Vale of Blackmoor from Woolland Hill.
15199 Oborne Fault, Oborne.
15200 Duncliffe Hill.
15201 View from Ridge Hill, Buckland Newton, towards Dogbury Hill.
15202 Dungeon Hill from Sharnhill Green, Buckland Newton.
15203 Forest Marble scarp below Charlton Gorse from the dipslope of the Inferior Oolite.
15204 River Terrace Deposits along Caundle Brook, near Lydlinch.
15205 Forest Marble scarp near Stowell
15206 Scarp at the base of the Forest Marble near Henstridge Bowden.
15207–08 The Cleeve, Charlton Horethorne.
15209 Highmore's Hill and Poyntington Fault.
15210 Middle Jurassic sequences in the Milborne Port district.
15211 The Haydon Fault, Haydon.
15212 Gold Hill, Shaftesbury.
15213 View westward from the Poyntington Fault.
15214 West block of the Poyntington Fault, Pinford.

FOSSIL INVENTORY

Latinised names only

GENERAL INDEX

BRITISH GEOLOGICAL SURVEY

Keyworth, Nottingham NG12 5GG
0115-936 3100

Murchison House, West Mains Road, Edinburgh
EH9 3LA 0131-667 1000

London Information Office, Natural History Museum
Earth Galleries, Exhibition Road, London SW7 2DE
0171-589 4090

The full range of Survey publications is available through the Sales Desks at Keyworth and at Murchison House, Edinburgh, and in the BGS London Information Office in the Natural History Museum (Earth Galleries). The adjacent bookshop stocks the more popular books for sale over the counter. Most BGS books and reports can be bought from HMSO and through HMSO agents and retailers. Maps are listed in the BGS Map Catalogue, and can be bought together with books and reports through BGS-approved stockists and agents as well as direct from BGS.

The British Geological Survey carries out the geological survey of Great Britain and Northern Ireland (the latter as an agency service for the government of Northern Ireland), and of the surrounding continental shelf, as well as its basic research projects. It also undertakes programmes of British technical aid in geology in developing countries as arranged by the Overseas Development Administration.

The British Geological Survey is a component body of the Natural Environment Research Council.

HMSO publications are available from:

HMSO Publications Centre
(Mail, fax and telephone orders only)
PO Box 276, London SW8 5DT
Telephone orders 0171-873 9090
General enquiries 0171-873 0011
Queuing system in operation for both numbers
Fax orders 0171-873 8200

HMSO Bookshops
49 High Holborn, London WC1V 6HB
(counter service only)
0171-873 0011 Fax 0171-831 1326
68–69 Bull Street, Birmingham B4 6AD
0121-236 9696 Fax 0121-236 9699
33 Wine Street, Bristol BS1 2BQ
0117-9264306 Fax 0117-9294515
9 Princess Street, Manchester M60 8AS
0161-834 7201 Fax 0161-833 0634
16 Arthur Street, Belfast BT1 4GD
01232-238451 Fax 01232-235401
71 Lothian Road, Edinburgh EH3 9AZ
0131-228 4181 Fax 0131-229 2734

HMSO's Accredited Agents
(see Yellow Pages)

And through good booksellers